D1141786

Strategic Decision Making

Analysis with Spreadsheets

WITHDRAWN

NAPIER UNIVERSITY LIBRARY

Strategic Decision Making
Multiobjective Decision Analysis with Spreadsheets

Craig W. Kirkwood
College of Business
Arizona State University

Duxbury Press
An Imprint of Wadsworth Publishing Company
I(T)P® An International Thomson Publishing Company

Belmont • Albany • Boston • Cincinnati • Detroit • London • Madrid • Melbourne
Mexico City • New York • Paris • San Francisco • Singapore • Tokyo • Toronto • Washington

Editor: Curt Hinrichs
Assistant Editor: Julie McDonald
Editorial Assistant: Cynthia Mazow
Advertising Project Manager: Joseph Jodar
Project Editor: Gary Mcdonald
Production: Ruth Cottrell
Print Buyer: Barbara Britton
Permissions Editor: Peggy Meehan
Copy Editor: Charles Cox
Cover Design: Mira Roytman, Christine Pramuk
Cover Photo: ©1997 TIB/Jeffrey M. Spielman
Printer: Quebecor Printing/Fairfield

COPYRIGHT © 1997 by Wadsworth Publishing Company
A Division of Thomson International Publishing Inc.

I(T)P The ITP logo is a registered trademark under license.
Duxbury Press and the leaf logo are trademarks used under license.

Microsoft and Visual Basic are registered trademarks, and Windows is a trademark of
Microsoft Corporation in the United States of America and other countries. DPL and
Decision Programming Language are trademarks of ADA Decision Systems. Logical
Decisions is a registered trademark of Logical Decisions. PCTEX for Windows is a
registered trademark of Personal TEX, Inc.

Printed in the United States of America
1 2 3 4 5 6 7 8 9 10
For more information, contact Duxbury Press at Wadsworth Publishing Company:

Wadsworth Publishing Company
10 Davis Drive
Belmont, California 94002, USA

International Thomson Editores
Campos Eliseos 385, Piso 7
Col. Polanco
11560 México D. F. México

International Thomson Publishing Europe
Berkshire House 168-173
High Holborn
London, WC1V 7AA, England

International Thomson Publishing GmbH
Königswinterer Strasse 418
53227 Bonn, Germany

Thomas Nelson Australia
102 Dodds Street
South Melbourne 3205
Victoria, Australia

International Thomson Publishing Asia
221 Henderson Road
#05-10 Henderson Building
Singapore 0315

Nelson Canada
1120 Birchmount Road
Scarborough, Ontario
Canada M1K 5G4

International Thomson Publishing - Japan
Hirakawacho Kyowa Building, 3F
2-2-1 Hirakawacho
Chioda-ku, Tokyo 102, Japan

All rights reserved. No part of this work covered by the copyright hereon may be
reproduced or used in any form or by any means—graphic, electronic, or mechanical, including photocopying, recording, taping, or information storage and retrieval
systems—without the written permission of the publisher.

Library of Congress Cataloging-in-Publication Data

Kirkwood, Craig W.
    Strategic decision making : multiobjective decision analysis with
spreadsheets / Craig W. Kirkwood.
        p.   cm.
    Includes bibliographical references and index.
    ISBN 0-534-51692-0
    1. Strategic planning—Decision making. 2. Strategic planning—
Data processing.   I. Title.
HD30.28.K57   1996
658.4'012—dc20   96-20271

To Lisa and Susan

Contents

Preface

This book presents methods for strategically making decisions using quantitative, spreadsheet-based, decision analysis methods. The intended audience is anyone responsible for decision making in an organizational setting, and the book provides a framework for thinking about decisions strategically, as well as practical tools that the reader can immediately apply. The book is suitable for use in classes on decision making, as well as for self-study.

Rules of thumb, intuition, tradition, and simple financial analysis are often no longer sufficient for addressing such common decisions as make-versus-buy, facility site selection, and process redesign. In general, the forces of competition are imposing a need for more effective decision making at all levels in organizations. The ongoing restructuring of businesses and other organizations increases the usefulness of the material in this book for a wide range of managers, analysts, and engineers. Traditionally, strategic decisions involving multiple competing objectives and significant uncertainties have been considered primarily the concern of top executives. However, with the current emphasis on downsizing and flattening organizations, individuals at lower levels in organizations must be concerned with such tradeoffs as cost versus quality, cost versus timeliness, or market share versus short-term return on investment.

The methods in this book have been applied for over twenty-five years, and they have a demonstrated capability to improve decision making. The methods have traditionally been considered advanced, in part because early presentations were framed in a mathematical terminology that is not familiar to many managers, and in part because early implementations of the methods required specialized software. This book brings the methods to a broader audience by explaining the intuitive basis for the methods, as well as how to implement them using spreadsheets.

Purpose   I wrote this book to make strategic decision making methods accessible for those who are making the decisions. The approaches in this book can help you to improve your decision making processes. The decision structuring methods in the first three chapters provide a framework for thinking about virtually all decisions ranging from those that are relatively tactical to corporate strategy. Numerous examples are included to give you a starting point for

your own decision analyses. Later chapters of the book present detailed procedures for quantitatively analyzing decisions using spreadsheets. The methods
are based on intuitively appealing principles, which are discussed as the methods
are presented. There are no mysterious procedures. You can understand, implement, and explain these methods without the need for specialized consultants or
software.

Focus and approach   The focus is on decisions where there are multiple
competing objectives that require consideration of tradeoffs among these objectives. This book brings the tools to analyze these decisions to a wider range of
decision makers through the use of spreadsheet methods. The approach is to
provide a structured, quantitative process for making such decisions by using
spreadsheet analysis methods. We take the view that decisions should be made
strategically, that is, in a skillful manner that is adapted to the ends that the
decision maker wishes to achieve. While there is probably little argument that
such an approach is desirable, the methods to actually carry out this type of
analysis have often been considered beyond the resources of managers below the
top level in large organizations.

Audience and prerequisites   The only prerequisite for much of the
book is an understanding of elementary algebra. The sections that consider
computer-based computational procedures also require an elementary understanding of electronic spreadsheets. Microsoft Excel, Version 5.0 or later, is used
in the book, including some use of Visual Basic, Applications Edition, which
was first included with Excel in Version 5.0. However, the reader does not need
to be familiar with Visual Basic or to learn anything about that programming
language to apply the methods.

Classes using the material in this book have included students in business
administration, engineering, health policy and administration, and public policy.
Most of these students have had practical experience in business, government,
or the not-for-profit sector. The quantitative and computer backgrounds of the
students in each class have generally varied substantially.

Organization   The book can be used in a variety of ways for either
class instruction or self-study. It is self-contained and can be used for a first
course in decision making which focuses on decisions with multiple objectives.
It can be used in conjunction with a text such as Robert T. Clemen's *Making
Hard Decisions: An Introduction to Decision Analysis*, Second Edition, Duxbury
Press, Belmont, California, 1996, in a first decision analysis course which places
less emphasis on decisions with multiple objectives. It can also be used as the
text for a course in multiobjective decision analysis which follows a traditional
first decision analysis course.

An instructor's manual is available with solutions to the exercises.

Chapters 1, 2, and 3 address decision problem structuring (formulating).
Chapter 4 addresses evaluation of alternatives with multiple objectives and no
uncertainty. Chapters 5, 6, and 7 address evaluation of alternatives with uncertainty. Chapter 8 reviews procedures for analyzing resource allocation decisions

with multiple objectives in the face of budget or other constraints. Chapter 9 presents theory that underlies the methods in earlier chapters. You do not have to understand this chapter to make practical use of the methods in the earlier chapters.

The appendices provide supplemental and background material. Appendix A presents a case that illustrates the application of the methods in Chapters 2, 3, and 4. Appendix B presents scenario planning approaches to analyzing decisions with uncertainty. These approaches can be used in conjunction with, or as an alternative to, the probability analysis methods in Chapters 5, 6, and 7. Appendix C presents a transcript of a probability elicitation session with a senior executive. Appendix D presents basic concepts of conditional probabilities, including use of influence diagrams and decision trees.

Chapter 9 is more abstract and requires somewhat more technical background than the other chapters and appendices. It can be assigned as optional reading or used as the primary reading for a more theoretical course. The material in Chapter 8 has also traditionally been considered more advanced, but with the current general availability of spreadsheet optimization features, these methods are now more widely used.

The latter parts of Chapters 4, 6, and 7, as well as Chapter 8, make use of electronic spreadsheets, and readers studying this material need to be familiar with elementary spreadsheet concepts. Some specific features of Excel, Version 5.0 or later, are used, although the approaches can be translated to other advanced spreadsheets. The Appendix A case study does not specifically require spreadsheet computations, but the calculations needed to complete the case will be tedious without a spreadsheet. A reader wishing to understand the methods presented in this book but not how to implement them can skip the spreadsheet material without loss of continuity. Some readers may wish to use specialized multiobjective decision analysis software, and an appropriate package is *Logical Decisions for Windows*, available from Logical Decisions, 1014 Wood Lily Drive, Golden, Colorado 80401.

All chapters include both Review Questions and Exercises. While either of these can be assigned as homework, the review questions are more open-ended and sometimes do not have a unique answer. Thus, these questions may be more suitable for in-class review purposes.

The diagram following this preface shows the primary precedence relationships among the chapters and appendices. In addition, Section 8.4 requires background provided in Section 7.7.

For readers who are familiar with multiobjective decision analysis methods, note that this book uses multiattribute value and utility analysis methods, including probabilistic analysis of decisions with uncertainty. A measurable value approach is taken, although an instructor who is so inclined can easily take a more traditional approach.

Acknowledgments   Special thanks to Curt Hinrichs at Duxbury Press for supporting this project. Thanks are also due to Donald L. Keefer and Jeffrey S. Stonebraker for many helpful comments on drafts of much of the book, and to Nancy A. Houston for reviewing several chapters. James L. Corner also provided helpful comments. I owe a substantial debt to the reviewers, whose comments helped me greatly improve the quality of the book. These include Richard H. Bernhard, North Carolina State University; Alan J. Brothers, Battelle, Pacific Northwest Laboratories; Robert T. Clemen, Duke University; L. Robin Keller, University of California, Irvine; Don N. Kleinmuntz, University of Illinois at Urbana-Champaign; Abu S. M. Masud, Wichita State University; H. V. Ravinder, University of New Mexico; Rakesh K. Sarin, University of California, Los Angeles; James E. Smith, Duke University; Donald N. Stengel, California State University, Fresno; L. James Valverde A., Jr., Massachusetts Institute of Technology; James N. Vedder, Syracuse University; and Robert L. Winkler, Duke University. I also wish to thank Ruth Cottrell for patiently guiding me through the preparation of this book and Charles Cox for correcting my sometimes fractured grammar. Thanks also to Lura Harrison for catching some longstanding errors. This book was typeset by the author using PCTEX for Windows.

Please write, phone, or e-mail me if you have questions, corrections, or ideas for improvements to this book.

Craig W. Kirkwood (602-965-6354; e-mail craig.kirkwood@asu.edu)
Department of Management
Arizona State University
Tempe, AZ 85287-4006

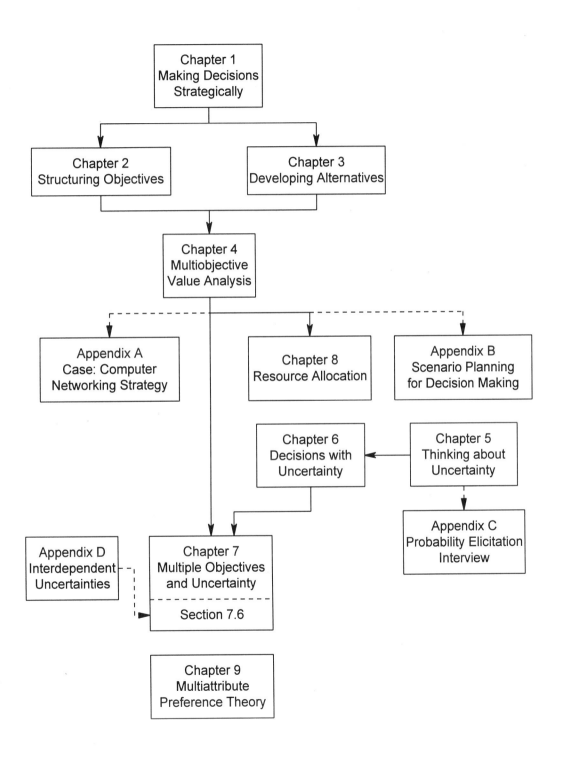

Strategic Decision Making

Multiobjective Decision Analysis with Spreadsheets

CHAPTER 1

Making Decisions Strategically

Personal and management decision making can be complicated and confusing. The future of your organization and the progress of your career can be profoundly affected by what you decide, and yet most people receive little instruction in decision making. Decision making does not have to be all intuition and "gut feel." Over the last several decades, a philosophy and a body of techniques have been developed to assist in analyzing decisions, and this approach has been used successfully in a wide variety of situations.

In this book, we concentrate on decisions where there are multiple competing objectives that require consideration of tradeoffs among these objectives. Such decisions are common in both management and personal decision making, particularly for tactical or strategic decisions that are above the purely routine operational day-to-day level. Business examples include tradeoffs between short-term financial return and market share when pricing products or services, between cost and control of quality when deciding make-versus-buy, and between efficiency and environmental impact when selecting certain manufacturing processes. At a more personal level, examples of decisions with multiple objectives include evaluation of career and employment opportunities, major purchases such as a home or an automobile, and decisions regarding marriage and methods of raising children.

The underlying assumption governing the methods presented here is that many significant decisions in modern organizations are made through a process involving technical staff and management at various levels in the organization as well as interested outsiders. Therefore, a key to good decision making is to provide a structured method for incorporating the information, opinions, and preferences of the various relevant people into the decision making process.

Do you participate in some form of athletic activity—for example, golf, tennis, or handball? Have you received coaching in that athletic activity? I have asked these questions of many groups. Usually, a large fraction of the group participates in some athletic activity, and most of those who do have received coaching. Then I ask how many people in the group make decisions, and how many have received coaching in how to make better decisions. Usually everyone agrees that they make decisions, and yet few have received coaching in making decisions.

It is clear that the ability to make good decisions is more important to the professional advancement of most people than their ability to play tennis well. Furthermore, it is possible to improve your decision making skills in the same way that you can improve your tennis game. Different people have differing natural talents for playing tennis or making decisions, but there are very few people who cannot benefit from some coaching in either tennis or decision making. Just as many fine athletic coaches do not have as much natural talent as the players they coach, I may not have as much natural decision making talent as you do. However, just as many athletic coaches know about the common mistakes in their sport and how to overcome them, this book presents methods to overcome the common decision making mistakes and improve your decision making "game."

The approach works. Better decision are made with less stress, and it is easier to explain the reasons for the decision that was made. Because of these factors, it is easier to get everyone working to make the selected alternative a success.

1.1 The Elements of a Decision

What is a decision? The one essential element of a decision is the existence of *alternatives*. That is, you must have a choice to make between at least two different things, only one of which you can select. If you don't have alternatives, then you may have a problem, but it isn't a decision problem. (Of course, some of the most vexing decisions are those where you do not seem to have any *good* alternatives.)

Most significant decisions involve situations where the various alternatives can lead to *differing consequences* or *outcomes*. If the results of all the alternatives are the same, then the decision problem doesn't warrant much analysis, but in most decisions this is not true. A primary difficulty in many decisions is figuring out what things are important in evaluating the consequences of decisions. This is considered in Chapter 2. Following this, Chapter 3 presents methods to help develop better alternatives for your decision. Another difficulty in decision making is determining the relative importance of different aspects of the consequences. Alas, in many realistic decisions we can't have it all—there are tradeoffs. If we want higher quality, we have to pay more. If we want faster service, we have to pay more or accept lower quality. Chapter 4 considers this difficulty.

Finally, many significant decisions involve *uncertainty about what consequence will result from each alternative*. The newspapers are full of stories about things that went wrong, and also about lucky people who did silly things but it all worked out okay. (Most experts agree that buying lottery tickets is not a wise use of your money, but some people do win the lottery!) We investigate decision making procedures for situations with uncertainty in Chapters 5, 6, and 7. Chapter 8 considers the important situation where scarce resources must be allocated among competing projects or activities, and Chapter 9 presents the theory that underlies the methods used in earlier chapters. Several appendices present supplemental material.

1.2 Thinking Strategically about Decisions

This book takes the view that decisions should be made *strategically*. That is, we should make decisions skillfully in a way that is adapted to the ends we wish to achieve. To make decisions strategically almost of necessity requires that we take a structured approach following a formal decision making process. Otherwise, it will be difficult to be sure that we have considered all the key aspects of the decision.

The approach taken is quantitative. We will see that using some numbers often clarifies the elements of a decision, and it forces us to be explicit about our reasoning. This improves decision making, and it also aids communication about the basis for a decision. It is typical in modern business and other organizational decision making that a variety of stakeholders need to understand and help implement a decision. Clarity of communications can be vitally important for this.

With the wide availability of spreadsheet programs, most business decision makers now understand that quantitative analysis can aid decision making. The methods in this book build on this understanding by using spreadsheets as the tool for the quantitative analysis methods. Specifically, we use Version 5.0 of the Excel spreadsheet. Excel is particularly useful because starting with Version 5.0 it now contains the Visual Basic programming language. This allows the easy addition of some straightforward analysis functions that simplify the methods we will use.

Readers who are not familiar with Visual Basic can easily enter these functions and use them without learning Visual Basic. The new functions just join such familiar functions as SUM in your Excel spreadsheet, and they allow simpler application of the methods in this book than would otherwise be possible.

A strategic approach to decision making includes the following five steps:

1 Specify objectives and scales for measuring achievement with respect to these objectives.
2 Develop alternatives that potentially might achieve the objectives.
3 Determine how well each alternative achieves each objective.
4 Consider tradeoffs among the objectives.
5 Select the alternative that, on balance, best achieves the objectives, taking into account uncertainties.

This process is just common sense, but following it systematically can yield big dividends in terms of improved decision making. The level of detail with which the process should be implemented depends on the complexity and importance of the decision. In the next two sections, a "report card" approach to this process is reviewed. The remainder of the book presents more detailed procedures.

1.3 CASE: Energy Computing International

This introductory case[1] illustrates some of the elements of the approach taken in this book. Section 1.4 discusses the issues that should be addressed in an analysis of this decision.

Coronado Power Company, a regulated public utility, seeks higher growth potential through expansion into nonregulated activities. To support this endeavor, Red Sand Holdings, Inc., a nonregulated holding company, has been formed with Coronado as a wholly owned subsidiary. While investigating potential new activities, an opportunity was identified related to specialized computer software that Coronado has developed to assist with its power system planning. This software has certain capabilities not provided by commercially available products, and preliminary investigation indicates that there might be a significant market for the software, both within the United States and in other countries.

Coronado has licensed a marketing agent to sell the software for several years, but sales under this arrangement have been limited. The current arrangement is unattractive for potential customers because they are unclear about the long-term prospects for support and development of the software. Also, Coronado's support for the software users has been less than ideal because there is little direct incentive to any group within Coronado to advance the sales of the software. To better exploit this opportunity, Red Sand is forming Energy Computing International (ECI) as a wholly owned subsidiary separate from Coronado Power Company.

Potential Alternative Organizations for the New Subsidiary

The specific goals for Energy Computing International, as stated in its business plan, are to market software and expertise developed over several decades by Coronado Power Company. The formation of ECI is to allow for more efficient use of Coronado Power Company assets to enhance shareholder return, provide cost benefits to ratepayers, and enrich work for employees. ECI wishes to organize itself in such a way as to maximize its potential to achieve these objectives.

Four strategic questions need to be resolved:

- **Location of software assets.** Where does ownership of the software reside?
- **Type of product.** In addition to the software itself, what types of services should ECI provide?
- **Structure.** To what extent should ECI directly provide services, versus serving as a general contractor and subbing these out?
- **Sources of contract personnel.** If ECI chooses to subcontract for services, where should the personnel for these contract services be obtained?

---

[1] This case is adapted from work conducted by D. G. Brooks and C. W. Kirkwood for the Decision Systems Research Center, Arizona State University.

Figure 1.1 shows the primary strategic options available for each of these aspects of the new organization.

From Figure 1.1, we see that there are potentially $3 \times 3 \times 3 \times 8 = 216$ different alternatives to be investigated. However, some of these are not physically possible, and others can be eliminated as clearly inferior after informal analysis. After an initial examination, ECI selected the eight alternatives shown in Figure 1.2 for further investigation.

Evaluation Considerations and Evaluation Measures

The alternative organizations shown in Figure 1.2 each have different advantages and disadvantages. To evaluate these advantages and disadvantages, five areas of evaluation concern were identified, with two specific evaluation considerations under each of these, as shown in Figure 1.3.

It was decided to use an A, B, C, D, E scale to evaluate the alternatives, where A represents an alternative that is "attractive" with respect to a specified evaluation consideration, C represents an alternative of "average merit," and E represents an "unattractive" alternative. A score of B represents an alternative that is intermediate between A and C scores, and a score of D represents an alternative that is intermediate between C and E scores.

The evaluation scores for the eight alternatives are shown in Figure 1.4. While evaluating the alternatives, the decision makers realized that there was significant uncertainty about the scores for some of the alternatives. This is represented by a "*" entry in the evaluation table.

1.4 Analyzing the Energy Computing International Case

This case illustrates the primary elements in a systematic evaluation of a decision. First, there are alternatives, which are specified in Figures 1.1 and 1.2. Second, there are multiple evaluation considerations, which are specified in Figure 1.3. There is a measuring scale for the relative performance of each alternative with respect to each evaluation consideration, namely the A, B, C, D, and E scale. There are scores (grades) for each alternative on this scale, as shown in Figure 1.4. There is also uncertainty about the scores for some of the alternatives.

Figure 1.4 illustrates a key dilemma in many decisions where there are multiple evaluation considerations: Alternatives that perform well with respect to some considerations perform less well with respect to others. There is no clear winner among the alternatives, and it is necessary to consider *tradeoffs* among the evaluation considerations to determine which alternative is preferred. Another way of saying this is that the preferred alternative depends on the *weights* that are assigned to the various evaluation considerations.

The issue of weighting evaluation considerations is analogous to the weighting that is often applied to grades in school to arrive at a grade point average. Usually grades are weighted by the number of credit hours for each course when a grade point average is calculated. Thus, a course with more credit hours,

- **Location of software assets**
1 The computer software remains with Coronado Power Company, and ECI is charged for its use.
2 The computer software is transferred to Red Sand Holdings, Coronado receives some form of credit, and ECI is charged by Red Sand for the use of the software.
3 The computer software is transferred to ECI, Coronado is credited, and ECI begins with its own product and a large amount of debt.

- **Type of product**
1 ECI confines its operations to marketing the software only, with minimal installation support.
2 ECI provides software, installation, and technical support, which is strictly software related.
3 ECI offers a full range of consulting services in addition to software support, including analysis of engineering, operations, and telecommunication problems.

- **Structure**
1 ECI functions as a general contractor with few employees, subcontracting out all installation, technical support, and consulting operations.
2 ECI employs its own technicians for software installation and basic support, but contracts out all consulting services.
3 ECI employs a full contingent of technicians and technical consultants to provide complete computer and consulting support.

- **Sources of contract personnel**
1 ECI continues to license marketing agents, either exclusively or nonexclusively, to market the software and relies on these agents for contract personnel.
2 ECI relies on Coronado personnel for contract services.
3 ECI relies on consulting firms for contract services.
4 ECI relies on a mix of marketing agent and Coronado personnel for contract services.
5 ECI relies on a mix of marketing agent and consulting firm personnel for contract services.
6 ECI relies on a mix of Coronado and consulting firm personnel for contract services.
7 ECI relies on a mix of marketing agent, Coronado, and consulting firm personnel for contract services.
8 ECI does not use contract personnel.

Figure 1.1 *Primary strategic options*

Alternative	Location of Software	Product	Structure	Contract Personnel Sources
1. Marketing	Coronado	Software only	Contract only	Coronado
2. Marketing support	Coronado	Software and support	Contract only	Marketing agents
3. Full marketing	Coronado	Software and support	ECI and contract	Coronado
4. Technical support	Coronado	Software and support	ECI and contract	Marketing agents and Coronado
5. Utility support	Red Sand	Software and support	ECI only	None
6. Consulting service	Coronado	Software, support, and consulting	ECI and contract	Coronado and consultants
7. Full consulting service	Red Sand	Software, support, and consulting	ECI and contract	Consultants
8. Independent subsidiary	ECI	Software, support, and consulting	ECI only	None

Figure 1.2   *Alternative organizations*

- Ratepayer cost benefits
  - Short-term benefits
  - Long-term benefits

- Shareholder return on investment
  - Short-term benefits
  - Long-term benefits

- Employee benefits
  - Rewarding and stimulating work environment
  - Pay and remuneration for innovation

- Managerial issues
  - Organizational structure changes
  - New managerial skill requirements

- High-level issues
  - Government/legislature relations, including public utilities commission
  - Public relations

Figure 1.3   *Evaluation considerations*

Alternative	Ratepayers		Shareholders		Employees		Management		High Level	
	Short	Long	Short	Long	Work	Pay	Org.	Skills	Gov't	Public
1. Marketing	B	B	B	B	B	B	A	A	A	A
2. Marketing support	B	B	B	B	B	B	B	B	B	B
3. Full marketing	C	A	C	A	A	A	A	B	B	B
4. Technical support	B	B	B	B	A	A	B	C	B	C
5. Utility support	D	*	D	*	B	B	B	B	*	*
6. Consulting service	C	A	B	A	A	A	B	C	B	B
7. Full consulting	E	*	E	*	B	*	D	D	E	E
8. Independent subsidiary	E	*	E	*	C	*	D	D	E	E

Figure 1.4   *Evaluation table*

which presumably requires more effort, receives greater weight in the calculation of a grade point average. The concept of weighting evaluation considerations to calculate an average rating, or value, for an alternative is similar.

Exercises 1.3 through 1.8 explore these issues further. Note that Exercise 1.3 requires some concepts that are presented in Chapter 3. Therefore, you may wish to return to this exercise after reading that chapter.

1.5 Review Questions

R1-1   What is the one essential element of a decision, and what are the two usual additional elements?

R1-2   Describe the steps presented in this chapter for a strategic decision making process.

R1-3   Give advantages and disadvantages of a quantitative approach to decision making.

1.6 Exercises

1.1 Project (Part A). This is the first part of a project that continues in Chapters 2, 3, 4, and 7. This project uses the methods presented in this book to analyze a business decision with multiple objectives and significant uncertainties. Examples of appropriate decisions include process improvement or re-engineering, facility siting, new ventures, new products/services, acquisitions, divestments, capital expenditures, lease-buy, make-buy, personnel planning, technology choice, and research/development planning.

The analysis must use the methods covered in this book. At minimum, the decision must have (1) at least three alternatives, (2) at least three evaluation concerns, and (3) significant uncertainty about at least one of the evaluation concerns. You must consult at least two outside expert data sources for information, and these sources must include both written material and an expert. While these requirements are minimums, the real requirement is that the decision problem be defensibly analyzed. That is likely to require more extensive analysis than the minimums.

For this part of the project assignment, prepare a proposal for the project. This should include (1) a summary of the decision to be analyzed, (2) a preliminary list of alternatives to be considered, (3) a preliminary list of evaluation considerations, and (4) proposed expert data sources to be consulted.

1.2 Project (Alternate Part A). This is an alternate to the Project (Part A) assignment above. This assignment is identical to the Part A assignment except that a personal decision is addressed. Examples of appropriate personal decisions include whether to change jobs or start a new career, what car to buy, and what house to buy.

Note: The exercises below relate to the Energy Computing International case presented in Section 1.3.

1.3 Construct a strategy generation table to summarize the information on primary strategic options in Figure 1.1. Illustrate the use of this strategy generation table to generate alternatives by showing alternative 3 (Full marketing) on the table that you have constructed.

1.4 For this exercise, consider only the first four alternatives in the evaluation table given in Figure 1.4. Using only the information in that table, determine whether it is possible to eliminate any of the four alternatives as a contender for the most preferred alternative. Specifically, show why you can or cannot eliminate each of the four alternatives as a contender for most preferred alternative.

1.5 For this exercise, consider only the first four alternatives in the evaluation table given in Figure 1.4. In report cards for school, it is common practice to calculate an average grade by assigning a numerical equivalent to each letter grade. In answering this exercise, assign numerical equivalents as follows: Assign 0 (zero) to E, 1 to D, 2 to C, 3 to B, and 4 to A. Determine the average grade for each of the four alternatives.

1.6 For this exercise, consider only the first four alternatives in the evaluation table given in Figure 1.4. In some grade averaging procedures—particularly in college grading—before being averaged, the grades are *weighted* by some measure of the effort required in the course (for example, credit hours). Similarly, in some evaluation procedures, weights are assigned to each evaluation consideration to indicate the "relative importance" of that consideration. Suppose that each of the six evaluation considerations in the ratepayer, shareholder, and employee categories receives the same weight. Suppose further that each of the four evaluation considerations in the management and high-level categories also receives the same weight, but that this weight is different from that received by the first six considerations. Finally, suppose that the "public relations" evaluation consideration receives half the weight of the "long-term ratepayer benefits" evaluation consideration.

  (i) Determine weights for each of the ten evaluation considerations that are consistent with this information and that will result in an alternative with all "A" scores receiving a weighted score of 4.0, while an alternative with all "E" scores receives a weighted score of 0.0.

  (ii) Determine the weighted scores and ranking of the four alternatives using these weights.

1.7 Assume the same situation as in the preceding exercise, except that we now wish to assign weights so that an alternative that has all "A" scores will receive a weighted score of 1.0, while an alternative that has all "E" scores will receive a weighted score of 0.0.

  (i) Determine the weights to be used for each evaluation consideration, and determine the weighted scores and ranking of the four alternatives.

  (ii) Determine an equation to convert the weighted scores on the 0 to 4 scale that you used in the preceding exercise into the weighted scores you found in part (i) of this exercise. [*Hint:* The two sets of scores are related by a straight line equation.]

1.8 Assume the same situation as in the preceding exercise, except that we now wish to consider all eight of the alternatives in Figure 1.4. Show that we can eliminate the three alternatives with "*" entries in that figure as possible contenders for the most preferred alternative, even though we do not know the scores for these alternatives with regard to some of the evaluation measures. [*Hint:* Consider what would happen if every missing score was an A. The unknown scores cannot be better than that.]

CHAPTER 2

Structuring Objectives

$\mathbf{A}$s the saying goes, "If you don't know where you're going, any road will do." Without clear objectives, it is difficult to make defensible decisions. Do we want quick payback, high long-term return on investment, or low capital requirements? Are we trying to improve our public image or better our union relations? Do we want to be first on the market with the electronic whiz-bang? Perhaps we want all these things at once.

An example of failure to specify objectives involved an electronics company with a small but profitable greeting-card business. The company received an offer to buy the subsidiary at a low price and sold it because greeting cards "did not fit" with its basic business in electronics. However, the subsidiary did fit very well with the objectives for growth of earnings and return on capital; in fact, it fit better than any other business in the electronics company's portfolio. These financial objectives were not clearly addressed and weighed against the "business fit" consideration. As a result, the low offering price looked attractive, and a below-market sale price resulted.

This chapter presents methods for developing and organizing objectives to use in analyzing decisions or evaluating alternatives. We first consider ways to develop a *qualitative* structure for your objectives, and then we examine ways to develop *quantitative* scales, which measure the degree of attainment of your objectives. The chapter concludes with several examples from actual applications.

2.1 Organizing and Presenting Objectives

The basic issues that must be addressed in dealing with objectives for a decision analysis are (1) determining what is important, and (2) figuring out how to measure how well the various decision alternatives perform with respect to the "important things." To address these issues, it is useful to define some terminology.

Evaluation consideration Any matter that is significant enough to be taken into account while evaluating alternatives should be an evaluation consideration. For example, evaluation considerations for a person examining different job offers might include salary, location, and prospects for advancement. Evaluation considerations for a company contemplating closing a plant that is the major employer in a town might include financial implications for the company, socioeconomic impacts on the town, and implications for the company's public image. Other terms sometimes used for evaluation considerations are **evaluation concern** or **area of concern**.

Objective   An objective is the preferred direction of movement with respect to an evaluation consideration. Thus, a job seeker may find a *higher* salary more desirable. The company contemplating closing a plant may wish to have a *more favorable* impact on its public image. This definition for objective implicitly assumes that preferences display monotonic behavior with respect to each evaluation consideration. That is, either "more is better" or "less is better" with respect to each evaluation consideration. There are some situations where this is not true. For example, suppose you are faced with a decision concerning a medical treatment that affects your blood pressure. There is a most preferred blood pressure level, and levels that are either higher or lower than this level are less preferable. However, such situations are unusual in the management decisions that are our focus.

Goal   A goal is a threshold of achievement with respect to an evaluation consideration which is either attained or not by any alternative that is being evaluated. For example, a job seeker might have a goal of attaining a salary of $50,000 per year. It should be noted that the use of the terms "objective" and "goal" is not completely standard in the literature. Some writers reverse the definitions from what are given here and use *goal* to mean what we have called an *objective* and vice versa.

Evaluation measure   A measuring scale for the degree of attainment of an objective is an evaluation measure. Thus, "annual salary in dollars" might be the evaluation measure for the job seeker's objective of finding a higher salary. Other terms that are sometimes used for an evaluation measure are **measure of effectiveness, attribute, performance measure,** or **metric.**

Level or **score**   The specific numerical rating for a particular alternative with respect to a specified evaluation measure constitutes its level (score).

Value structure   The value structure encompasses the entire set of evaluation considerations, objectives, and evaluation measures for a particular decision analysis.

Value hierarchy or **value tree**   A value structure with a hierarchical or "treelike" structure is called a value tree or value hierarchy. The remainder of this chapter will focus on value hierarchies, and the meaning of "tree-like" structure will become clearer as we proceed.

An example of a hierarchical organization of evaluation considerations is given in Figure 2.1. This value hierarchy was developed to assist a manufacturing firm that had decided to standardize on the word processing software used by all employees. This figure makes clearer the source of the term "value tree" to refer to such a structure. The figure looks somewhat like an upside-down tree with its root at the top of the diagram.

Layer or **tier**   The evaluation considerations at the same distance from the top of a value hierarchy constitute a layer or tier. In the Figure 2.1 value hierarchy, the first layer or tier below the topmost consideration of "purchase best value software" consists of the considerations "cost" and "suitability for use." Similarly, the next lower layer or tier of evaluation considerations consists of "software outlay"; "training, maintenance, and upgrades"; "hardware outlay"; "production, R&D, and engineering"; "finance and administration"; and "marketing." This example shows that the number of layers or tiers may not be uniform across a value hierarchy. There is an additional tier of considerations under the three "suitability for use" considerations ("production, R&D, and engineering"; "finance and administration"; and "marketing"), but there is not an additional tier under the three "cost" considerations ("software outlay"; "training, maintenance, and upgrades"; and "hardware outlay").

A properly organized value structure will be *hierarchical*. That is, an evaluation consideration closer to the root of the tree consists of the considerations "below" it (that is, in the next layer) that emanate from the specified consideration. For example, the overall evaluation consideration shown in Figure 2.1 is "purchase best value software." By proceeding to the next tier in the hierarchy, we obtain a more detailed understanding of what "purchase best value software" means in the context of this software selection decision. The second-tier evaluation considerations are "cost" and "suitability for use," and this means that the overall evaluation consideration "purchase best value software" is made up of "cost" and "suitability for use." That is, if we know how well an alternative performs with respect to these two second-tier evaluation considerations, then we know everything we need in order to evaluate the alternative with respect to the overall evaluation consideration "purchase best value software."

In a similar manner, some of the second-tier evaluation considerations are subdivided into a number of third-tier evaluation considerations. For example, "marketing" consists of "marketing graphics," "printing and mail merge," and "interface." As was true at the second tier of the hierarchy, it is true that knowing how well an alternative performs with respect to "marketing graphics," "printing and mail merge," and "interface" provides a complete description of how well the alternative performs with respect to "marketing."

Alternative Ways of Presenting Value Hierarchies

Figure 2.1 shows the tree-like structure of a value hierarchy, but the diagram does not fit well on the usual vertically oriented page of a book or paper. The diagram has to be turned on its side, as it is in that figure. An alternative way of presenting the value hierarchy is shown in Figure 2.2. This has the tree turned on its side with the root at the left-hand side of the page.

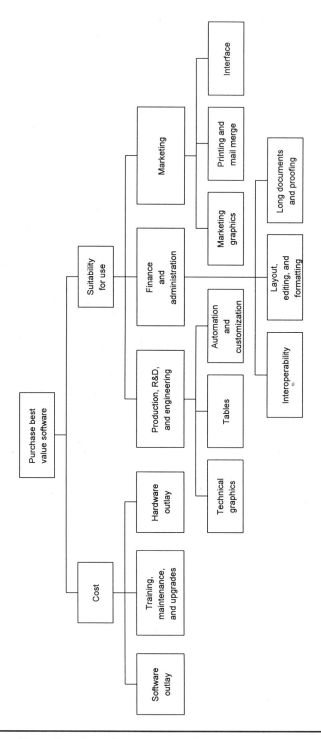

Figure 2.1   *Hierarchy of word processor evaluation considerations*

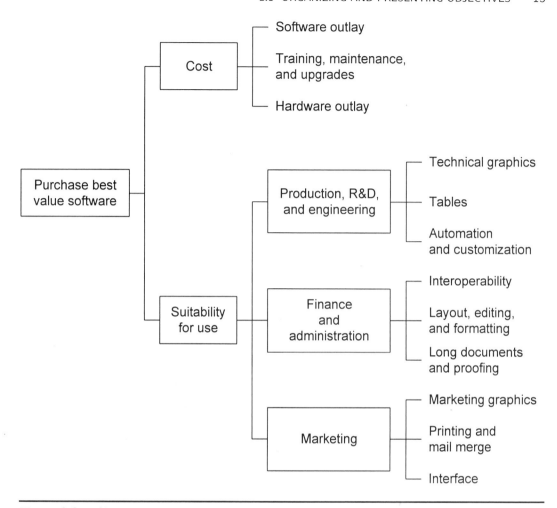

Figure 2.2 *Alternative hierarchy of word processor evaluation considerations*

Either of the two presentations in Figure 2.1 and Figure 2.2 allows the reader to quickly visualize the structure of the evaluation considerations for a decision analysis. However, they require either that the figures be drawn by hand or that some type of graphics program be used. An alternative way of displaying a value hierarchy, which can be constructed easily with a standard word processor, is shown in Figure 2.3. Note that with this form, the overall evaluation consideration ("purchase best value software") is not explicitly shown in the figure.

Here is a small technical point about Figure 2.1 through Figure 2.3. These are not complete value hierarchies as this term was defined above. According to the definition, a value hierarchy includes evaluation considerations, objectives, and evaluation measures. The figures include only evaluation considerations, and not objectives or evaluation measures. Usually, it is not practical to include all of the information about a value hierarchy in a single diagram because it would be too complex for easy comprehension. Often the evaluation considerations are

1. Cost
   a. Software outlay
   b. Training, maintenance, and upgrades
   c. Hardware outlay

2. Suitability for use
   a. Production, R&D, and engineering
      i. Technical graphics
      ii. Tables
      iii. Automation and customization
   b. Finance and administration
      i. Interoperability
      ii. Layout, editing, and formatting
      iii. Long documents and proofing

   c. Marketing
      i. Marketing graphics
      ii. Printing and mail merge
      iii. Interface

Figure 2.3   *Outline form for the word processor evaluation considerations*

presented in one diagram, and objectives and evaluation measures are presented separately. Examples are shown later in this chapter.

Desirable Properties of Value Hierarchies

Desirable properties for a value hierarchy include completeness, nonredundancy, decomposability, operability, and small size. Each of these properties is discussed in this section.

Completeness   For a value hierarchy to be complete, the evaluation considerations at each layer (tier) in the hierarchy, taken together as a group, must adequately cover all concerns necessary to evaluate the overall objective of the decision. Thus, if the value tree shown in Figure 2.1 through Figure 2.3 is complete, it is necessary to know only how well an alternative performs with respect to the lowest-tier evaluation considerations to know how well it performs with respect to the overall evaluation consideration of "purchase best value software." (Note that there are twelve lowest-tier evaluation considerations.)

There is a second requirement for a value hierarchy to be complete. In addition to the lowest-tier evaluation considerations adequately covering all evaluation concerns, it must also be true that the evaluation measures for those lowest-tier evaluation considerations adequately measure the degree of attainment of their associated objectives. Developing evaluation measures is considered in Section 2.4.

Nonredundancy   In addition to being complete, a value hierarchy should be nonredundant. This means that no two evaluation considerations in the same layer or tier of the hierarchy should overlap. This follows directly from the definition of a hierarchy—each layer in a hierarchy "divides up" the layer above it into more detailed pieces. Thus, in Figure 2.1, the "cost" evaluation consideration is divided up into the three categories "software outlays"; "training, maintenance, and upgrades"; and "hardware outlay." If this is a nonredundant division, then every cost can be assigned to exactly one of these three categories. That is, no cost can fit into more than one of these categories.

The two properties of completeness and nonredundancy are sometimes expressed by saying that the evaluation considerations in each layer of a value hierarchy must be "collectively exhaustive and mutually exclusive." That is, the evaluation considerations in each layer, taken as a whole, must include everything needed to evaluate the decision alternatives (collectively exhaustive), and nothing necessary to do the evaluation can be included in more than one evaluation consideration (mutually exclusive).

The reasons for requiring completeness and nonredundancy are somewhat intuitive from the hierarchical way in which we are displaying evaluation considerations, but a careful reader may wonder why these properties are necessary. Giving a completely defensible reason for requiring completeness and nonredundancy requires detailed study of decision theory (as presented by, for example, Keeney and Raiffa 1976). We will restrict ourselves to a few informal remarks here.

The need for completeness is fairly easy to see. If not all important evaluation considerations are included, then an evaluation of alternatives may not be able to distinguish between alternatives that are, in fact, of different preferability. For example, suppose that you are evaluating possible automobiles for purchase. Your evaluation might or might not need to consider the color of the automobile to be complete. This is a significant evaluation consideration for some people and not for others.

The need for nonredundancy becomes clearer when you start to work with evaluation measures for different objectives and want to combine these evaluation measures together to reach an overall evaluation of each alternative. Often this is done by a procedure that assigns a numerical weight to each evaluation measure and then uses a weighted sum of the scores on the various evaluation measures to determine the overall rating for alternatives. If more than one evaluation measure indicates the degree of attainment for a particular objective (that is, the evaluation measures are redundant), then that objective will probably receive more weight than was intended when the weights were assigned to the various evaluation measures. (Sometimes this is expressed by saying that an evaluation consideration is "double counted.")

Decomposability or **independence**   As with completeness and nonredundancy, a detailed discussion of decomposability requires a study of decision theory. However, the basic issue can be illustrated with an example. Suppose that a job seeker has as an evaluation consideration "economic issues," and has proposed as lower-tier evaluation considerations for economic issues the following: "salary," "pension benefits," and "medical coverage." These are nonredundant because no part of economic issues is counted in more than one of the lower-tier considerations, but they may not be decomposable. This is because the value attached to variations in the level (score) of the evaluation measure for any of the lower-tier considerations probably depends on the levels of the other two lower-tier considerations. For example, if there are very good pension benefits, then the value of an additional $5,000 in salary may be less than if the pension plan is poor and the job seeker will need to provide for his or her retirement out of the salary.

Lack of decomposability causes difficulties when attempting to develop a procedure to combine evaluation measures to determine the overall preferability of alternatives. Without decomposability, the required procedures can be too complicated for practical use in many situations.

For this economic issues example, the difficulty can be overcome by assigning equivalent dollar values to salary, pension benefits, and medical coverage. (This may require taking into account tax issues, as well as the time value of money.) Then a single economic issues evaluation measure can be used, which consists of the sum of the equivalent dollar values for the three lower-tier considerations.

Operability   An operable value hierarchy is one that is understandable for the persons who must use it. Clearly, operability is in the eye of the beholder. For the economic issues example above, using the discounted present value of a pension might be very meaningful (and hence operable) for a graduating MBA student, but not for a person who is less sophisticated about financial matters.

The issue of operability often comes up when technical specialists interact with the public. For example, during the Three Mile Island nuclear power plant incident, technical specialists had difficulty presenting an assessment of risks in a manner that was understandable to journalists or the general public. The evaluation measures that the specialists were using were not operable for the intended audience. Similar difficulties arise with use of terms like *sulfur dioxide concentration* or *particulate size* when specialists discuss air pollution. While these measures may be operable for the specialists who use them in their work, they have little meaning for most other people, and hence are not operable for nonspecialists.

In practice, it may be necessary to compromise with respect to some of the other desirable characteristics in order to use evaluation measures that are operable. In the economic issues example above, it might be necessary to use the three lower-tier evaluation considerations rather than combining them as discussed above in order to have evaluation measures that a particular audience views as complete, but still understandable (that is, operable). Determining when such compromises are likely to cause analysis difficulties is part of the art of developing a value hierarchy.

Small size   Other things being equal, it is desirable to have a smaller value hierarchy. A smaller hierarchy can be communicated more easily to interested parties and requires fewer resources to estimate the performance of alternatives with respect to the various evaluation measures.

In many practical business decision problems at the operational level it is necessary to consider only a single evaluation measure—often some indicator of financial performance. In some situations—for example, capital budgeting or research and development decisions—cash flow streams over time must be considered but the cash flow can be reduced to a single discounted cash flow number that can be used to evaluate alternatives. For business decisions that are more strategic, it can be necessary to include additional evaluation considerations. Often, quality and timeliness issues are important in addition to purely financial matters. Other issues that can be relevant at the strategic level include public image, government regulations, environmental impacts, and socioeconomic impacts.

In government and not-for-profit organizations, multiple evaluation considerations are often central to decision making at even low levels. Quality and timeliness of service are usually important in addition to financial performance. (In fact, at the operational level in government and not-for-profits, the crux of many decisions is a tradeoff between cost and quality/timeliness of service.) A variety of issues related to distribution of impacts over different groups and the views of various stakeholder groups may also be important.

There is a tendency to keep adding evaluation considerations until the value hierarchy becomes so complex that any analysis using the hierarchy will be difficult to conduct and interpret. In practical applications, the quest for completeness and fine detail must be balanced against the need to finish an analysis within a realistic time frame and budget. Part of the art of conducting such analyses is making this balance in such a way that the analysis is defensible while still being practical.

As we will see when we discuss development of evaluation measures below, there is usually a decision to be made during an analysis about how many layers to have in a value hierarchy. Often with more layers it is easier to develop evaluation measures but more time consuming to conduct the analysis of alternatives, while with fewer layers it is more difficult to develop evaluation measures but less time consuming to conduct the analysis.

The "test of importance" (Keeney and Raiffa 1976, Section 2.3.2) provides a rough indication of whether to include a particular evaluation consideration in a value hierarchy. This test states that an evaluation consideration should be included in a value hierarchy only if possible variations among the alternatives with respect to the proposed evaluation consideration could change the preferred alternative. Of course, it is generally not possible to answer this question for certain until a detailed analysis including the evaluation consideration in question is completed, but often a more informal review will make the answer clear. For example, I don't have much preference among colors for automobiles except that I will not buy one that is green. Almost all autos come in at least two colors, so that if I ignore color, I will not change my evaluation of possible autos to purchase. Thus, color does not pass the test of importance in my auto purchase decision, and hence does not have to be included as an evaluation consideration.

2.2 Approaches to Structuring Value Hierarchies

The preceding section discussed how value hierarchies can be displayed and also reviewed desirable characteristics of value hierarchies. This section considers how such hierarchies are developed. Keeney and Raiffa (1976), Buede (1986), and Keeney (1988, 1992) discuss this process in more detail.

Generally, it is necessary to develop a value hierarchy that is specific to a particular decision. This is because a general-purpose value hierarchy that would be applicable to a wide range of decisions would be so complex as to be impractical. However, in the several decades that value hierarchies have been used to structure objectives, a large number of different decision problems have been analyzed, and you may be able to find a value hierarchy for a previously analyzed decision problem that shares some characteristics in common with your decision. If so, you will not have to start from scratch in developing the value hierarchy for your decision problem. For example, I have worked on selection of sites for various types of energy facilities (Bennedsen and Kirkwood 1982; Kirkwood 1982). While coal-fired, nuclear, and pumped-storage hydroelectric plants differ in substantial ways, they also have many features in common. Thus, the value hierarchy for siting one of these types of facilities serves as a starting point in developing a value hierarchy for siting another type of facility.

Bottom-up versus Top-down Value Structuring

An appropriate approach to developing a value hierarchy differs somewhat depending on whether or not the alternatives to be evaluated are known at the time the hierarchy is being developed. If the alternatives are known, then a *bottom-up* or *alternatives-driven* approach may be appropriate. With this approach, the alternatives are examined to determine the ways in which they differ. Evaluation measures are developed to measure the evaluation considerations for which the alternatives differ, and these evaluation measures are grouped together to form higher layers in the value hierarchy. Thus, this approach to developing a value hierarchy starts with the alternatives ("alternatives-driven"), develops the bottom layer of the value hierarchy, and constructs the rest of the hierarchy on top of this bottom layer ("bottom-up").

Golabi, Kirkwood, and Sicherman (1981) present an application of this approach. This involved developing a procedure for the U.S. Department of Energy to use in evaluating proposals submitted in response to a Request for Proposals to design and build solar energy demonstration projects. In this evaluation, the general characteristics of the projects were specified in the Request for Proposals. Thus, the task was first to determine how the submitted proposals might differ, and then to use this as a basis for developing a value hierarchy. In such a situation, a bottom-up approach was appropriate.

In situations where the alternatives are not well specified at the start of the analysis, an approach starting with the overall objective and successively subdividing objectives is more appropriate. For example, Brooks and Kirkwood (1988) worked on a decision analysis to determine a strategy for a company to follow with regard to microcomputer networking. In this case, the possible alternatives were unclear at the beginning of the analysis, and in fact one of the purposes of the analysis was to identify potential alternatives. An approach was used that started with the overall objective and subdivided this overall objective to develop the evaluation considerations in successively greater detail. Such an approach is referred to as *top-down* or *objectives-driven*.

Sources of Evaluation Considerations and Measures

Review relevant literature   As noted above, it is usually necessary to develop a value hierarchy that is specific to a particular decision problem. However, many decision situations are sufficiently similar to previously studied decisions that obtaining information about how these previous decisions were addressed will help in developing a value hierarchy. Even if informal methods were used in the earlier decisions so that a value hierarchy was not developed, learning what the people considered while analyzing the decision will help in developing a value hierarchy for your decision.

Within the decision analysis literature, works by Keeney and Raiffa (1976), Buede (1986), and Keeney (1988, 1992) include examples of value hierarchies, as well as more advice on how to develop such hierarchies. The review article by Corner and Kirkwood (1991) includes a variety of applications classified by application area, as well as references to additional sources of applications. Gustafson, Cats-Baril, and Alemi (1992) discuss a variety of medical-oriented applications. Outside of the decision analysis literature, many books in particular applications areas include discussions of value structures. While many of these structures were informally developed and may violate some of the desirable characteristics presented earlier, they can serve as useful starting points for developing a value structure addressing the decision problem of interest to you.

Perform casual empiricism   Even if there is substantial literature about previous similar decision situations, it is still usually necessary to talk with the relevant stakeholders for the decision problem of interest. This process, sometimes called *casual empiricism*, is necessary to ensure that the specific evaluation considerations of interest to the stakeholders in the decision problem are included in the value hierarchy and also to obtain "buy-in" from the relevant stakeholders.

For decisions being made on relatively routine operational matters in a business organization, there may be only a small number of persons directly interested in the decision. However, there are often others with valuable experience on similar decision problems. Consulting these persons can help ensure that no important elements are missed.

For more strategic business decisions, or for many decisions within government or not-for-profit organizations, there may be many stakeholders who have an interest in the decision, and whose agreement on a value hierarchy is useful or necessary to obtain later implementation of the analysis results. In such situations, developing the value hierarchy can be an involved process that requires structured interview approaches. Keeney (1988, 1992) discusses such approaches in more detail.

Ends Objectives versus Means Objectives

One specific issue arises sufficiently often during the structuring of value hierarchies that it is worth some discussion. This is the distinction between *ends* objectives and *means* objectives. This concept can be introduced with an example. Consider once again our job seeker who wishes to include financial issues in an evaluation of possible jobs. Salary, pension benefits, and company-provided medical coverage were discussed as possible lower-tier evaluation considerations for financial issues. That discussion also presented possible difficulties with respect to decomposability for this set of evaluation considerations. There is another possible difficulty—whether these evaluation considerations really address the ultimate *ends* that are important to the job seeker, or are just *means* to the ultimate ends.

Suppose it is true that the ultimate financial ends of interest to the job seeker are greater short-term and long-term financial security. If this is so, then there may be other ways to achieve these ends than more income (salary and pension) and better medical coverage. For example, selecting a job in a low-cost area would mean that a lower salary would go further. Changing lifestyle to require lower expenditures is another way. Stopping smoking or other unhealthy activities to lower medical risks, and hence required medical coverage, is another possible approach.

You should pause from time to time as you develop a value hierarchy to consider whether the hierarchy you are developing is really measuring your ultimate objectives (*ends*) or the *means* to achieve these ultimate objectives. By putting more focus on ends, you may be able to develop more creative alternatives that better meet your objectives.

2.3 Uses for Value Hierarchies

A value hierarchy, either with or without evaluation measures (which are discussed in the next section), has several uses.

Guide to information collection   When you know what is important to you, then you can be sure to collect information about the important evaluation considerations and not waste time collecting information about considerations that are not important. Both the "one more study" and "nobody knows this business like I do" syndromes are common in analyzing decisions. Sometimes we think that if we just do *one more study*, then we will know what is the best decision. Good information is important for good decision making, but if you always seem to be collecting more information and never making the decision, then you should question the relevance of the information. As the saying goes, "No choice is also a choice." Understanding what considerations are important to the decision, as specified by a value hierarchy, will help to avoid the "one more study" phenomenon.

On the other hand, it is also common to collect too little information before making a decision. It seems that the more important the decision is, the more likely we are to commit this error. When it comes to strategic decisions affecting the entire future of an organization, some top managers think *nobody knows this business like I do* and fail to get relevant information from the most knowledgeable people. Having a value hierarchy that specifies the important evaluation considerations in a decision can make it clearer what additional information is required.

Help to identify alternatives   In a decision situation where the alternatives are not prespecified, a value hierarchy provides a basis for designing good alternatives. When you know what you are trying to accomplish (that is, your objectives), then you can attempt to identify alternatives that address these objectives. Many people do not cast their nets wide enough when they consider alternatives. Failure to consider all the alternatives is encouraged by the technical complexity of many modern management decisions. You must rely on specialists for information that is important to the decision, but these specialists sometimes analyze the options they know how to solve, rather than the preferable ones. Having a value hierarchy helps everyone involved in the decision to better understand the full breadth of considerations that are important in evaluating alternatives.

Facilitate communications   Many decisions that are significant enough to justify formal analysis involve multiple *stakeholders* (that is, people or groups with an interest in the decision and its outcome). A value hierarchy helps to facilitate communications among the stakeholders about what are important evaluation considerations for making the decision. In situations with controversy, a common understanding about what are important evaluation considerations may provide a better basis for compromise and/or consensus with regard to selecting alternatives.

Evaluate alternatives   An important use of value hierarchies is to provide a basis for evaluating alternatives. Remember the quote at the beginning of this chapter: "If you don't know where you're going, any road will do." Often simply developing a value hierarchy will make clearer the relative desirability of various alternatives. In other cases, it is useful to apply formal methods to arrive at a quantitative evaluation of the various alternatives. Such formal methods start with a value hierarchy and use a mathematical function to combine the evaluation measures from the hierarchy to rank alternatives. (See Chapter 4.)

2.4 Developing Evaluation Measures

A qualitative value hierarchy—for example, the one shown in Figure 2.1 through Figure 2.3—can be useful in a decision analysis for the reasons discussed earlier. Sometimes it is also useful to go beyond this qualitative value hierarchy and specify evaluation measures (also called measures of effectiveness, attributes, or metrics) for the degree of attainment of objectives. These evaluation measures allow an unambiguous rating of how well an alternative does with respect to each objective. For example, stating that a particular alternative will return a net present value profit of one million dollars is less ambiguous than stating that it will return a "significant profit." Keeney (1981, 1992) discusses the development of evaluation measures in detail, and this section reviews the process.

Types of Evaluation Measure Scales

Evaluation measure scales can be classified as either *natural* or *constructed*, and also as either *direct* or *proxy*. A *natural* scale is one that is in general use with a common interpretation by everyone. Thus, "number of fatalities" is a natural scale for evaluating alternatives with risks of death. Profit in dollars is a natural scale for many business decisions.

A *constructed* scale is one that is developed for a particular decision problem to measure the degree of attainment of an objective. Examples of constructed scales are presented below. Constructed scales are used in a variety of situations where natural scales are not appropriate. One such situation is when there is no existing natural scale that can be used. For the word processor standardization value hierarchy shown in Figure 2.1 through Figure 2.3, there were not established natural scales for many of the evaluation considerations. Thus, constructed scales were needed for these.

A *direct* scale directly measures the degree of attainment of an objective, while a *proxy* scale reflects the degree of attainment of its associated objective, but does not directly measure this. Profit in dollars is usually a direct scale. Gross national product is a proxy scale for the economic well-being of the country.

It is possible to have direct scales that are either natural or constructed, and similarly, it is possible to have proxy scales that are either natural or constructed. Furthermore, the sharp distinctions "natural versus constructed" and "direct versus proxy" actually represent the extremes of a range of possibilities. For example, is the scale "net present value profit" natural or constructed? To some extent, this may depend on whom you talk to. When this scale was originally developed to represent the time value of money, it would probably have met our definition of a constructed scale. However, it has now been in use long enough that for financially literate people it has become a natural scale. On the other hand, there are many people with an interest in money who are not financially literate. Many of them probably view net present value as an exotic constructed scale.

A similar situation exists with respect to the direct/proxy distinction. Is gross national product a direct or proxy scale? When it was first developed, it was a (constructed) proxy scale for the economic well-being of a country. Now it is in general use, and it has become in some quarters the accepted direct measure of economic well-being. Furthermore, it has come into such widespread use that it may meet the definition of a natural scale. Similar things have happened with such scales as the consumer price index, the Dow Jones industrial average for stock prices, the Richter scale for earthquake intensity, and the decibel scale for loudness.

Developing Evaluation Measure Scales

The difficulty of selecting or developing a scale to measure attainment of a particular objective can vary greatly. At one extreme, using net present value profit as a measure of the objective "maximize profit" is probably not controversial (although even in this case there are contenders such as payback period and internal rate of return). Perhaps at the other extreme is developing a scale to measure the visual aesthetic impact of a coal-fired power plant. How do you measure beauty, particularly for a power plant?

Several questions often arise while developing evaluation measures:

1 Should we use a natural scale that is a proxy, or should we develop a constructed direct scale?

2 Should we subdivide an evaluation consideration into more detailed subconsiderations for which natural scales might exist, or should we construct a scale to measure the evaluation consideration without subdividing it further?

3 Should we use a natural scale that is precise, but uses technical jargon, or should we use a constructed scale that may not be as precise but that may be more understandable to some stakeholders in the decision process?

4 How carefully should we specify the scale definition for a constructed scale?

We now consider each of these questions in more detail.

Natural-proxy versus constructed-direct scales   Natural scales
have some nice properties: You do not have to spend the time to develop the scale definition. Their use may be less controversial because they are in general use. The difficulty is that natural scales are not all that easy to come by, and you may have to use a proxy scale in order to find a natural scale for your evaluation consideration.

For example, consider a decision to select the number and type of elevators for a proposed high-rise building. One objective might be to minimize delays faced by elevator users. A natural scale for this could be the expected number of minutes that an elevator user has to wait before an elevator arrives. However, "delay" is really a subjective quantity. The delay experienced by an elevator user depends on not only the actual time until the elevator arrives, but also such things as how pleasant the surroundings are and what activities are available. (Installing mirrors seems to reduce the perceived delay, for example.) Capturing these aspects of delay may require using one or more constructed scales.

**How much should we subdivide an evaluation considera-
tion?**   This issue is illustrated by the constructed evaluation measure scale
shown in Figure 2.4 to measure ecological impacts of construction of a proposed
power plant. Examining this scale, we see that it considers several different aspects of ecological impact. Grassland, shrubland, pinyon-juniper, riparian, and
wetland habitat are all considered. Raptor habitat is considered, as are habitats
for threatened, endangered, or otherwise unique species. If the ecological impact
evaluation consideration were to be subdivided into lower-tier considerations addressing each of these aspects, it might not be necessary to develop a constructed
scale. Instead, the impact on each habitat could be directly measured—for example, in square miles of each type of habitat.

However, there are two potential difficulties with this. First, it may be that
ecological impact is a small part of the overall evaluation of power plant sites
(perhaps because sites with significant ecological problems were screened from
consideration before beginning the evaluation of potential sites). Using several
evaluation measure scales could give a misleading indication of the relative importance of ecological impacts in the analysis.

The second potential difficulty with subdividing the ecological impact scale is
that it could require more effort than is warranted to obtain scores for the several
lower-tier evaluation measure scales, and hence the resulting scores could give an
unwarranted indication of accuracy. The constructed scale in Figure 2.4 has only
thirteen different levels, and thus it is clear that a relatively coarse assessment
of ecological impact is being done when this scale is used. Suppose, on the other
hand, that five or ten separate scales were used. Then the resulting scores would
either require a more detailed assessment (so that impact on each type of habitat
could be determined more accurately), or use of rounded scores (for example,
to the nearest 100 acres of productive wetlands or the nearest 10 acres of rare
species habitat), where the level of accuracy would not be immediately apparent
from looking at the scale.

This example also illustrates a typical characteristic of constructed scales—
the scale level definitions represent only a small number of all the possible actual
situations that might be encountered with actual decision alternatives. Note that
in Figure 2.4 the only values used for percentage of different types of habitat lost
are 10 and 25. Thus, a biologist assessing a particular power plant site will need
to use his or her judgment to determine which of the defined categories is most
like the situation at the actual site. Suppose, for example, that a particular site
has a loss of 20 percent riparian or wetland habitat and 15 percent of actual
or potential habitat for threatened, endangered, or otherwise unique species.
This doesn't exactly meet the definition of any level on the scale. Thus, some
judgment is needed in assigning it a level. Perhaps this situation should even
have a fractional score that is between two of the defined levels.

**Should we use a precise natural scale that is technical or a
more operable constructed scale that is less precise?**   Suppose
that the smell and visual haze associated with sulfur dioxide is a concern for
some industrial process. This can be measured precisely by using the natural
scale "sulfur dioxide concentration," but this may not be meaningful for some

0 Removal of 6 square miles having > 25 percent of cultivated agricultural use.

1 Removal of 6 square miles having ≤ 25 percent of cultivated agricultural use with remaining area grass and shrubland or pinyon-juniper.

2 Removal of 6 square miles of grassland habitat, shrubland habitat, or pinyon-juniper habitat with no important or unique species or habitats present.

3 Removal of 6 square miles of pinyon-juniper/ponderosa pine habitat with no important or unique species or habitats present.

4 Removal of 6 square miles of grassland, shrubland, or pinyon-juniper habitat that includes ≤ 10 percent riparian or wetland habitat.

5 Removal of 6 square miles of grassland, shrubland, or pinyon-juniper habitat that is within 1 mile of significant actual or potential raptor habitat.

6 Removal of 6 square miles of grassland, shrubland, or pinyon-juniper habitat that includes $\leq$ 10 percent riparian or wetland habitat and is within 1 mile of significant actual or potential raptor habitat.

7 Removal of 6 square miles of grassland, shrubland, or pinyon-juniper habitat, of which $\leq$ 25 percent is actual or potential habitat for threatened, endangered, or otherwise unique species.

8 Removal of 6 square miles of grassland, shrubland, or pinyon-juniper habitat within 1 mile of significant actual or potential raptor habitat, and of which ≤ 25 percent is actual or potential habitat for threatened, endangered, or otherwise unique species.

9 Removal of 6 square miles of grassland, shrubland, or pinyon-juniper habitat within 1 mile of significant actual or potential raptor habitat, and including ≤ 10 percent riparian or wetland habitat and ≤ 25 percent actual or potential habitat for threatened, endangered, or otherwise unique species.

10 Removal of 6 square miles of grassland, shrubland, or pinyon-juniper habitat including > 25 percent actual or potential habitat for threatened, endangered, or otherwise unique species.

11 Removal of 6 square miles of grassland, shrubland, or pinyon-juniper habitat within 1 mile of significant actual or potential raptor habitat, and of which > 25 percent is actual or potential habitat for threatened, endangered, or otherwise unique species.

12 Removal of 6 square miles of grassland, shrubland, or pinyon-juniper habitat within 1 mile of significant actual or potential raptor habitat and including ≤ 10 percent riparian or wetland habitat and > 25 percent actual or potential habitat for threatened, endangered, or otherwise unique species.

Note: Important or unique species include federally listed endangered and threatened species, as well species proposed for listing and those classified as status undetermined. The category also includes species listed by the state and other reputable organizations as endangered, threatened, or otherwise unique and protected.

Significant raptors are defined as those that are considered important by the state or federal government (i.e., threatened and endangered species, etc.) but do not include common, widespread species such as red-tailed hawks.

Figure 2.4 *Evaluation measure scale for ecological impacts at plant site*

of the decision stakeholders (for example, a city council that has to approve a permit for the facility). A constructed scale that refers to rotten egg smell or uses pictures of visual haze in its definition may be more operable for the decision makers, even though it may be less precise than using sulfur dioxide concentration.

How carefully should we define the scale levels?   We have all seen the results of public opinion surveys that asked people to assess their confidence about the future using a scale with the following levels: very pessimistic, pessimistic, neutral, optimistic, or very optimistic. This five-point scale is not well defined. Suppose one person answers that she is "optimistic" while another answers that he is "pessimistic." Did their different ratings occur because they have a different assessment of what is likely to happen or because they assign different meanings to the different levels? Perhaps one of the people is naturally pessimistic, and it would require the promise of eternal bliss before he would assign an "optimistic" rating, while the other would assign "optimistic" if it looks like unemployment will drop.

Ambiguous scales impede communications. The ideal is to have scales that pass the *clairvoyance test*. That is, if a clairvoyant were available who could foresee the future with no uncertainty, would this clairvoyant be able to unambiguously assign a score to the outcome from each alternative in a decision problem? Most natural scales pass the clairvoyance test, but it is more difficult to develop constructed scales that do this. The scale in Figure 2.4 is fine as long as the alternatives meet the specific conditions of one of the defined scale levels. However, for a situation that does not exactly meet one of the definitions, some judgment is called for to determine a score. Thus, the scale is somewhat lacking with respect to the clairvoyance test. The example evaluation measure scales shown in the next section vary in how well they meet the clairvoyance test.

The analysis tradeoff is between effort spent to develop the scale, and ease of assessing alternatives and communicating the results of the analysis. Study of the examples in the next section will help you in thinking about what is involved in specifying constructed evaluation measure scales that are less ambiguous.

2.5 Illustrative Value Structures

This section presents several value structures taken from actual decision analysis applications. As you examine these, consider the degree to which each of them meets the various desirable criteria presented above.

Microcomputer Networking Strategies

This example is adapted from Brooks and Kirkwood (1988), who developed a multiobjective decision analysis procedure to assist a public utility company in selecting a strategy for dealing with microcomputer networking. The decision analysis successfully addressed key management issues in selecting a microcomputer networking strategy. Based on a literature review and discussion with the managers involved and computer professionals, key evaluation considerations were identified for selecting a microcomputer networking strategy for the public utility company as follows:

1 *Impact on Microcomputer Users:* The degree to which addition of a specified microcomputer network changes the work environment of microcomputer users. Since the impact may be different for different groups of users, each group is considered separately.

    a. *Productivity Enhancement:* Changes in the quantity and quality of work output due to the network.

    b. *User Satisfaction:* The degree to which network users perceive their work environment has been improved or degraded due to the network.

2 *Impact on Mainframe Capacity:* Changes in utilization of the organization's mainframes due to the network.

3 *Costs:* Capital and operating costs to support the network.

4 *Upward Compatibility of the Network:* Ability of the network to adapt to future changes in the environment.

    a. Compatibility with potential future industry networking standards.

    b. Compatibility with potential expanded network uses.

5 *Impacts on Organizational Structure:* The degree and type of changes in organizational structure required for successful network operation.

6 *Risks:* The potential for difficulties in maintaining or upgrading the network, and in maintaining data and software integrity and security.

The remainder of this section presents the definitions of the scales that were used to assess the performance of proposed networking strategies with respect to the various evaluation considerations given above. All of these evaluation measure scales are constructed scales except the one for cost. To keep the presentation compact, the scale definitions are presented in paragraph form. For example, the scale for Productivity Enhancement has four possible levels (scores): -1, 0, 1, and 2. The definition of level -1 is "User group productivity is diminished sufficiently that noticeably longer time or more resources are required to provide the same level of service." The definition for level 0 is "No perceivable change in user group productivity," and so on for the other two levels.

Here are the evaluation measure scales:

1 *Impact on Microcomputer Users*

Productivity Enhancement. −1) User group productivity is diminished sufficiently that noticeably longer time or more resources are required to provide the same level of service. 0) There is no perceivable change in user group productivity. 1) User group productivity is enhanced to the extent that group members are perceived by their clients to be providing better service, or somewhat fewer resources are required to provide service at the same level as before the network was installed. 2) There is a significant and easily perceived increase in user group productivity. Indicators of this could include significant reduction in staffing level required to carry out user group activities or considerable improvement in financial performance of the group.

User Satisfaction. −1) A significant number of user group members do not accept use of the network, or they feel the network is a detraction from their work environment. 0) There is no noticeable change in user group satisfaction with their microcomputer resources. 1) Many user group members believe the addition of the network has enhanced their work environment. 2) Virtually all user group members believe the addition of the network has moderately or significantly enhanced their work environment.

2 *Impact on Mainframe Capacity*

Impact on Mainframe Hardware or Software Necessitated or Made Possible by Addition of the Network. −1) The addition of the network causes a major increase in system load, necessitating an acceleration of the system capacity expansion schedule. 0) There is either no change or a noticeable but minor change in the system and mainframe load due to implementation of the network. 1) The addition of the network causes a major decrease in system load by supporting a significant amount of the computing load that would otherwise have been on the system.

3 *Costs*

Net Present Value of Capital and Operating Costs for the Network.

4 *Upward Compatibility of the Network*

Degree of Compatibility with Potential Emerging Networking Standards. 0) The specified network configuration appears unlikely to become or remain a widely accepted approach to networking microcomputers. 1) The configuration will likely be a widely accepted approach.

Degree of Compatibility with Potential Future Expanded Uses for the Network. 0) While the network meets current needs, it has little capability for expanding to meet other needs. 1) The network either currently has the capability of serving all needs currently projected for the network or can be expanded to serve these needs in a straightforward manner. 2) The network could be expanded in a straightforward manner to meet all potential expanded needs, including future uses that do not seem likely but are possible.

5 *Impacts on Organizational Structure*

Degree of Impact on Organizational Structure. −1) Significant potentially disruptive and detrimental changes in organizational structure are necessary or desirable to effectively utilize the specified network. 0) No changes in organizational structure are necessary or desirable to effectively utilize the specified network. 1) While some changes in organizational structure may be necessary or desirable to effectively utilize the specified network, these changes are of a minor nature or are generally beneficial, and can be made without disruptive dislocations.

6 *Risks*

Vendor Viability and the Continuing Availability of Support (Hardware and Software). 0) No continuing support is available, and the network becomes a "throwaway" investment. 1) Limited support for the network is available from a small number of vendors. 2) Full support is available, but no network enhancements are available. 3) Full support and network enhancements are readily available from the primary vendor and other industry sources.

Integrity and Security of System Data and Software. −2) The addition of the network causes a potentially serious decrease in system control and security for the use of data or software. −1) There is a noticeable but acceptable diminishing of system control and security. 0) There is no detectable change in system control or security. 1) System control or security is enhanced by addition of a network.

Planning for a Professional Services Firm

During the early 1970s, Woodward-Clyde Consultants, a professional services consulting firm, conducted an analysis of its objectives to aid in long-range planning. This analysis is presented in detail in Keeney and Raiffa (1976), Section 7.4, and two figures adapted from the analysis are presented here. Figure 2.5 shows the structure of the evaluation considerations developed during the original 1972 assessment by Woodward-Clyde Consultants management, and Figure 2.6 shows the evaluation measure scales used for the twelve lowest-tier evaluation considerations shown in Figure 2.5.

Frozen Blood

Keeney and Raiffa (1976), Section 7.7.2, present a value hierarchy for a decision by a hospital about whether to invest in expensive blood freezing equipment. An initial value hierarchy for this decision is shown in Figure 2.7.

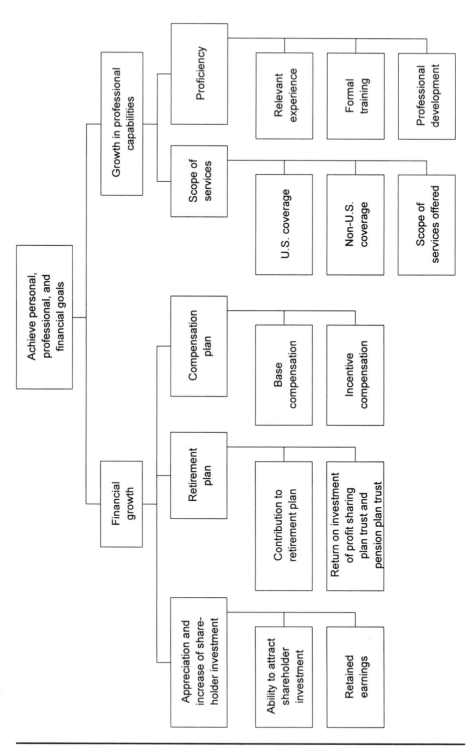

Figure 2.5 *Value hierarchy for Woodward-Clyde Consultants*

Evaluation Measure	Measurement Scale
Ability to attract shareholder investment	$\dfrac{\text{Number of shares requested}}{\text{Fees}}$
Retained earnings	Percent of fees
Contribution to retirement plan	Percent of fees
Return on investment for retirement plan	Percent of fees
Base compensation	Percent annual increase
Incentive compensation	Percent of fees
U.S. coverage	$\dfrac{\text{Geographic centers adequately covered}}{\text{Centers where relevant work can be generated}}$
Non-U.S. coverage	$\dfrac{\text{Geographic centers adequately covered}}{\text{Centers where relevant work can be generated}}$
Scope of services offered	$\dfrac{\text{Number of disciplines having threshold capability}}{\text{Number of synergistic disciplines required by society}}$
Relevant experience	$\dfrac{\text{Existing man-years experience}}{\text{Required man-years experience}}$
Formal training	Number of degrees per professional staff member
Professional development	Percent of fees

Figure 2.6 *Evaluation measures for Woodward-Clyde Consultants*

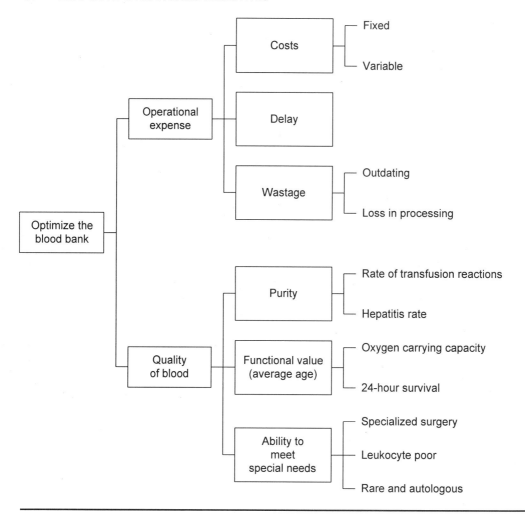

Figure 2.7   *Value hierarchy for frozen blood decision*

Job Selection

Figure 2.8 and Figure 2.9 present two value hierarchies for job selection. The one in Figure 2.8 is adapted from Keeney and Raiffa (1976), Section 7.7.4, and it was applied to a situation where a graduate student needed to evaluate employment offers immediately following graduation. The hierarchy in Figure 2.9 was developed by a midcareer professional considering a job change that might also involve a change of career. Note that in this case all the alternatives being considered were in the same general geographical area.

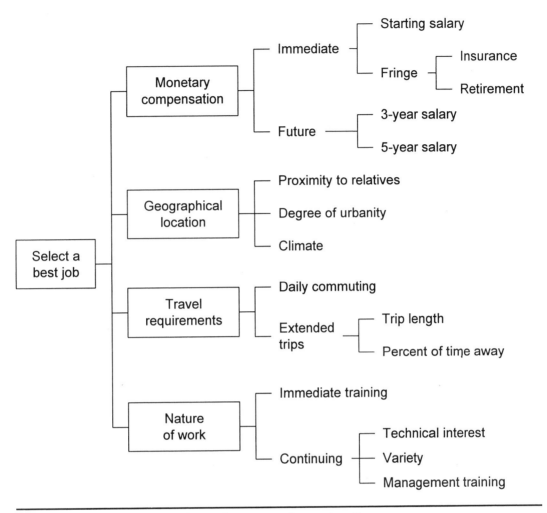

Figure 2.8   *Value hierarchy for evaluating employment options*

Nuclear Power Plant Siting

This section summarizes the value hierarchy used in a decision analysis to select a site for a nuclear power plant (Kirkwood 1982). The proposed sites to be evaluated were located in a semi-arid region, so that considerations related to water use were particularly significant. Because of the limited availability of water, sites were considered that did not have sufficient water locally available. Thus, it was necessary to consider impacts at both the plant site location and also the water source location. In addition, the impacts of an electrical transmission line carrying electrical power from the plant, as well as a water transmission pipeline between the water source and plant site, were of concern.

Figure 2.10 shows the evaluation considerations, objectives, and evaluation measures for this decision problem. The three environmental effects evaluation

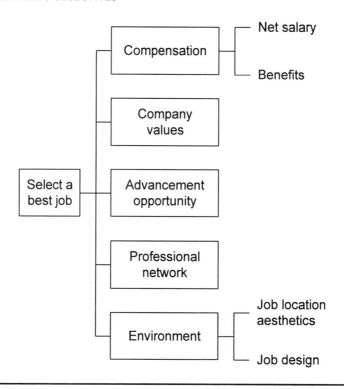

Figure 2.9   *Midcareer considerations for evaluation of jobs*

measures had constructed scales. The scale used to measure the ecological impacts at the plant site was discussed earlier, and it is shown in Figure 2.4. (Note that sites having "fatal flaws" with respect to ecological impacts were removed from consideration before the sites were evaluated in detail.) A constructed scale was used, which incorporated several different types of ecological impacts into a single scale. This was done because the relatively moderate degree of ecological impact at the various sites did not justify developing several different scales and doing the data collection needed to determine the levels for several scales.

As shown in Figure 2.10, the evaluation measure scale used to measure environmental impacts of electrical transmission and water supply pipeline corridors was "equivalent miles of corridor through unpopulated rangeland." From the standpoint of environmental impacts, unpopulated rangeland is the least important type of terrain, and thus having the electrical transmission line or water supply pipeline traversing a different type of terrain results in a more serious environmental impact. This was accounted for by multiplying the actual mileage of transmission line or pipeline through other types of terrain to arrive at a number of miles of unpopulated rangeland that was judged to have an equivalent environmental impact.

The multipliers that were used for electrical transmission line mileage are shown in Figure 2.11. Examine the entry for a multiplier of 5.0. This means, for example, that a transmission line with 8 miles of "route traversing pristine undisturbed areas of extraordinary scenic, recreational, historic, scientific, or

Evaluation Consideration	Objective	Evaluation Measure
System economics	Minimize overall project cost	Levelized system cost, in 1989 dollars
Health and safety	Minimize potential radiological exposure to nearby population	Site population factor
Environmental effects	Minimize ecological impacts at plant site	Thirteen-point constructed scale
	Minimize ecological impacts at water source	Three-point constructed scale
	Minimize environmental impacts of electrical transmission and water supply pipeline corridors	Equivalent miles of corridor through unpopulated rangeland
Socioeconomic effects	Minimize socioeconomic impacts at plant site	Maximum annual population growth rate, in percent
	Minimize short-term socioeconomic impacts due to water diversion	Annual crop value foregone due to withdrawal of irrigation, in 1989 dollars
	Minimize long-term socioeconomic impacts due to water diversion	Regional personal income potential foregone due to diversion of water, in levelized 1989 dollars
Land use	Minimize loss of irrigable land due to water diversion	Equivalent square miles of irrigable land preempted from agricultural use

Figure 2.10   *Evaluation considerations for nuclear power plant siting decision*

cultural value" was considered to have an equivalent environmental impact to a line traversing $8 \times 5.0 = 40.0$ miles of unpopulated rangeland. The total equivalent mileage through different types of terrain was summed to arrive at the overall score on this evaluation measure.

Integrated Circuit Tester

Keeney and Lilien (1987) discuss a study conducted for a firm considering introducing a new integrated circuit tester. The firm had developed a technical breakthrough that they felt would give them a significant cost advantage in manufacturing test equipment for very large scale integrated circuits. The firm was interested in how prospective customers would evaluate a potential entry in this highly competitive market, and how they should design their product. Figure 2.12 shows the evaluation considerations and evaluation measures used in that study.

Multiplier	Criteria
1.0	Route traversing unpopulated rangeland; not visible* from highways or high-use roadway. Route not affecting any known endangered species or important limited habitats; does not intrude on a "pristine," historic, or culturally significant area.
1.5	Route traversing sparsely populated area with no planned urban expansion and minimal scenic intrusion, not utilizing existing corridor. Route in area of low occurrence of known archaeologic and paleontologic resources.
2.0	Route traversing populated areas. Route traversing Bureau of Land Management or public lands, not utilizing an existing corridor. Route having aesthetic intrusion on primary highways and high-use roadways (parallel to and/or visible* from highway); or aesthetic intrusion on a national or state monument or park.
2.5	Route traversing state or federal forested lands, wildlife management, or critical habitat areas. Route traversing lands having high known archaeologic or paleontologic resource density.
5.0	Route traversing pristine undisturbed areas of extraordinary scenic, recreational, historic, scientific, or cultural value.
10.0	Route traversing state or national parks or monuments, military bases, or military research areas. Route traversing ecologically sensitive wetlands and migratory wildfowl refuges; or habitats containing unusual or unique communities, endangered species, or introduced game species.

* Visible determined as being within 3 miles of highway.

Figure 2.11 *Conversion multipliers for electrical transmission line mileage*

Evaluation Consideration	Evaluation Measure
Technical	
Pin capacity	Quantity
Vector depth	Memory size (megabits)
Data rate	MHz
Timing accuracy	Picoseconds
Pin capacitance	Picofarads
Programmable measurement units	Number
Economic	
Price	Total cost
Uptime	Percent
Delivery time	Months
Software	
Software translator	Percent conversion
Networking: communications	Yes/no
Networking: open	Yes/no
Development time	Mean time (months)
Data analysis software	Yes/no
Vendor support	
Vendor service	Time until system works (hours)
Vendor performance	Time until response (hours)
Customer applications	Yes/no

Figure 2.12   *Evaluation considerations for integrated circuit tester*

2.6 References

M. S. Bennedsen and C. W. Kirkwood, "Selecting Sites for Coal-fired Power Plants," *Journal of the Energy Division, Proceedings of the American Society of Civil Engineers,* Vol. 198, No. EY2, pp. 69–78 (1982).

D. G. Brooks and C. W. Kirkwood, "Decision Analysis to Select a Microcomputer Networking Strategy: A Procedure and a Case Study," *Journal of the Operational Research Society,* Vol. 39, pp. 23–32 (1988).

D. M. Buede, "Structuring Value Attributes," *Interfaces,* Vol. 16, No. 2, pp. 52–62 (March-April 1986).

J. L. Corner and C. W. Kirkwood, "Decision Analysis Applications in the Operations Research Literature, 1970–1989," *Operations Research,* Vol. 39, pp. 206–219 (1991).

K. Golabi, C. W. Kirkwood, and A. Sicherman, "Selecting a Portfolio of Solar Energy Projects Using Multiattribute Preference Theory," *Management Science,* Vol. 27, pp. 174–189 (1981).

D. H. Gustafson, W. L. Cats-Baril, and F. Alemi, *Systems to Support Health Policy Analysis: Theory, Models, and Uses,* Health Administration Press, Ann Arbor, Michigan, 1992.

R. L. Keeney, "Measurement Scales for Quantifying Attributes," *Behavioral Science,* Vol. 26, pp. 29–36 (1981).

R. L. Keeney, "Structuring Objectives for Problems of Public Interest," *Operations Research,* Vol. 36, pp. 396–405 (1988).

R. L. Keeney, *Value-Focused Thinking: A Path to Creative Decisionmaking,* Harvard University Press, Cambridge, Massachusetts, 1992.

R. L. Keeney and G. L. Lilien, "New Industrial Product Design and Evaluation Using Multiattribute Value Analysis," *Journal of Product Innovation Management,* Vol. 4, pp. 185–198 (1987).

R. L. Keeney and H. Raiffa, *Decisions with Multiple Objectives: Preferences and Value Tradeoffs,* Wiley, New York, 1976.

C. W. Kirkwood, "A Case History of Nuclear Power Plant Site Selection," *Journal of the Operational Research Society,* Vol. 33, pp. 353–363 (1982).

2.7 Review Questions

R2-1   Define the following: evaluation consideration, objective, goal, evaluation measure (attribute), level (score), and value structure.

R2-2   What special characteristics are required for a value structure to be a value hierarchy (value tree)?

R2-3   State and define the five desirable properties of a value hierarchy.

R2-4   What is the "test of importance"?

R2-5   Describe the two general approaches to structuring value hierarchies. Can these be combined?

R2-6   What are the two usual sources of information for developing value hierarchies?

R2-7   Describe four uses for value hierarchies.

R2-8   Compare natural versus constructed evaluation measure scales; direct versus proxy scales. What combinations of these different types of evaluation measure scales are possible?

R2-9   Discuss the significance of the clairvoyance test and the test of importance as these relate to evaluation measures.

2.8 Exercises

2.1   Summarize desirable properties for an individual evaluation measure and for a set of evaluation measures. Explain why each of these properties is desirable.

2.2   In accidents that result in deaths (for example, an airplane crash), the issue of placing a value on life often comes up when there are lawsuits. One possible way to value a life is to use expected lifetime earnings with a suitable discount rate. Other possible evaluation measures are current earnings or expected remaining years of lifetime. Suppose there is a fixed total settlement to be distributed among the heirs of all the accident victims. Briefly comment on how using each of these three different evaluation measures might change how this fixed total settlement is distributed among the heirs. Discuss why one of these evaluation measures (or another that you might propose) is preferable to the others. Based on your discussion, comment on the following statement: "The selection of evaluation measures can require significant value judgments."

2.3   Select one of the constructed evaluation measure scales presented in this chapter and critique it with respect to how well it meets the clairvoyance test.

2.4   **Project (Part B).** This is a continuation of the project that you started in Chapter 1. Develop evaluation considerations and evaluation measures for your decision. Describe the process used to determine these evaluation considerations and evaluation measures, including a discussion of other evaluation considerations that seem relevant and the reasons that they are not included. Present a value hierarchy, and completely describe the final set of evaluation measures. Thus, completely present any constructed evaluation measure scales.

CHAPTER 3

Developing Alternatives

This chapter addresses the specification of alternatives for a decision. Many people do not cast their nets wide enough when they consider alternatives for a perplexing decision. For example, suppose you are trying to figure out *where* to build that new plant to produce whiz-bangs. Maybe you should also be looking at *whether* to build it. There are other options: how about buying a plant that someone else is closing down, or closing down your mechanical clunker factory and converting it to whiz-bangs? Maybe you could lease a plant or license someone else to do production. Perhaps production could be done more cheaply offshore.

Failure to consider all the alternatives is encouraged by the technical complexity of many management decisions. You must rely on engineers and financial analysts for information that is important to the decision. However, these specialists sometimes analyze the options they know how to solve, rather than the preferable ones. This chapter discusses the reasons that we often have difficulty coming up with good alternatives, and then presents some procedures to help with this.

There is no substitute for a good alternative. The most complete analysis of decision alternatives can show you only the best of the identified alternatives. If none of the alternatives is very good, then the best alternative will be only the best of a poor lot. Following the procedures in this chapter will help you to identify better alternatives.

3.1 Why We Have Difficulty Identifying Good Alternatives

The root of our difficulty in identifying good alternatives seems to be in basic human reasoning processes. Dawes (1988, p. 68) notes that since John Locke (1632–1704) there has been increasing support for the view that thought is primarily an *associative process.* That is, we think about a new situation by making mental associations with previous situations that seem relevant, and, furthermore, these associations occur with relatively little conscious control on our part. Something "pops into our mind" that seems relevant, and we use this as a basis for structuring our consideration of the new situation.

Associative reasoning processes are efficient and effective in a stable environment with little uncertainty. As you gain more experience, you are able to quickly reach accurate conclusions about a wider variety of situations. ("There's no substitute for experience.") However, when you face a new situation that is outside your previous experience, associative reasoning can be dangerous. In such situations, the real-world associations between your current situation and previous situations you have faced are weak. Nevertheless, associations still form in your mind anyway, and you have limited conscious control of this process. A convincing "story" shapes itself in your mind. You remember previous situations that seem to be related to the current one, and you quickly recall alternatives that worked well in those previous situations. You "lock on" to those alternatives and fail to consider others that might be better.

You can demonstrate associative reasoning to yourself by considering, for example, the color blue. What pops into your mind? One summer day in Phoenix I was sitting with my wife at a restaurant table that had a blue tablecloth. This reminded me of my blue-eyed daughter, and I pictured her with her blond hair. This led to a thought of the yellow sun, which reminded me of how hot it is in Phoenix in the summer. From there my thoughts rapidly moved to considering a vacation in cool San Diego with its blue ocean. (Back to blue again!) I said to my wife, totally out of the blue as far as she was concerned, "We ought to consider spending a few days in San Diego."

All of this happened in a few seconds with no conscious effort or control. Associative reasoning processes work almost like magic. Sometimes these processes can generate amazingly creative thoughts, but often they quickly lock us into a premature conclusion that we have identified the best alternatives for our decision problem. The remainder of this chapter reviews ways to improve your reasoning about alternatives.

3.2 Too Many Alternatives

In some decisions we are faced with too many alternatives to thoroughly analyze, while in other situations we have too few alternatives. We will consider situations with too many alternatives in this section. Howard (1988) reports that his son drew his attention to an advertisement of a fast food hamburger chain that boasted that you could get 1,024 different hamburgers at their establishments. This claim came from observing that there were ten different possible ingredients, each of which could either be included or not. While it is true that this leads to $2^{10} = 1,024$ different possible hamburgers, some of these alternatives are not very interesting—what about the one that contains *none* of the ten ingredients (the "null burger")? Or the one that contains only lettuce?

In a more serious vein, Golabi, Kirkwood, and Sicherman (1981) developed a computer-based decision support system to aid in selecting proposals for funding from those submitted to the U.S. Department of Energy in response to a Request for Proposals to design solar energy demonstration projects. It was anticipated that about seventy-five proposals would be submitted, of which about six would be funded. It turns out that there are over 200 million different possible combinations of six proposals that can be formed from the seventy-five proposals. How do you analyze all these possibilities? (This decision is discussed further in Section 8.3.)

For an even more extreme example, consider the decision about where to locate a nuclear power plant analyzed by Kirkwood (1982). The possible sites included an entire state in the western United States. There were literally an infinite number of possible sites.

In some situations with a large number of alternatives, the primary difficulty is organizing the information about the alternatives, while in other situations the primary difficulty is that it is not possible to collect the required information about all the potential alternatives. In the hamburger example there is no difficulty figuring out the characteristics of each of the 1,024 possible alternatives. The difficulty is working through all the cases before you lose your appetite. Similarly, for the solar energy decision, there were only seventy-five different proposals that needed to be assessed, which was time consuming but feasible. The difficulty came in considering the 200 million ways these could be combined. In contrast, for the nuclear power plant site selection, it was not feasible to collect detailed data about every location in the state. In that type of situation, we need a way to identify a smaller number of alternatives for more detailed study.

Combinatorial Problems

Often, when data about the alternatives are available, but there are too many alternatives to analyze by routine methods, it is because the decision alternatives are *combinations* of simpler elements. In the hamburger example, the possible hamburgers are combinations of bun, meat, cheese, lettuce, etc. In the solar energy example, the alternatives are combinations of proposals, where the number of proposals that can be funded is constrained by the available budget, as well as programmatic considerations.

For such situations, the tools of *mathematical programming* (also called *optimization*) can be helpful for considering all of the many alternatives (Winston 1994). In the solar energy decision, a particular type of mathematical programming called *0-1 linear integer programming* was used to identify the most preferred combination of proposals. The various mathematical programming techniques are able to determine the best alternative without having to consider all possibilities (which would probably take too much time to be feasible). While the theory underlying this approach is complex, you do not need to understand this theory in order to use mathematical programming. Specialized software to do the required calculations is available for personal computers, and spreadsheet programs like Excel now also include the necessary capabilities. This approach is considered further in Chapter 8.

Issue	Consideration	Measure	Criterion
HEALTH/ SAFETY	Radiological release due to surface fault rupture	Distance to faults	Areas greater than 5 miles from faults
	Radiological release due to accident at nearby hazardous facility	Distance to airports	Areas greater than 5 miles from major airports
ENVIRONMENTAL IMPACT	Land use conflict	Location with respect to designated land use areas	Outside of designated land use areas larger than 1,000 acres
SYSTEM RELIABILITY AND COST	Rugged terrain	Slope	Areas with less than 10 percent slope

Figure 3.1   *Example of screening criteria for nuclear power plant sites*

Data Collection Problems

In the nuclear power plant site selection decision, it was not feasible to collect enough data about the entire state to evaluate the desirability of every spot in the state. Thus, some method was needed to reduce the number of sites under consideration until this number was small enough that data could be collected for the sites. In another decision where data cannot be obtained for all possible alternatives, McNamee and Celona (1990) consider a case where a brokerage firm owns a small share of the office complex where its offices are located. The majority owner of the complex offers current shareholders the opportunity to purchase an additional interest in the complex. The brokerage firm must consider this offer, as well as the possibility of moving out of the complex. The decision is made more complex by a variety of uncertainties, as well as the necessity to consider a range of possible investment levels. The decision must be made rapidly, and there is not time to collect and analyze data on all possible relevant alternatives.

For the nuclear power plant site selection, *screening criteria* were established to reduce the number of sites that needed to be considered. Figure 3.1 shows some of the criteria that were used. Only sites meeting all of the criteria (and there were many more criteria than those shown in this figure) were retained. The use of screening criteria involves an approximation. For example, a site that is 4.9 miles from a fault is not infinitely worse than a site that is 5.1 miles from a fault, but the first one will be removed from consideration, while the second will be retained. Thus, it is appropriate to select screening criteria that are relatively loose—that is, we anticipate that the most preferred alternative, once it is identified, will meet all the criteria with ease. For example, we anticipate that the best site will be much farther than 5 miles from a fault. If loose screening criteria are not used, then an alternative might be eliminated that is desirable

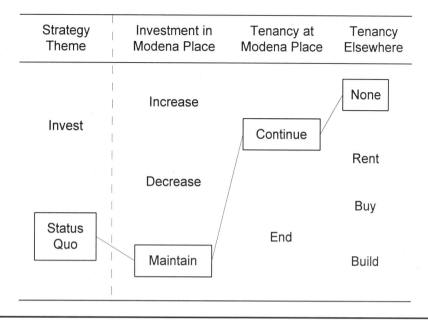

Strategy Theme	Investment in Modena Place	Tenancy at Modena Place	Tenancy Elsewhere
			None
Invest	Increase	Continue	Rent
	Decrease		Buy
Status Quo		End	
	Maintain		Build

Figure 3.2   *Example of a strategy generation table*

enough with respect to other criteria to overcome its poor rating on the screening criterion that eliminated it.

In the office complex investment decision, the *strategy-generation table* shown in Figure 3.2 (adapted from McNamee and Celona 1990, Figure 6-2) was helpful for identifying desirable alternatives to analyze in detail. In this figure, all of the columns except the first one (which is to the left of the dashed vertical line) represent different aspects of the decision alternatives. The name of the office complex is Modena Place, and Figure 3.2 shows columns to the right of the dashed vertical line for the level of investment in Modena Place, whether the brokerage firm should continue its tenancy at Modena Place, and what type of tenancy, if any, the firm should take at another location.

An alternative is constructed by selecting one entry from each of the columns to the right of the dashed vertical line. The column to the left of the dashed vertical line is used to give shorthand names to each strategic alternative that is selected for further analysis. For example, an alternative labeled "Status Quo" is marked on the figure. This alternative includes the entry in each column that is boxed, and the combination of entries is connected with lines. Different alternatives are indicated by different combinations of "investment in Modena Place," "tenancy at Modena Place," and "tenancy elsewhere."

Note that in this strategy generation table some of the columns show only a few of the total possible variations under the column. For example, the column "investment in Modena Place" shows only the three possibilities "increase," "decrease," and "maintain." In reality, there are many possible levels of investment for either the "increase" or "decrease" cases. A strategy generation table helps to sort out what types of alternatives make sense to analyze in more detail, but often doesn't specify exact alternatives. This is the reason that the leftmost column is labeled "strategy theme" rather than "alternative." Once the most promising strategy themes have been identified, then specific alternatives can be defined for each theme.

A strategy generation table can be useful when alternatives are made up of "bundles" of characteristics like the office complex investment example, and where a structured procedure is needed to sort out reasonable alternatives to study in more detail. By working through the various possibilities, it may be possible to quickly eliminate some.

A strategy generation table may be less useful in situations where many alternatives make sense, and it is not possible to quickly eliminate some. In the hamburger example discussed earlier, a strategy generation table could be constructed, but it would consist of ten columns to the right of the dashed vertical line (one for each possible component of a hamburger), and each of these columns would have only two entries: "include this component" and "don't include it." Similarly, for the solar energy decision, there would be seventy-five columns (one for each proposal), and once again each of these would only have "include" or "don't include" entries. In both the hamburger and solar energy decisions, the strategy generation table doesn't provide much insight, but for the office complex investment example it is more useful.

3.3 Too Few Alternatives

When there are too few alternatives—especially if none of these appears to be very good—the task is to develop additional alternatives that are better. Your associative reasoning processes can be both a help and a hindrance in this process. These reasoning processes can help you because they may generate some ideas that do not seem at first to be relevant (like the blue tablecloth to San Diego vacation example above) but which turn out to be useful. (I remembered about a vacation before the summer was over!) However, associative reasoning processes can also be a hindrance because they quickly build a "good story" about why you already have all the alternatives that are possible. Thus, we often have a tendency to "rush to judgment" and select an alternative before we have given careful consideration to other possibilities.

The basic issue is how to be creative and come up with additional good alternatives. There is a large literature on creativity, and we will make only a few general remarks here. Clemen (1996), Chapter 6, summarizes some of the creativity-enhancing methods that are most relevant for decision analysis. Keller and Ho (1988) also examine methods for developing alternatives. Several of their methods start either with existing alternatives or with a list of objectives for the decision, and attempt to develop new alternatives that are more attractive.

A strategy generation table can be useful for this purpose. If you develop the columns for a strategy generation table that represents your existing alternatives, then the table may suggest other entries within some of the columns that result in better alternatives. Or perhaps it will suggest other columns that ought to be in the table, and hence additional desirable characteristics for alternatives.

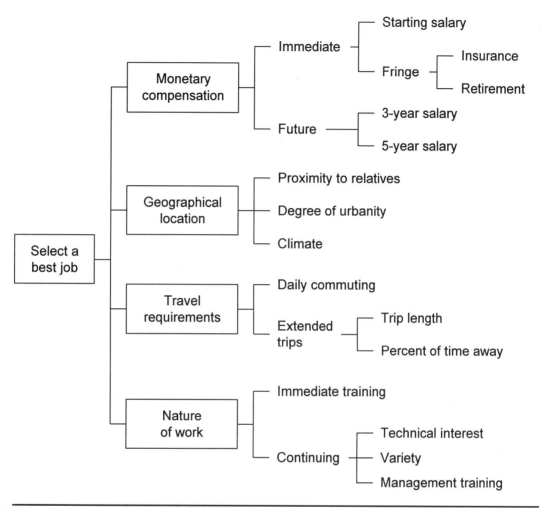

Figure 3.3   *Value hierarchy for evaluating employment alternatives*

More Alternatives from Your Objectives

As Chapter 2 discusses, a value hierarchy for a decision problem summarizes your objectives for the decision, and thus for an alternative to be good it must perform well with respect to at least some of the considerations in the value hierarchy. If you are having difficulty developing good alternatives, then you should develop a value hierarchy if you have not already done so. This may give you ideas for new alternatives. Keeney (1992), Chapters 7–8, focuses on developing alternatives from decision problem objectives.

Figure 3.3 repeats the value hierarchy for evaluating employment alternatives that was presented in Figure 2.8. You can use this to stimulate thinking about additional alternatives in several ways. First, consider the evaluation considerations at the lowest tier of the hierarchy one at a time. Try to develop alternatives that do as well as possible with respect to each evaluation consideration without taking the other evaluation considerations into account. While each of the alternatives that you develop may be too "one-dimensional" to be attractive by itself, there may be feasible alternatives that combine the strong points of the (possibly) extreme alternatives that you have developed by this procedure.

A second approach is to maximize a particular objective at a higher tier in the hierarchy. In Figure 3.3, examine the second tier of the hierarchy (monetary compensation/geographical location/travel requirements/nature of work). Work to develop alternatives that do well with respect to one of these. This is likely to result in alternatives that are somewhat more balanced than if you focus on only one lowest-tier objective.

Another approach is to directly concentrate on developing alternatives that provide balance across all the evaluation considerations.

3.4 Developing Alternatives When There Is Uncertainty

The methods presented above for developing alternatives are relevant whether or not there is uncertainty about the outcome of a selected alternative. However, there are some special issues that arise in developing alternatives for decisions under uncertainty. These are illustrated by the decision problem in Crawford, Huntzinger, and Kirkwood (1978). This decision addresses the selection of a transmission line size and tower configuration for a high voltage electrical transmission line. The basic tradeoff is between the capital cost to build the original transmission line and the operating cost of the line. By spending more on construction (higher capital cost), you can build a larger transmission line that has a lower operating cost because it will have lower electrical losses due to heating.

If there is no uncertainty, then determining the best alternative is a simple calculation—select a discount rate and find the alternative with the lowest net present value. However, when there is uncertainty, the situation is not so simple. In the transmission line decision, the projected operating cost is a function of the cost of the oil that will be used to generate the electricity being transported over the line. At the time the analysis was done, there was substantial uncertainty about the cost of oil over the projected fifty-year lifetime of the transmission line.

In this situation, some alternatives that are not relevant under certainty can make sense, and these should be considered as possibilities. You might want to *hedge* against the uncertainty by building a medium-size transmission line. This might not be the cheapest alternative with either low or high future oil prices, but it will not do too badly in either case. By selecting this hedging alternative, you ensure that you will not do very poorly regardless of what happens. Of course, you also give up the opportunity of doing very well if you happen to guess right about the future.

It may also make sense to select an alternative that allows you to *sequence* your decisions. In the conductor selection problem, an alternative was investigated where large towers would be built but a small transmission line would be installed. Thus, if the price of oil was later high, it would be possible to install larger transmission lines without having to build new towers. In this case, as in many other sequential decisions, you would have to spend some additional money up front (to build larger towers) to preserve the option of responding when the uncertainty is later resolved.

Another possible approach is to *risk share* by taking a partner. When you do this, you lose the opportunity to make all the profit if the uncertainty turns out well, but you also share the loss if things don't turn out so well. Finally, there may be the opportunity to take *insurance* against the risks posed by the uncertainties. While this was not considered in the transmission line decision, it might have been possible to find someone willing to build the line and rent capacity on it to the electric utility for a fixed cost. Usually when you buy insurance, you do this because you do not want to assume the risk for an event that is unlikely but that could leave you in dire straits if it occurred. (For example, young breadearners often take out life insurance. They don't expect to die, but if they do, their family could be left in serious trouble.)

Hedging, sequencing, risk sharing, and insuring are all approaches to consider in decisions where there is uncertainty. Sometimes one of these approaches can substantially reduce the risk of an undesirable outcome while still maintaining the opportunity to take advantage of attractive outcomes.

3.5 References

R. T. Clemen, *Making Hard Decisions: An Introduction to Decision Analysis,* Second Edition, Duxbury Press, Belmont, California, 1996

D. M. Crawford, B. C. Huntzinger, and C. W. Kirkwood, "Multiobjective Decision Analysis for Transmission Conductor Selection," *Management Science,* Vol. 24, pp. 1700–1709 (1978).

R. M. Dawes, *Rational Choice in an Uncertain World,* Harcourt Brace Jovanovich, San Diego, 1988.

K. Golabi, C. W. Kirkwood, and A. Sicherman, "Selecting a Portfolio of Solar Energy Projects Using Multiattribute Preference Theory," *Management Science,* Vol. 27, pp. 174–189 (1981).

R. A. Howard, "Decision Analysis: Practice and Promise," *Management Science,* Vol. 34, pp. 679–695 (1988).

R. L. Keeney, *Value-Focused Thinking: A Path to Creative Decisionmaking,* Harvard University Press, Cambridge, Massachusetts, 1992.

R. L. Keeney and H. Raiffa, *Decisions with Multiple Objectives: Preferences and Value Tradeoffs,* Wiley, New York, 1976.

L. R. Keller and J. L. Ho, "Decision Problem Structuring: Generating Options," *IEEE Transactions on Systems, Man, and Cybernetics,* Vol. 18, pp. 715–728 (1988).

C. W. Kirkwood, "A Case History of Nuclear Power Plant Site Selection," *Journal of the Operational Research Society*, Vol. 33, pp. 353–363 (1982).

P. McNamee and J. Celona, *Decision Analysis with Supertree, Second Edition*, The Scientific Press, South San Francisco, California, 1990.

W. L. Winston, *Operations Research: Applications and Algorithms, Third Edition*, PWS-Kent, Boston, 1994.

3.6 Review Questions

R3-1   Describe the significance of associative reasoning processes for the development of decision alternatives.

R3-2   Describe two procedures that are used to reduce the number of alternatives considered in a decision analysis. Give advantages and disadvantages of each procedure.

R3-3   Describe two approaches to developing more alternatives in a decision analysis.

3.7 Exercises

3.1   Consider the three concepts of *ends objectives*, *means objectives*, and *alternatives*. Discuss the relationship among these concepts in a decision problem. Specifically, (i) define each of the concepts, including an example of each, (ii) review how each of these is different from the others, and (iii) discuss why all three concepts are important in systematically analyzing a decision.

3.2   We have discussed the use of a hierarchical ("tree") structure for organizing objectives and evaluation measures, and also an associative structure as a model for how people's reasoning processes work. Explain how an associative structure differs from a hierarchy. As part of your explanation, include a definition of each. Examples may also help to clarify your presentation.

3.3   Assume that human reasoning processes work by association. Discuss one reason that this might make it difficult for people to develop innovative alternatives for a decision problem.

3.4   **Project (Part C).** This is a continuation of the project from Chapters 1 and 2. Specify alternatives for your project. Describe the process used to determine alternatives, including a discussion of other alternatives that seem relevant and the reasons these are not included. If appropriate, include a strategy generation table. If this is not appropriate, discuss why it is not appropriate. Describe the final set of alternatives to be used in your decision analysis.

CHAPTER 4

Multiobjective Value Analysis

This chapter presents a multiobjective value analysis procedure to rank alternatives and select the most preferred alternative. This procedure is appropriate when there are multiple, conflicting objectives and no uncertainty about the outcome of each alternative. The presentation assumes that evaluation measures and decision alternatives have been specified as discussed in Chapters 2 and 3, and also that the evaluation measure scores (levels) have been determined for each alternative.

To conduct a multiobjective value analysis, it is necessary to determine a *value function*, which combines the multiple evaluation measures into a single measure of the overall value of each evaluation alternative. The form of this function that is used here is a weighted sum of functions over each individual evaluation measure. Thus, determining a value function requires that

a *Single dimensional value functions* be specified for each evaluation measure. (Single dimensional value functions are also called *single attribute value functions*.)

b *Weights* be specified for each single dimensional value function. (Weights are also called *scaling constants* or *swing weights*.)

The multiobjective value analysis procedure will be demonstrated with an example based on an actual application of the method to the selection of a strategy for networking personal computers within an organization. (An expanded version of this application is included as a case study in Appendix A.) The calculations needed for a multiobjective value analysis can be carried out using a spreadsheet, and the process for doing this is presented.

The procedure presented in this chapter is based on the Simple Multiattribute Rating Technique using Swings (SMARTS). Further discussion of the SMARTS procedure is presented in Edwards and Barron (1994). See also Edwards (1977), Edwards and Newman (1986), Edwards, von Winterfeldt, and Moody (1988), and von Winterfeldt and Edwards (1986, Chapter 8). This method of analysis is based on measurable value theory, which is presented in Dyer and Sarin (1979) and reviewed in Chapter 9.

4.1 An Example: Selecting a Networking Strategy

The example discussed here is simplified from a networking strategy decision presented in Brooks and Kirkwood (1988). A company was deciding what strategy to follow with respect to networking its personal computers. (In the original application by Brooks and Kirkwood, the personal computers were not connected together at all, but a similar approach could be used in a situation where an organization is considering upgrading an existing network.) The company believed that the productivity of its design engineers might potentially be enhanced by networking its computers. However, the design work conducted by those engineers was both complex and proprietary, and the company was concerned about the security and integrity of this work if the computers were networked. In addition, the cost of the network was of concern.

For the purposes of this chapter, assume that the company has selected three evaluation measures: Productivity Enhancement, Cost Increase, and Security. Cost Increase is the net present value of the increased cost for an alternative relative to the current situation (measured in thousands of dollars), while Productivity Enhancement and Security each have constructed scales. Specifically, Productivity Enhancement has the following four-point scale:

−1 User group productivity is diminished sufficiently that noticeably longer time or more resources are required to provide the same level of service.

0 No change in user group productivity is perceived.

1 User group productivity is enhanced to the extent that group members are perceived by their clients to be providing better service, or somewhat fewer resources are required to provide service at the same level as before the network was installed.

2 There is significant and easily perceived increase in user group productivity. Indicators of this could include a significant reduction in the staffing level required to carry out user group activities or considerable improvement in the financial performance of the group.

Security has the following four-point scale:

−2 The addition of the network causes a potentially serious decrease in system control and security for the use of data or software.

−1 There is a noticeable but acceptable diminishing of system control and security.

0 There is no detectable change in system control or security.

1 System control or security is enhanced by addition of a network.

Note that the scales are defined so that the *status quo* (that is, not adding a network) has a score of zero on each scale.

Assume that four alternative strategies are under consideration: Status Quo, High Quality/High Cost, Medium Quality/Medium Cost, and Low Quality/Low

	Evaluation Measure		
Alternative	Productivity Enhancement (X_p)	Cost Increase (X_c)	Security (X_s)
Status Quo	0	0	0
High Quality/High Cost	2	125	0.5
Medium Quality/Medium Cost	1	95	0
Low Quality/Low Cost	0.5	65	-1

Table 4.1 *Evaluation measure scores for networking alternatives*

Cost. The scores of the four alternatives with respect to each of the three evaluation measures are shown in Table 4.1.

The remainder of this chapter reviews a procedure to obtain the value function over the three evaluation measures and to rank the four alternatives using this value function. During this presentation, it is necessary to keep straight the distinction between (1) the particular level or score that an alternative receives on an evaluation measure and (2) the rating or value that is attached to having that score. We will use the terms *score* or *level* to refer to the number that an alternative receives on a particular evaluation measure, while we will use the terms *rating* or *value* to refer to the number that is used to show the "goodness" that is assigned to receiving a particular score on an evaluation measure. For example, in the table above, the Status Quo alternative receives scores of zero on all three of the evaluation measures. However, the rating or value that this alternative receives on the three evaluation measures will differ depending on the particular rating procedure that is used. Several different procedures are discussed below, and then a preferred procedure is presented.

Note that higher scores (levels) on an evaluation measure are not necessarily better. For the Productivity Enhancement and Security evaluation measures, higher scores are more preferred, but for the Cost Increase evaluation measure, a higher score is less preferred. However, values (ratings) are defined so that higher-value numbers are always more preferred.

4.2 The Multiobjective Value Function

Table 4.1 shows that there is no clear "winner" among the four alternatives. This is because the alternatives that are cheaper (and hence more desirable with respect to Cost Increase) have worse performance with respect to Productivity Enhancement and/or Security than the more expensive alternatives. Thus, *tradeoffs* among the three evaluation measures must be considered to determine which alternative is most preferred. That is, the four alternatives can be ranked only if some procedure is used to combine the three evaluation measures into a single index of the overall desirability of an alternative. This section discusses the issues that need to be addressed by a procedure that combines the evaluation measures. This is done by examining some intuitively reasonable combination procedures and showing the difficulties with these procedures. Finally, a procedure is presented that overcomes these difficulties.

The discussion below gives intuitive arguments for the particular procedure that is recommended. A more detailed presentation of the theory underlying this procedure is given in Section 9.2. It is not necessary to study that theory in order to apply the procedure.

Using Simple Averaging

A simple procedure to combine the evaluation measures is to take an average of the scores on each evaluation measure. Some thought shows that to use this procedure, it is necessary to make allowances for whether it is better to have more of a particular evaluation measure or to have less. In our example, it is better to have higher scores for Productivity Enhancement and Security, but it is better to have lower scores for Cost Increase. To account for this, we will *subtract* the scores of the Cost Increase evaluation measure, but we will *add* the scores of the other two evaluation measures. Using this procedure, the ratings for the four alternatives are

> Status Quo: $(0 - 0 + 0)/3 = 0$
> High Quality/High Cost: $(2 - 125 + 0.5)/3 = -40.83$
> Medium Quality/Medium Cost: $(1 - 95 + 0)/3 = -31.33$
> Low Quality/Low Cost: $(0.5 - 65 - 1)/3 = -21.83$

Thus, with this procedure, the Status Quo alternative is most preferred, the Low Quality/Low Cost alternative is second most preferred, and the Medium Quality/Medium Cost and High Quality/High Cost alternatives are successively less preferred.

The Problem of Units for Evaluation Measures

While simple averaging is indeed simple, a little thought shows that it has a flaw. Specifically, the relative ratings of the alternatives can change depending on what *units* are used for an evaluation measure. For example, suppose that rather than measuring Cost Increase in thousands of dollars, it is measured in millions of dollars. Then the ratings for the four alternatives become

> Status Quo: $(0 - 0 + 0)/3 = 0$
> High Quality/High Cost: $(2 - 0.125 + 0.5)/3 = 0.79$
> Medium Quality/Medium Cost: $(1 - 0.095 + 0)/3 = 0.30$
> Low Quality/Low Cost: $(0.5 - 0.065 - 1)/3 = -0.19$

Thus, with this change in units, the High Quality/High Cost alternative becomes the most highly rated alternative. However, there has not been a change in the alternative—only the measuring units have changed! This is like changing from measuring temperature in Fahrenheit degrees to Celsius degrees. When this change is made, the physical temperature doesn't change. In the same way, the actual characteristics of the alternatives do not change when the units are changed for Cost Increase. Thus, it is not reasonable for the relative ratings of the alternatives to change merely because the units have changed for one of the evaluation measures.

This difficulty can be addressed by *normalizing* the ratings on each evaluation measure scale. This is done by using as the measure of performance for each evaluation measure, not the actual evaluation measure score that each alternative receives, but rather the *proportion of the way along the allowed range* of that evaluation measure scale where the score for the alternative lies. With this procedure, the ratings become

$$\text{Status Quo: } \frac{1}{3} \times \left\{ \frac{0 - (-1)}{2 - (-1)} + \frac{125 - 0}{125 - 0} + \frac{0 - (-2)}{1 - (-2)} \right\} = 0.67$$

$$\text{High Quality/High Cost: } \frac{1}{3} \times \left\{ \frac{2 - (-1)}{2 - (-1)} + \frac{(125 - 125)}{125 - 0} + \frac{0.5 - (-2)}{1 - (-2)} \right\} = 0.61$$

$$\text{Medium Quality/Medium Cost: } \frac{1}{3} \times \left\{ \frac{1 - (-1)}{2 - (-1)} + \frac{125 - 95}{125 - 0} + \frac{0 - (-2)}{1 - (-2)} \right\} = 0.52$$

$$\text{Low Quality/Low Cost: } \frac{1}{3} \times \left\{ \frac{0.5 - (-1)}{2 - (-1)} + \frac{125 - 65}{125 - 0} + \frac{-1 - (-2)}{1 - (-2)} \right\} = 0.44$$

This set of ratings requires some study to understand. Take the Status Quo alternative as an example. For the Productivity Enhancement evaluation measure, the total range for this evaluation measure is from -1 to 2. Thus, the length of this range is $2 - (-1) = 3$. The score for the Status Quo alternative on this evaluation measure is 0, which is $0 - (-1) = 1$ unit along the scale from the least preferred end of the scale. Therefore, as a proportion of the total range for this scale, the rating of the Status Quo alternative on the Productivity Enhancement evaluation measure is $[0 - (-1)]/[2 - (-1)] = 0.33$. The same procedure shows that on the Security evaluation measure, the rating for the Status Quo alternative is $[0 - (-2)]/[1 - (-2)] = 0.67$.

Thus, in general, for an evaluation measure where higher scores are more preferred, use the following formula to calculate the rating:

$$\text{Rating} = \frac{\text{Score} - \text{Lowest Level}}{\text{Highest Level} - \text{Lowest Level}}$$

The situation is different in two ways for the Cost Increase evaluation measure. First, there is no range of values given for this scale, and second, higher costs are less preferred rather than more preferred as was true with the other two evaluation measures. For the moment, let's use the range of scores for Cost Increase that actually occurs among the four alternatives (from 0 to 125 thousand dollars) as the range for this evaluation measure. Then the length of the range is $125 - 0 = 125$. A little thought shows that the preference decrease for higher Cost Increase scores can be accounted for by using the proportion of the distance that an alternative score is from the *highest* end of the Cost Increase scale, rather than the *lowest* end as was done with the other two evaluation measures. Thus, the rating for the Status Quo alternative on the Cost Increase evaluation measure is $(125 - 0)/(125 - 0) = 1$. In general, for an evaluation measure where higher scores are less preferred, use the following formula to calculate the rating:

$$\text{Rating} = \frac{\text{Highest Level} - \text{Score}}{\text{Highest Level} - \text{Lowest Level}}$$

This procedure of using the proportion of the total evaluation measure range as a rating for the evaluation measure corrects the problem with ratings changing if the units for a particular evaluation measure are changed. Ratings will no longer change when evaluation measure units change because both the numerator and denominator for a particular evaluation measure rating will be multiplied by the same factor when units are changed, and hence the ratings will stay the same.

However, the issue of exactly what range of values should be used for each evaluation measure is still troublesome. In the ratings above, this issue came up for the Cost Increase evaluation measure. It seemed natural to use the range of values that show up in the actual scores for the four decision alternatives (that is, from 0 to 125). However, this is not consistent with how the Security evaluation measure was treated. The scores on that evaluation measure that show up in the four decision alternatives are from -1 to 0.5, but the total specified range of scores for that scale (from -2 to 1) was used in the evaluation procedure. If the Security evaluation measure had been treated consistently with the Cost Increase measure, a range from -1 to 0.5 would have been used, rather than from -2 to 1. The solution to this problem is now discussed.

The Problem of Ranges for Evaluation Measures

Some experimentation with different ranges will quickly show that it is possible to change the rankings of the alternatives by changing the range that is used for each evaluation measure. Thus, while changing the *units* that are used for each evaluation measure no longer affects the ratings, changing the *range* of scores that is used for each evaluation measure can affect the rankings. This does not seem reasonable. Using the temperature measuring analogy that was discussed above, this would be like having the physical temperature change if the range of temperatures covered by a thermometer changes!

Even without this difficulty, there is another problem with the procedure given above, which is that it treats variations over the specified range of each evaluation measure as having equal importance. To see this, consider two new alternatives, which each have a value of 1 on the Security evaluation measure, but which differ on the Productivity Enhancement and Cost Increase evaluation measures. Specifically, assume that Alternative 1 has a Productivity Enhancement score of 2 and a Cost Increase of 125, while Alternative 2 has scores of -1 and 0, respectively, for these evaluation measures. That is, Alternative 1 has the best possible score on Productivity Enhancement and the worst possible score on Cost Increase, while Alternative 2 has the worst possible score on Productivity Enhancement and the best possible score on Cost Increase. The overall ratings for these two alternatives using the procedure discussed above are

$$\text{Alternative 1: } \frac{1}{3} \times \left\{ \frac{2-(-1)}{2-(-1)} + \frac{125-125}{125-0} + \frac{1-(-2)}{1-(-2)} \right\} = 0.67$$

$$\text{Alternative 2: } \frac{1}{3} \times \left\{ \frac{-1-(-1)}{2-(-1)} + \frac{125-0}{125-0} + \frac{1-(-2)}{1-(-2)} \right\} = 0.67$$

Thus, these two alternatives are equally preferred. Is this reasonable? Perhaps, but it implicitly assumes that equal importance is attached to improving Productivity Enhancement from its worst to best score as is attached to improving Cost Increase from its worst to its best score. It is easy to visualize many situations when this would not be reasonable.

Thus, there are two problems with the evaluation procedure presented above:

1 The evaluation results depend on the range of variation that is specified for each evaluation measure, and
2 The procedure assumes that variations on each evaluation measure from the worst to the best specified score are of equal importance.

Using Weights for Evaluation Measures

The solution to both these problems is to allow different *weights* for each evaluation measure. In the procedure given above, each term in the sum of evaluation measure ratings is being multiplied by 1/3. If this multiplying factor, or weight, is allowed to be different for different evaluation measures, then it is possible to account for both (1) changes in the range of variation for each evaluation measure and (2) different degrees of importance being attached to these ranges of variation.

As an example, suppose that for the ranges of variation given above for each evaluation measure, a weight of 0.5 is attached to Productivity Enhancement, a weight of 0.3 is attached to Cost Increase, and a weight of 0.2 is attached to Security. Then the ratings for Alternatives 1 and 2 discussed above become

$$\text{Alternative 1: } 0.5 \times \frac{2-(-1)}{2-(-1)} + 0.3 \times \frac{125-125}{125-0} + 0.2 \times \frac{1-(-2)}{1-(-2)} = 0.70$$

$$\text{Alternative 2: } 0.5 \times \frac{-1-(-1)}{2-(-1)} + 0.3 \times \frac{125-0}{125-0} + 0.2 \times \frac{1-(-2)}{1-(-2)} = 0.50$$

Thus, for this set of weights, the desirable Productivity Enhancement score for Alternative 1 outweighs the desirable Cost Increase score for Alternative 2, and hence Alternative 1 is more highly rated overall than Alternative 2. That is, variations over the range for Productivity Enhancement are *more important* than variations over the range for Cost Increase when using this set of weights.

The use of weights solves the problems associated with the somewhat arbitrary range of variation specified for each evaluation measure, but it has introduced another problem—determining the weights. Clearly, there is some subjectivity to this. Different individuals or organizations might have different weights for Productivity Enhancement versus Cost Increase. We will address this issue below after first considering one final difficulty in combining evaluation measure scores.

Analyzing Returns to Scale Effects

The final issue that needs to be addressed in combining evaluation measure scores is more subtle than the weighting issue. This issue is called "returns to scale." The difficulty can be demonstrated by examining the definition of the Security evaluation measure scale that was given earlier:

−2 The addition of the network causes a potentially serious decrease in system control and security for the use of data or software.

−1 There is a noticeable but acceptable diminishing of system control and security.

0 There is no detectable change in system control or security.

1 System control or security is enhanced by addition of a network.

Intuitively, a score of −2 seems to imply a significant potential for some security problem, while a score of −1 doesn't seem very serious. Thus, it seems that the increase in value that results when a score of −2 is improved to a score of −1 is greater than the increase in value that results when a score of −1 is improved to a score of 0. When the increase in value that results from moving to each successively more preferable score on the evaluation measure scale is less, then it is said that there are "decreasing returns to scale" for the evaluation measure. While this is the most common situation, there are also cases where each successive unit of increase in preferability has a higher value. This case is referred to as "increasing returns to scale." It is useful to have the capability of including these returns-to-scale effects in a multiobjective value analysis. This could be done by putting a *function* over each evaluation measure that accounts for the returns to scale before combining the evaluation measure scores. Such a function is called a *single dimensional value function* or a *single attribute value function.*

While this capability seems useful, it introduces another complexity—the necessity of determining the single dimensional value function over each evaluation measure. This is addressed below after considering how the resulting single dimensional value functions are combined to obtain an overall rating or value for each alternative.

The Value Function

With the weights and single dimensional value functions, the final value function becomes

$$v(X_p, X_c, X_s) = w_p v_p(X_p) + w_c v_c(X_c) + w_s v_s(X_s)$$

where X_p, X_c, and X_s are the evaluation measures for Productivity Enhancement, Cost Increase, and Security, respectively; w_p, w_c, and w_s are the weights on the three evaluation measures; and $v_p(X_p)$, $v_c(X_c)$, and $v_s(X_s)$ are the single dimensional value functions over each of the three evaluation measures.

Typically, the weights are determined so that they sum to one, and hence $1 = w_p + w_c + w_s$. Furthermore, it is usual to have each single dimensional value function vary between 0 and 1 over the range of scores that is of interest. With these conventions, the least preferred score being considered for a particular evaluation measure will have a single dimensional value of zero, and the most preferred score will have a single dimensional value of one. Furthermore, an alternative that has the least preferred scores for all of the evaluation measures will have an overall value of zero, and an alternative that has the most preferred scores for all of the evaluation measures will have an overall value of one.

4.3 Determining the Single Dimensional Value Functions

In this section, procedures are presented for determining single dimensional value functions. Two different procedures are presented that are often used in practical applications. One of these procedures results in a single dimensional value function that is made up of segments of straight lines that are joined together into a *piecewise linear* function, while the other procedure uses a specific mathematical form called the *exponential* for the single dimensional value function.

In practical applications, it is often possible to use either of these forms of single dimensional value functions for a particular evaluation measure. While the use of one form versus the other may result in a somewhat different specific shape for the value function, the difference is often not of practical significance. However, it is generally most natural to use a piecewise linear single dimensional value function when the evaluation measure being considered has a small number of possible different scoring levels, as is the case for Productivity Enhancement (X_p) and Security (X_s) in the networking strategy. Thus, the procedure to determine a piecewise linear function will be illustrated by considering the two evaluation measures X_p and X_s.

Piecewise Linear Single Dimensional Value Functions

The procedure for determining a piecewise linear single dimensional value function requires that the relative value increments be specified between each of the possible evaluation measure scores. This information is then used to specify the value function.

This procedure can be demonstrated using the Productivity Enhancement evaluation measure. Suppose that the value increment between $X_p = 0$ and $X_p = 1$ is the smallest increment between any of the two neighboring scores on the Productivity Enhancement scale, and that this value increment is the same as that between $X_p = 1$ and $X_p = 2$. Furthermore, the value increment between $X_p = -1$ and $X_p = 0$ is greater than that between $X_p = 0$ and $X_p = 1$. Specifically, this value increment is twice as great as that between $X_p = 0$ and $X_p = 1$.

To determine the single dimensional value function over the Productivity Enhancement evaluation measure, it is necessary to find values for each of the different levels of that evaluation measure that result in the value increments given in the preceding paragraph. This can be done using the fact that the total value increment between the lowest possible level of Productivity Enhancement (which is -1) and the highest possible level (which is 2) is 1. The fact that this value increment is the sum of the increments going from -1 to 0, 0 to 1, and 1 to 2 can be used to determine the necessary values with a small amount of algebra. To do this, use the symbol x to represent the smallest value increment (that is, the increment going from 0 to 1). Then the increment going from 1 to 2 is also x, and the increment going from -1 to 0 is $2x$. Thus the sum of all the value increments is $2x + x + x$. Hence, $2x + x + x = 1$, or $4x = 1$. Thus, $x = 1/4 = 0.25$.

It follows from this that the increment in value going from 0 to 1 is 0.25, since this increment is equal to x. The increment in value going from 1 to 2 is also 0.25, since this increment is also equal to x. Finally, the increment in value going from -1 to 0 is 0.50, since this increment is equal to $2x$. To obtain the values for each of the defined levels of the Productivity Enhancement evaluation measure, add up the value increments between the lowest possible level and the level of interest. Thus, the values are

$$v_p(-1) = 0.00 \text{ (since this is the least preferred level)}$$
$$v_p(0) = 0.00 + 2x = 0.00 + 2 \times 0.25 = 0.50$$
$$v_p(1) = 0.00 + 2x + x = 0.00 + 2 \times 0.25 + 0.25 = 0.75$$
$$v_p(2) = 0.00 + 2x + x + x = 0.00 + 2 \times 0.25 + 0.25 + 0.25 = 1.00$$

Note that this procedure has a built-in check on whether the arithmetic has been done correctly, since the value for the highest level must be 1.

A similar procedure works for the Security evaluation measure. Suppose that the value increment between $X_s = 0$ and $X_s = 1$ is the smallest, followed by that between $X_s = -1$ and $X_s = 0$, and then that between $X_s = -2$ and $X_s = -1$. Specifically, suppose the value increment between $X_s = -1$ and $X_s = 0$ is twice as large as that between $X_s = 0$ and $X_s = 1$, and the value increment

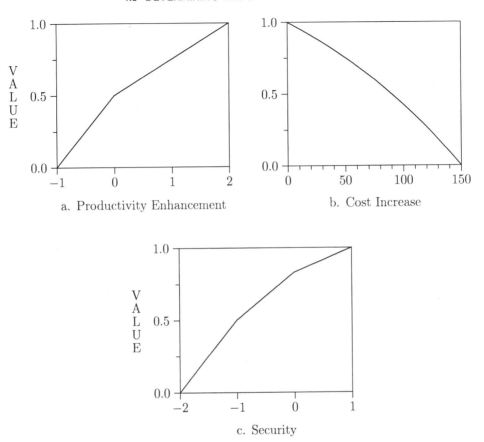

a. Productivity Enhancement   b. Cost Increase

c. Security

Figure 4.1   *Single dimensional value functions for networking decision*

between $X_s = -2$ and $X_s = -1$ is three times as large as that between $X_s = 0$ and $X_s = 1$. Then, since the value increments must sum to 1, it is true that $3v + 2v + v = 6v = 1$, where v represents the value increment between $X_s = 0$ and $X_s = 1$. Hence, $v = 1/6$, and thus

$$v_s(-2) = 0.00$$
$$v_s(-1) = 0.00 + 3v = 0.00 + 3 \times (1/6) = 0.50$$
$$v_s(0) = 0.00 + 3v + 2v = 0.00 + 3 \times (1/6) + 2 \times (1/6) = 0.83$$
$$v_s(1) = 0.00 + 3v + 2v + v = 0.00 + 3 \times (1/6) + 2 \times (1/6) + (1/6) = 1.00$$

The single dimensional value functions that were just determined for Productivity Enhancement and Security are shown in Figure 4.1a and Figure 4.1c. In these figures, straight lines are drawn between each of the single dimensional values that we just determined above, and the resulting functions are called *piecewise linear single dimensional value functions*. The straight lines are drawn so that values can be estimated for levels of the evaluation measure that lie between the levels for which values were found. It is sometimes useful to determine such intermediate values in order to make fine distinctions among the alternatives. For example, Table 4.1 shows that the High Quality/High Cost alternative has a score of 0.5 for the Security evaluation measure, and the Low Quality/Low Cost alternative has a score of 0.5 for the Productivity Enhancement evaluation measure. These levels are between the defined scores on those evaluation measures, and hence show that the alternatives have characteristics that lie between those of the defined scores.

To summarize, here are the steps to determine a piecewise linear single dimensional value function:

1 For an evaluation measure where higher scores are more preferred, consider the increments in value that result from each successive increase in the score (level) of the evaluation measure, and place these increments in order of successively increasing value increments.[1] For an evaluation measure where higher scores on the measure are less preferred, consider the increase in value for each successive *decrease* in the score on the evaluation measure, and place these in order of successively increasing value increments.

2 Quantitatively scale the value increments as multiples of the smallest value increment.

3 Set the smallest value increment so that the total of all the increments is 1.

4 Use the result of Step 3 to determine the single dimensional value for each possible score of the evaluation measure as illustrated by the examples presented above from the networking decision.

Exponential Single Dimensional Value Functions

The procedure just presented for determining the single dimensional value functions over the Productivity Enhancement and Security evaluation measures does not work for the Cost Increase evaluation measure. This is because Cost Increase can take on essentially an infinite number of different levels. Thus, it would be necessary to find a very large number of value increments to determine a piecewise linear single dimensional value function.

An approximate piecewise linear function can be determined for Cost Increase by selecting a small number of equally spaced scores that cover the range of possible scores, and then applying the procedure covered above to determine a piecewise linear function through this small number of scores. For example, Cost Increase scores of 0, 50, 100, and 150 might be considered, and the relative value increments determined between these and used to specify a piecewise linear function.

---

[1] The phrase "increment in value," or "value increment," refers to the degree to which the decision maker prefers the higher score level to the lower. Thus, if the value increment between $X_p = -1$ and $X_p = 0$ is greater than the value increment between $X_p = 0$ and $X_p = 1$, this means that increasing X_p from -1 to 0 is more preferable than increasing X_p from 0 to 1.

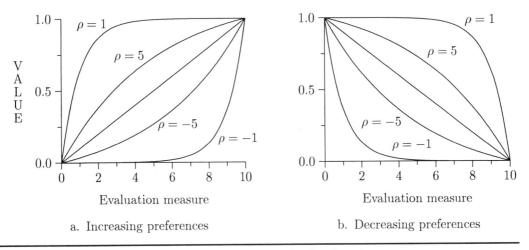

Figure 4.2   *Example exponential single dimensional value functions*

An alternate approach that is sometimes simpler is to use a particular type of mathematical function called the *exponential*. Kirkwood and Sarin (1980) have shown that this form is often reasonable, and this theory is presented in Section 9.2, particularly in Theorem 9.16. It is not necessary to understand this theory to apply the procedure below.

The equations for the exponential value function have a particular form, which depends on the range of the evaluation measure and a constant, which is signified by the Greek letter ρ (rho), and which is called the *exponential constant*. Examples of possible exponential single dimensional value functions are shown in Figure 4.2. As this figure shows, the specific shape of the exponential single dimensional value function depends on ρ. The curves in Figure 4.2 show that for smaller values of ρ, the value function is more curved. As the value of ρ increases, the graph becomes less curved until finally, when ρ is very large, the value function becomes almost a straight line. In fact, when ρ is infinitely large, the value function is exactly a straight line. That case is shown in the center of Figure 4.2a and Figure 4.2b.

For those who are interested, here are the equations for the exponential single dimensional value function. (Spreadsheet methods to do computations with the exponential are given in Section 4.7, and when these methods are used, it is not necessary to work directly with the exponential function.) To simplify the presentation, the subscript that indicates the specific evaluation measure of interest is dropped in this paragraph. If preferences are *monotonically increasing* over an evaluation measure x (that is, higher amounts of x are preferred to lower amounts as shown in Figure 4.2a), then the exponential single dimensional value function $v(x)$ can be written

$$v(x) = \begin{cases} \dfrac{1 - \exp\left[-(x - \text{Low})/\rho\right]}{1 - \exp\left[-(\text{High} - \text{Low})/\rho\right]}, & \rho \neq \text{Infinity} \\[2em] \dfrac{x - \text{Low}}{\text{High} - \text{Low}}, & \text{otherwise} \end{cases} \qquad (4.1)$$

and if preferences are *monotonically decreasing* over x (that is, lower amounts of x are preferred to higher amounts as shown in Figure 4.2b), then

$$v(x) = \begin{cases} \dfrac{1 - \exp\left[-(\text{High} - x)/\rho\right]}{1 - \exp\left[-(\text{High} - \text{Low})/\rho\right]}, & \rho \neq \text{Infinity} \\[2ex] \dfrac{\text{High} - x}{\text{High} - \text{Low}}, & \text{otherwise} \end{cases} \tag{4.2}$$

where "Low" is the lowest level of x that is of interest, "High" is the highest level of interest, and ρ is the *exponential constant* for the value function. The value function is scaled so that it varies between 0 and 1 over the range from $x = $ Low to $x = $ High. That is, for monotonically increasing preferences, $v(\text{Low}) = 0$ and $v(\text{High}) = 1$, while for monotonically decreasing preferences $v(\text{Low}) = 1$ and $v(\text{High}) = 0$. In these equations, "$\exp(x)$" represents the exponential function. An alternative notation for this is e^x, where e has a value of $2.7182\ldots$ Financial and scientific calculators have an exponential function, and the button for this function is usually marked either exp or e^x.

A study of the equations for the exponential value function given in the preceding paragraph shows that appropriate values for ρ depend on the range of possible scores for the evaluation measure. In particular, realistic values of ρ will generally have a magnitude greater than one-tenth of the range of possible scores for the evaluation measure. For the example in Figure 4.2, realistic values of ρ will have a magnitude greater than one-tenth of the range from 0 to 10; thus, ρ will be greater than 1 if it is positive or will be less than -1 if it is negative. In fact, this figure shows that for $\rho = 1$ or $\rho = -1$, the value functions are very curved. Notice further that when ρ is greater than zero, the single dimensional value functions are bowed upward, while when ρ is less than zero, the value functions are bowed downward. This is a general property of exponential value functions.

There is no upper limit to the realistic values for the magnitude of ρ. However, if the magnitude is greater than ten times the range of possible evaluation measure scores, then the value function curve will be almost a straight line. Therefore, it is rarely necessary to use values of ρ with a greater magnitude than this—just use a straight line. Thus, for the example in Figure 4.2, it would not be necessary to have positive values of ρ greater than 100, or negative values of ρ less than -100, since these correspond to functions that are essentially straight lines.

The procedure to determine the value of ρ for a specific exponential single dimensional value function depends on the concept of the *midvalue* for the range of evaluation measure scores that is of interest. The midvalue of a range is defined to be the score such that the difference in value between the lowest score in the range and the midvalue is the same as the difference in value between the midvalue and the highest score.

As discussed previously, the single dimensional value for one end of the range of scores being considered is zero, while the single dimensional value for the other end is 1. Thus, if the value differences between the midvalue and either end of the range are the same, it must be true that the single dimensional value for the midvalue is 0.5. Why is this? Because the increment in value between the least preferred score in the range and the midvalue must be equal to the increment in value between the midvalue and the most preferred score in the range. But from our earlier discussion of single dimensional value functions, we know that the sum of the value increments must be 1. Hence, since the two value increments must be equal, they must each be 0.5, and thus the single dimensional value for the midvalue is 0.5.

If the two endpoints for a range are known, along with the midvalue, then either equation 4.1 or 4.2 can be solved to determine the exponential constant. This is done by setting $v(x_m) = 0.5$ for the appropriate equation, where $v(x)$ is the appropriate one of equation 4.1 or 4.2 and x_m is the midvalue. Since everything will be known in the equation except ρ, the equation can be solved for ρ.

Unfortunately, there is no closed form solution to the resulting equation, and hence it must be solved numerically. Readers who are advanced Excel users may wish to experiment with using the Excel Goal Seek or Solver commands to determine ρ for a specified midvalue. However, it is generally quicker to use the procedure presented below based on Table 4.2. This table presents the exponential constants that correspond to various possible midvalues. Since there are an infinite number of different possible Low and High levels in equations 4.1 and 4.2, a table that included Low and High would be very large. To keep the table to a reasonable size, a user is required to do some conversions on the midvalue and exponential constant.

An examination of the curves in Figure 4.2 leads to some conclusions about the relationship between the midvalue and the value of ρ:

1 If the midvalue is equal to the *average* of the highest and lowest possible scores of the evaluation measure, then the value function is a straight line. Thus, for the example in Figure 4.2, if the midvalue is $(10 - 0)/2 = 5$, then the value function will be a straight line. This result holds in general, and it holds whether higher scores are more preferred (as in Figure 4.2a) or less preferred (as in Figure 4.2b).

2 For the situation where higher scores are more preferred (as in Figure 4.2a), if the midvalue is less than the average of the highest and lowest score in the range, then ρ will be greater than zero. If the midvalue is greater than the average of the highest and lowest score in the range, then ρ will be less than zero. This result holds in general for situations where higher scores are more preferred.

3 For situations where higher scores are less preferred (as in Figure 4.2b), if the midvalue is greater than the average of the highest and the lowest scores in the range, then ρ will be greater than zero. If the midvalue is less than the average of the highest and lowest scores in the range, then ρ will be less than zero. This result holds in general for situations where higher scores are less preferred.

Once the midvalue has been determined for some range of an evaluation measure, the value of the exponential constant ρ can be found using Table 4.2 as follows:

1 Calculate the *normalized midvalue* by taking the difference between the midvalue and the *less preferred* of the two ends of the range of interest and dividing this by the difference between the highest and lowest scores in the range. When doing this, take each of the two differences so that the result has a positive sign.

2 Look up the normalized midvalue in Table 4.2 under the column marked $z_{0.5}$, and find the *normalized exponential constant* R that corresponds to this.

3 The value of the exponential constant ρ that corresponds to this value of R is found by multiplying R by the distance between the highest and lowest scores in the range.

As an example, consider the Cost Increase evaluation measure in the networking strategy decision and assume that the single dimensional exponential value function is desired for the range from $X_c = 0$ to $X_c = 150$. If the midvalue of this range is 90, then the normalized midvalue is $(150 - 90)/(150 - 0) = 0.40$. (Remember that higher levels of X_c are less preferred; therefore, the less preferred end of the range from 0 to 150 is 150. As specified in Step 1 above, the differences are taken so that the result has a positive sign.)

Looking up 0.40 in Table 4.2 under the column marked $z_{0.5}$, we find that the corresponding normalized exponential constant is $R = 1.216$. Thus, $\rho = 1.216 \times (150 - 0) = 182.4$. With this value of ρ, it is then possible to use equation 4.2 to calculate the exponential value for any specified Cost Increase score. The exponential value function for Cost Increase is shown in Figure 4.1b.

4.4 Determining the Weights

As discussed earlier in this chapter, the value function for the networking strategy decision has the form

$$v(X_p, X_c, X_s) = w_p v_p(X_p) + w_c v_c(X_c) + w_s v_s(X_s)$$

In the preceding section, the three single dimensional value functions $v_p(x_p)$, $v_c(x_c)$, and $v_s(x_s)$ were determined. Thus, to complete the determination of the value function for the networking strategy example, it is necessary to find the weights w_p, w_c, and w_s for the three evaluation measures.

To understand the procedure for determining the weights, it is useful to review some of the properties of the value function. First, the single dimensional value functions have been specified so that each of them is equal to zero for the least preferred level that is being considered for the corresponding evaluation measure. Similarly, each of the single dimensional value functions has been specified so that it is equal to one for the most preferred level that is being considered for the corresponding evaluation measure.

From these properties of the single dimensional value functions, it follows that the weight for an evaluation measure is equal to the increment in value that is received from moving the score on that evaluation measure from its least preferred level to its most preferred level. This property provides a basis for a

This table presents pairs of numbers $z_{0.5}$ and R that solve the equation

$$0.5 = \frac{1 - \exp(-z_{0.5}/R)}{1 - \exp(-1/R)}$$

$z_{0.5}$	R	$z_{0.5}$	R	$z_{0.5}$	R	$z_{0.5}$	R
0.00	—	0.25	0.410	0.50	Infinity	0.75	-0.410
0.01	0.014	0.26	0.435	0.51	-12.497	0.76	-0.387
0.02	0.029	0.27	0.462	0.52	-6.243	0.77	-0.365
0.03	0.043	0.28	0.491	0.53	-4.157	0.78	-0.344
0.04	0.058	0.29	0.522	0.54	-3.112	0.79	-0.324
0.05	0.072	0.30	0.555	0.55	-2.483	0.80	-0.305
0.06	0.087	0.31	0.592	0.56	-2.063	0.81	-0.287
0.07	0.101	0.32	0.632	0.57	-1.762	0.82	-0.269
0.08	0.115	0.33	0.677	0.58	-1.536	0.83	-0.252
0.09	0.130	0.34	0.726	0.59	-1.359	0.84	-0.236
0.10	0.144	0.35	0.782	0.60	-1.216	0.85	-0.220
0.11	0.159	0.36	0.845	0.61	-1.099	0.86	-0.204
0.12	0.174	0.37	0.917	0.62	-1.001	0.87	-0.189
0.13	0.189	0.38	1.001	0.63	-0.917	0.88	-0.174
0.14	0.204	0.39	1.099	0.64	-0.845	0.89	-0.159
0.15	0.220	0.40	1.216	0.65	-0.782	0.90	-0.144
0.16	0.236	0.41	1.359	0.66	-0.726	0.91	-0.130
0.17	0.252	0.42	1.536	0.67	-0.677	0.92	-0.115
0.18	0.269	0.43	1.762	0.68	-0.632	0.93	-0.101
0.19	0.287	0.44	2.063	0.69	-0.592	0.94	-0.087
0.20	0.305	0.45	2.483	0.70	-0.555	0.95	-0.072
0.21	0.324	0.46	3.112	0.71	-0.522	0.96	-0.058
0.22	0.344	0.47	4.157	0.72	-0.491	0.97	-0.043
0.23	0.365	0.48	6.243	0.73	-0.462	0.98	-0.029
0.24	0.387	0.49	12.497	0.74	-0.435	0.99	-0.014

Table 4.2 *Calculating the exponential constant*

procedure to determine the weights. The procedure for determining weights is similar to that for finding the piecewise linear single dimensional value function. Specifically, the steps are as follows:

1 Consider the increments in value that would occur by increasing (or "swinging") each of the evaluation measures from the least preferred end of its range to the most preferred end, and place these increments in order of successively increasing value increments.

2 Quantitatively scale each of these value increments as a multiple of the smallest value increment.

3 Set the smallest value increment so that the total of all the increments is 1.

4 Use the results of Step 3 to determine the weights for all the evaluation measures.

This procedure will be illustrated with the networking strategy decision. Suppose that the swing over the total range for Productivity Enhancement from $X_p = -1$ to $X_p = 2$ has the smallest increment of value, followed by the swing over the total range for Cost Increase from $X_c = 150$ to $X_c = 0$, and lastly the swing over the total range for Security from $X_s = -2$ to $X_s = 1$. (That is, the swing over the Security evaluation measure gives the greatest increment in value.) Suppose further that the swing over Cost Increase from $X_c = 150$ to $X_c = 0$ has 1.5 times as great a value increment as the swing over Productivity Enhancement from $X_p = -1$ to $X_p = 2$, and the swing over Security from $X_s = -2$ to $X_s = 1$ has 1.25 times the value increment of the swing over Cost Increase from $X_c = 150$ to $X_c = 0$.

Since the increment in value from swinging an evaluation measure over its total range is equal to the weight for the evaluation measure, it follows that

$$w_c = 1.5w_p$$
$$w_s = 1.25w_c = 1.25 \times 1.5w_p$$

Since the weights must sum to 1,

$$1 = w_p + w_c + w_s = w_p + 1.5w_p + 1.25 \times 1.5w_p = w_p(1 + 1.5 + 1.25 \times 1.5)$$

Hence,

$$w_p = 1/(1 + 1.5 + 1.25 \times 1.5) = 0.23$$
$$w_c = 1.5 \times 0.23 = 0.35$$
$$w_s = 1.25 \times 1.5 \times 0.23 = 0.43$$

Note that these weights do not sum to exactly 1 due to roundoff error. In some of the procedures presented below, we will be checking to make sure that the weights sum to 1. Therefore, we will decrease the weight for Productivity Enhancement by 0.01 so that the weights sum to 1. Thus, in further work with this example, we will use the following weights:

$$w_p = 0.22, \ w_c = 0.35, \ \text{and} \ w_s = 0.43$$

An Alternate Approach to Determining the Weights

This section presents another approach that can be used to determine weights, or to check the weights that are determined using the approach just presented. This method is most easily illustrated using the specific example of the networking strategy decision.

Consider a hypothetical alternative that has the least preferred level for all three evaluation measures, namely a Productivity Enhancement of -1, a Cost Increase of 150, and a Security of -2. Now suppose that you could move one, and only one, of the evaluation measures from its least preferred level to its most preferred level. Which would you move?

If you are consistent with the answers given above, you should select Security, since you receive the greatest value increment from swinging this over its complete range from -2 to 1.

Suppose now that you could not change Security. Which of the two remaining evaluation measures would you choose to move from its least preferred level to its most preferred level if you could move only one? If you are consistent with the answers given above, you should select Cost Increase.

Now consider the evaluation measure that you most prefer to move from its least preferred level to its most preferred level—namely, Security—and compare it with the evaluation measure that you second most prefer to move from its least preferred level to its most preferred level—namely, Cost Increase. Suppose that you could either move Cost Increase all the way from its least preferred level (150) to its most preferred level (0), or you could move Security from its least preferred level to some intermediate level. Select the intermediate level for Security for which you would be indifferent between these two possibilities. (This question is usually easiest to answer by considering a specific intermediate level for Security, and then adjusting this level until indifference is established.)

Suppose that when you consider these two hypothetical situations, you select a level of 0 for Security as the level for which you are indifferent. Then it must be true that the value for a hypothetical alternative with Security = 0, Cost Increase = 150, and Productivity Enhancement = -1 is equal to the value for a hypothetical alternative with Security = -2, Cost Increase = 0, and Productivity Enhancement = -1, since these two situations are equally preferred. Therefore,

$$w_s v_s(0) + w_c v_c(150) + w_p v_p(-1) = w_s v_s(-2) + w_c v_c(0) + w_p v_p(-1)$$

But the earlier analysis to determine the single dimensional value functions showed that $v_s(0) = 0.83$, $v_c(150) = 0$, $v_p(-1) = 0$, $v_s(-2) = 0$, and $v_c(0) = 1$. Therefore, the equation above reduces to

$$w_s \times 0.83 + w_c \times 0 + w_p \times 0 = w_s \times 0 + w_c \times 1 + w_p \times 0$$

or $w_c = 0.83 w_s$.

In a similar manner, consider two hypothetical alternatives, one where you can move Productivity Enhancement all the way from its least preferred level (-1) to its most preferred level (2), and one where you can move Security from its least preferred level (-2) to an intermediate level. Select the intermediate level of Security for which you are indifferent between these two alternatives. Suppose that you select a level of -1 for this. Then, by reasoning analogous to that in the preceding paragraph, it is true that $w_p = w_s v_s(-1)$, or $w_p = w_s \times 0.50$.

However, since the weights must sum to 1, we can solve for the weights from the information in the two immediately preceding paragraphs as follows:

$$1 = w_s + w_c + w_p = w_s + 0.83w_s + 0.50w_s = 2.33w_s$$

and therefore $w_s = 1/2.33 = 0.43$. Thus, $w_c = 0.83 \times 0.43 = 0.36$, and $w_p = 0.50 \times 0.43 = 0.22$. These weights are close to those determined using the earlier assessment procedure; therefore, this assessment provides further support for these values.

The specification of the weights completes the determination of the value function over Productivity Enhancement, Cost Increase, and Security. This value function will now be used to find the overall value for each alternative.

4.5 Determining the Overall Values for the Alternatives

Finding the overall values for the alternative using the value function determined above is just arithmetic. This can be tedious, and in practical decision problems you will probably do these calculations with either an electronic spreadsheet or a specialized program. (Use of a spreadsheet is reviewed in Section 4.7.) However, in case you want to do the calculations by hand, the procedure is demonstrated for the networking strategy decision.

Use the following steps:

1 Use the single dimensional value function to determine the single dimensional values that correspond to the evaluation measure scores for each alternative.
2 Multiply each of the single dimensional values by the appropriate weight, and sum these weighted single dimensional values to determine the overall value for each alternative.

Determining the single dimensional values for Productivity Enhancement and Security is straightforward using the functions shown in Figure 4.1a and Figure 4.1c. The only potentially confusing point is that two of the alternatives have evaluation measure scores that lie between the levels for which we assessed values. These are the High Quality/High Cost alternative, which has a score of 0.5 for Security, and the Low Quality/Low Cost alternative, which has a score of 0.5 for Productivity Enhancement. In each of these cases, the score is halfway between two scores for which we have values—thus, the value is also halfway between the values that were assessed. Specifically, for Productivity Enhancement, the value for a Productivity Enhancement of 0 was determined earlier to be 0.50, and the value for a Productivity Enhancement of 1 was determined to be 0.75. Hence, the value for a Productivity Enhancement of 0.5 is $(0.50 + 0.75)/2 = 0.63$. In a similar manner, the value for a Security rating of 0.5 is $(0.83 + 1.00)/2 = 0.92$.

Alternative	Single Dimensional Values		
	Productivity Enhancement (X_p)	Cost Increase (X_c)	Security (X_s)
Status Quo	0.50	1.00	0.83
High Quality/High Cost	1.00	0.23	0.92
Medium Quality/Medium Cost	0.75	0.46	0.83
Low Quality/Low Cost	0.63	0.66	0.50

Table 4.3 *Single dimensional values for networking alternatives*

The single dimensional values for Cost Increase can be determined with a calculator or spreadsheet program using the equations for the exponential value function that were given in equations 4.1 and 4.2. Preferences for Cost Increase are decreasing, and therefore equation 4.2 applies. It was earlier determined that Low $= 0$, High $= 150$, and $\rho = 182.4$. The Status Quo alternative has a Cost Increase of 0, and hence has a value of 1. The High Quality/High Cost alternative has a Cost Increase of 125, and therefore, from equation 4.2, it follows that the value for this is

$$\frac{1 - \exp[-(150 - 125)/182.4]}{1 - \exp[-(150 - 0)/182.4]} = 0.23$$

The Medium Quality/Medium Cost alternative has a Cost Increase of 95, and therefore the value for this is

$$\frac{1 - \exp[-(150 - 95)/182.4]}{1 - \exp[-(150 - 0)/182.4]} = 0.46$$

Finally, for the Low Quality/Low Cost alternative, the Cost Increase is 65, and hence the value is
$$\frac{1 - \exp[-(150 - 65)/182.4]}{1 - \exp[-(150 - 0)/182.4]} = 0.66$$

The single dimensional values just determined for the four alternatives are as given in Table 4.3.

The overall values of the alternatives are obtained by multiplying the single dimensional values by the corresponding weights. Thus, the overall values are

Status Quo: $0.22 \times 0.50 + 0.35 \times 1.00 + 0.43 \times 0.83 = 0.82$

High Quality/High Cost: $0.22 \times 1.00 + 0.35 \times 0.23 + 0.43 \times 0.92 = 0.70$

Medium Quality/Medium Cost: $0.22 \times 0.75 + 0.35 \times 0.46 + 0.43 \times 0.83 = 0.68$

Low Quality/Low Cost: $0.22 \times 0.63 + 0.35 \times 0.66 + 0.43 \times 0.50 = 0.58$

Hence, the Status Quo alternative is most highly rated, followed by the High Quality/High Cost alternative, which is closely followed by the Medium Quality/Medium Cost alternative; and finally, the Low Quality/Low Cost alternative brings up the rear.

4.6 The Meaning of Value Numbers

In the preceding section, it was determined that for the networking decision problem the Status Quo, High Quality/High Cost, Medium Quality/Medium Cost, and Low Quality/Low Cost alternatives have values of 0.82, 0.70, 0.68, and 0.58, respectively. Higher values are more preferred, and therefore the Status Quo is the most preferred alternative. But what does the value number 0.82 for the Status Quo alternative mean?

The value determination procedure has been specified so that an alternative that has the least preferred score on all of the evaluation measures will have an overall value of zero. Similarly, an alternative that has the most preferred score on all of the evaluation measures will have an overall value of 1. In many decisions, there will not be any actual alternatives that are this good or bad, but the value numbers for the actual alternatives can be interpreted by comparing these actual alternatives with (possibly hypothetical) alternatives with overall values of zero and 1. The value number for a particular alternative gives the proportion of the distance, in a value sense, that the alternative is from the (possibly hypothetical) alternative with an overall value of zero to the (also possibly hypothetical) alternative with an overall value of 1.

In the networking decision problem, the least preferred levels being considered for the evaluation measures are -1 for Productivity Enhancement, \$150 for Cost Increase, and -2 for Security. Similarly, the most preferred levels being considered are 2 for Productivity Enhancement, \$0 for Cost Increase, and 1 for Security.

In this decision, there are no actual alternatives that are as bad or good as these two possibilities. Thus, consider a hypothetical alternative that has the least preferred levels for all the evaluation measures and a second hypothetical alternative that has the most preferred levels for all the evaluation measures. Then the value of 0.82 for the Status Quo alternative means that this alternative is 0.82 (82 percent) of the distance in a value sense from the hypothetical worst possible alternative to the hypothetical best possible alternative. Another way to express this is to say that by selecting the Status Quo alternative, you obtain 82 percent of the value improvement, relative to the hypothetical worst possible alternative, that you would obtain if you could exchange the worst possible alternative for the (equally hypothetical) best possible alternative.

This discussion shows that no specific meaning can be given to value numbers without knowing the ranges of the evaluation measures that are being used. We showed earlier that the ranges of the evaluation measures must be considered when determining weights for a value function. In a similar way, the ranges of the evaluation measures must be considered when interpreting the overall value numbers for alternatives.

4.7 Spreadsheet Analysis Methods

This section presents spreadsheet methods to carry out the calculations for a multiobjective value analysis. Using such spreadsheet methods is helpful because the hand calculation procedure presented in Section 4.5 to determine values can be tedious when there are many evaluation measures or alternatives. In addition, in many practical decision problems, it is desirable to investigate whether changes in assumptions about the data for the decision lead to changes in the ranking of alternatives. For example, there may be disagreements among the different parties to a decision about the weights for the evaluation measures. In such a situation, it can be useful to repeat the value calculations using different weights to determine whether the ranking of alternatives changes with different weights. Such an investigation of the impact of variations in assumptions is called a *sensitivity analysis*, and it is usually a part of the analysis of important decisions. However, sensitivity analysis can be tedious, and is subject to calculation errors, if it is done by hand. This is another reason to use a spreadsheet calculation procedure.

In the following presentation, the reader is assumed to be familiar with standard spreadsheet notation. Cell references are given in the usual column-row notation. That is, the cell reference B11 refers to column B, row 11, and the reference is relative to the current cell. Thus, if the cell reference B11 is used in a formula in cell D14, and if this formula is then copied to cell E15, then the cell reference B11 will become C12 in this new location. Similarly, the usual $ notation is used for absolute cell references. For example, if the cell reference B$11 is used in a formula in cell D14, and if this formula is copied to cell E15, then the cell reference B$11 will become C$11 in the new location.

The spreadsheet approach presented in this section uses Microsoft Excel, Version 5.0. Specifically, some use is made of Visual Basic, Applications Edition, which is included in Excel starting with Version 5.0. You do not need to know Visual Basic programming to apply the approach presented here, but you must have access to Version 5.0 or later of Excel.

Figure 4.3 shows the spreadsheet that is desired for the networking strategy decision, and we will now present a procedure to develop this spreadsheet. An analogous procedure can be used to calculate values for any decision problem, and thus this spreadsheet serves as an example, or *template*, of the spreadsheet needed for any decision.

There are four logical parts to this spreadsheet. The first two include the input data for the decision analysis:

1 The range A4:G11 specifies the value function for the decision.
2 The range A13:F17 specifies the scores (levels) for each alternative.

The last two parts of the spreadsheet show the analysis results, and these are the parts that the spreadsheet calculates:

3 The range A19:G24 shows the weighted single dimensional values and the overall value for each alternative.

4 The stacked bar chart at the bottom of the spreadsheet shows the weighted single dimensional values in graphical form.

Each of these parts of the spreadsheet will now be discussed in further detail.

Value function data   The single dimensional value function for each evaluation measure is specified in two columns: Columns A and B for Productivity Enhancement, columns C and D for Cost Increase, and columns E and F for Security. Since Productivity Enhancement and Security have piecewise linear single dimensional value functions, these value functions are specified by giving the endpoints for each straight line segment in the value function. Thus, in column A the evaluation measure levels are specified for the Productivity Enhancement single dimensional value function, and in column B the corresponding value is specified next to each evaluation measure level. (These values were determined in Section 4.3.) Similarly, the levels and corresponding values for the Security single dimensional value function are given in columns E and F.

As equations 4.1 and 4.2 show, an exponential single dimensional value function is specified by its Low and High levels, and its monotonicity ("increasing" or "decreasing") and exponential constant ("rho"). Columns C and D present this information for the Cost Increase evaluation measure, as this was determined in Section 4.3.

Finally, the weights are given in row 11 for the three evaluation measures, as these were determined in Section 4.4. (A check equation that adds up the weights is included in cell G11. By inspecting this, you can quickly catch common errors in entering weights.)

Scores for alternatives   Rows 14 through 17 of Figure 4.3 show the scores for the alternatives, as these are given in Table 4.1.

Weighted single dimensional value calculations   In order to fill in the table of weighted single dimensional values in range A19:F24, it is necessary to calculate the single dimensional values for each alternative, and then multiply each value by the correct weight, which is given in row 11. To have the spreadsheet automatically determine the single dimensional values, it is necessary to have functions available that can calculate exponential and piecewise linear values for specified examples of these types of value functions. Excel does not have such value functions built in, but it is possible to add them by using the Visual Basic programming language, which is included with Excel. The straightforward procedure to do this is given below after we discuss how such functions are used.

For the moment, assume that these functions are available. Then Figure 4.4 shows the equations needed to determine the weighted single dimensional values. (This figure shows the equations for the spreadsheet in Figure 4.3.) In this figure, the equations for the left portion of the spreadsheet are shown at the top, and equations for the right portion are shown at the bottom. Rows 21 through 24 show the desired equations. The piecewise linear single dimensional value function is called ValuePL, and the exponential single dimensional value function is called ValueE.

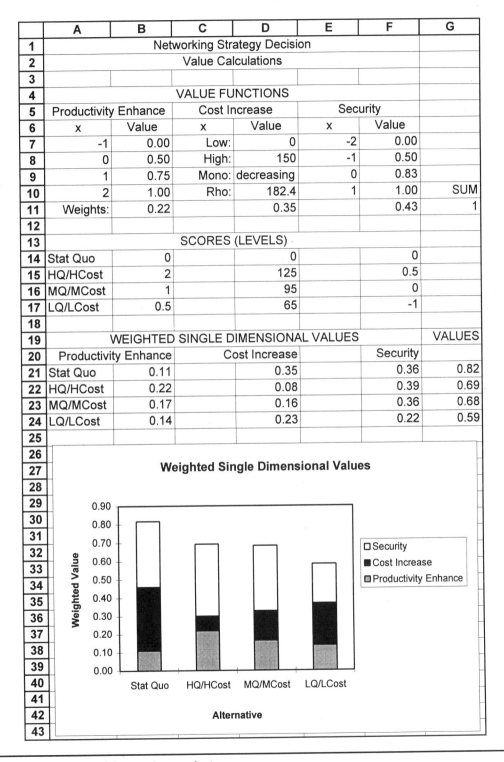

	A	B	C	D	E	F	G
1	Networking Strategy Decision						
2	Value Calculations						
3							
4	VALUE FUNCTIONS						
5	Productivity Enhance		Cost Increase		Security		
6	x	Value	x	Value	x	Value	
7	-1	0.00	Low:	0	-2	0.00	
8	0	0.50	High:	150	-1	0.50	
9	1	0.75	Mono:	decreasing	0	0.83	
10	2	1.00	Rho:	182.4	1	1.00	SUM
11	Weights:	0.22		0.35		0.43	1
12							
13	SCORES (LEVELS)						
14	Stat Quo	0		0		0	
15	HQ/HCost	2		125		0.5	
16	MQ/MCost	1		95		0	
17	LQ/LCost	0.5		65		-1	
18							
19	WEIGHTED SINGLE DIMENSIONAL VALUES						VALUES
20	Productivity Enhance		Cost Increase		Security		
21	Stat Quo	0.11		0.35		0.36	0.82
22	HQ/HCost	0.22		0.08		0.39	0.69
23	MQ/MCost	0.17		0.16		0.36	0.68
24	LQ/LCost	0.14		0.23		0.22	0.59
25							

Figure 4.3   *Spreadsheet value analysis*

The piecewise linear value function has three arguments as follows:

$$\text{ValuePL}(x, X\text{-list}, V\text{-list})$$

where x is the level (score) for which the value is needed, X-list is the list of evaluation measure levels that specifies the piecewise linear function, and V-list is the list of corresponding values for these levels.

The exponential value function has five arguments as follows:

$$\text{ValueE}(x, \text{Low}, \text{High}, \text{Monotonicity}, \text{Rho})$$

where x is the level (score) for which the value is needed, and the other four arguments specify the value function, as shown in equations 4.1 and 4.2.

The easiest way to learn how to use these functions is to study the equations in Figure 4.4 in the range B21:F24. For example, consider the equations in range B21:B24, which calculate the weighted single dimensional values for Productivity Enhancement. These equations are entered by first entering the correct equation into cell B21, and then filling this down into cells B22, B23, and B24.

The first factor in the cell B21 equation refers to cell B11, which contains the weight for Productivity Enhancement. Note that the row reference in this is absolute. (That is, B$11 is entered.) This is done so that when the equation is filled down into the range B22:B24, this factor will still refer to the cell that contains the Productivity Enhancement weight. The first argument of the function ValuePL in the cell B21 equation (that is, B14) refers to the Productivity Enhancement score for the Status Quo alternative. Note that this is a relative reference. (That is, it contains no dollar signs.) This is done so that when the equation is filled down into the range B22:B24, the reference will change to refer to the cells containing the Productivity Enhancement scores for each of the different alternatives.

The final two arguments for ValuePL in cell B21 (A$7:A$11 and B$7:B$10) refer to the lists of evaluation measure levels and values that specify the piecewise linear value function. Note that the row references in these are absolute, so that when the equation is filled down into range B22:B24, the resulting equations will still refer to the correct lists.

The equations for the weighted single dimensional values for Cost Increase (in range D21:D24) and Security (in range F21:F24) are entered in a similar manner. In each case, the correct equation is entered into the topmost cell of the range, and then filled down through the remainder of the range. This is straightforward if you remember that in each of these equations the reference to the cell containing the score for alternative should be relative, while all other cell references should have an absolute reference for the row number.

Overall value calculations

The equations to calculate the overall values for the alternatives are shown in range G21:G24 of Figure 4.4. Each of these is simply the sum of the three weighted single dimensional values to its left. The necessary equation is entered into cell G21 and then filled down into range G22:G24. (Note that a slight shortcut is used in this equation. Since the SUM function assumes that any blank cell has a zero in it, all of the cells in the

	A	B	C	D	E
1					
2					
3					
4					
5					
6	x	Value	x	Value	x
7	-1	0	Low:	0	-2
8	0	0.5	High:	150	-1
9	1	0.75	Mono:	decreasing	0
10	2	1	Rho:	182.4	1
11	Weights:	0.22		0.35	
12					
13					
14	Stat Quo	0		0	
15	HQ/HCost	2		125	
16	MQ/MCost	1		95	
17	LQ/LCost	0.5		65	
18					
19					
20		Productivity Enhance		Cost Increase	
21	Stat Quo	=B$11*ValuePL(B14,A$7:A$10,B$7:B$10)		=D$11*ValueE(D14,D$7,D$8,D$9,D$10)	
22	HQ/HCost	=B$11*ValuePL(B15,A$7:A$10,B$7:B$10)		=D$11*ValueE(D15,D$7,D$8,D$9,D$10)	
23	MQ/MCost	=B$11*ValuePL(B16,A$7:A$10,B$7:B$10)		=D$11*ValueE(D16,D$7,D$8,D$9,D$10)	
24	LQ/LCost	=B$11*ValuePL(B17,A$7:A$10,B$7:B$10)		=D$11*ValueE(D17,D$7,D$8,D$9,D$10)	

	F	G
1		
2		
3		
4		
5		
6	Value	
7	0	
8	0.5	
9	0.83	
10	1	SUM
11	0.43	=SUM(B11:F11)
12		
13		
14	0	
15	0.5	
16	0	
17	-1	
18		
19		VALUES
20	Security	
21	=F$11*ValuePL(F14,E$7:E$10,F$7:F$10)	=SUM(B21:F21)
22	=F$11*ValuePL(F15,E$7:E$10,F$7:F$10)	=SUM(B22:F22)
23	=F$11*ValuePL(F16,E$7:E$10,F$7:F$10)	=SUM(B23:F23)
24	=F$11*ValuePL(F17,E$7:E$10,F$7:F$10)	=SUM(B24:F24)

Figure 4.4 *Equations for spreadsheet value analysis*

weighted single dimensional value function table to the left of each overall value entry are included in the SUM, even though only three of these cells contain values.)

Weighted single dimensional value chart   The bottom portion of the Figure 4.3 spreadsheet shows the same information as is in the weighted single dimensional value table, but in graphical form. From this chart, it is easy to quickly determine the ranking of the alternatives, as well as the contributions of the various evaluation measures to the overall value of each alternative. This chart is constructed by applying the Excel Chart Wizard to the information in range A20:F24 and constructing a stacked bar graph.

Implementing ValuePL and ValueE

The approach above is straightforward if the functions ValuePL and ValueE are available. These can easily be added to Excel using Visual Basic. This section first presents the procedure to do this, and then discusses the logic of these functions. You do not need to understand the logic of the functions to use them. You just need to enter them into your Excel workbook. You do this by selecting the Insert item from the main Excel menu, then selecting the Macro item from the submenu that is displayed, and finally selecting the Module item from the third-level submenu that is displayed.

When you do this, a Visual Basic module will be created in your workbook. You can move between this module and your worksheets in the usual manner by clicking on the appropriate tab at the bottom of the Excel screen. To create the functions ValuePL and ValueE, enter the code shown in Figure 4.5 into the Visual Basic module you have created. The spacing in this is not important as long as you keep the material shown on each line in Figure 4.5 on separate lines. Note that there is an underscore (_) at the end of the sixth line of code in Figure 4.5. This must be preceded by at least one space. (This indicates that the expression on that line continues onto the next line.)

Once this code is entered into your Excel workbook, you can use the ValuePL and ValueE functions just as if these functions were built into Excel. These functions have been kept short so that they are easy to enter. As a result, the functions do not include any error checking. You must manually check to make sure that the data in your spreadsheet are correct. For example, if you enter evaluation measure scores that are outside the range you specify for a value function, you will obtain incorrect answers. Note in particular that the list of evaluation measure levels and corresponding values that specify a piecewise linear value function (for example, the lists in ranges A7:B10 and E7:F10 in Figure 4.3) must be entered in ascending order of the levels, or the ValuePL function will yield incorrect results.

```
Function ValuePL(x, Xi, Vi)
  i = 2
  Do While x > Xi(i)
    i = i + 1
  Loop
  ValuePL = Vi(i - 1) _
      + (Vi(i) - Vi(i - 1)) * (x - Xi(i - 1)) / (Xi(i) - Xi(i - 1))
End Function

Function ValueE(x, Low, High, Monotonicity, Rho)
  Select Case UCase(Monotonicity)
    Case "INCREASING"
      Difference = x - Low
    Case "DECREASING"
      Difference = High - x
  End Select
  If UCase(Rho) = "INFINITY" Then
    ValueE = Difference / (High - Low)
  Else
    ValueE = (1 - Exp(-Difference / Rho)) / (1 - Exp(-(High - Low) / Rho))
  End If
End Function
```

Figure 4.5 *Equations for value functions*

Logic of the ValuePL and ValueE Functions

It is not necessary to understand the material in this section to use the ValuePL and ValueE functions. Readers interested only in the practical application of these functions can skip ahead to the section on sensitivity analysis.

The ValuePL function is straightforward. First the Do While...Loop determines which straight line segment of the piecewise linear function applies, and then the value is calculated using the equation for that segment.

The ValueE function is also straightforward, but it uses one slight trick to shorten the code. First, the Select...Case determines which of the equations 4.1 or 4.2 applies. These equations differ from each other only in that where equation 4.1 has the term $x -$ Low, equation 4.2 has the term High $- x$. Thus, once ValueE determines which of these equations applies, it calculates the value of the appropriate of these two terms and stores it in a variable called Difference. Then ValueE uses an If...Else...End If to determine whether the value function is linear or exponential, and it calculates the value using the appropriate equation.

Sensitivity Analysis

Once a spreadsheet model like the one shown in Figure 4.3 has been developed, it can be used to conduct a *sensitivity analysis* to determine the impact on the ranking of alternatives of changes in various model assumptions. One sensitivity analysis that is often of interest is on the weights. These weights represent the relative importance that is attached to changes in the different evaluation measures, and this is sometimes a matter of disagreement among the various stakeholders for a particular decision. Figure 4.6 shows a spreadsheet to conduct a sensitivity analysis in the networking decision for the weight on Productivity Enhancement. This spreadsheet was constructed from the one in Figure 4.3 by making only a few changes. However, making these changes requires understanding some algebraic manipulations.

One difficulty in conducting a sensitivity analysis on the weight for Productivity Enhancement is that as this weight is changed, it is also necessary to change the weights for Cost Increase and Security. This is because the sum of all three weights must be 1, and therefore if one weight is changed, at least one of the other two weights must also be changed to keep the sum equal to 1. Of course, you can manually make changes in all of the weights as you conduct the sensitivity analysis, but this is error prone and time consuming if very many different cases are investigated. Therefore, it is often desirable to enter into the spreadsheet equations that automatically change the weights for Cost Increase and Security as the weight for Productivity Enhancement is changed. Once such equations are entered, changes can be made to the weight for Productivity Enhancement in cell B12 of the Figure 4.6 spreadsheet, and the weights for Cost Increase and Security in cells D12 and F12 will automatically change.

The process used to add this type of sensitivity analysis capability to the Figure 4.6 spreadsheet is as follows: First, a new row is inserted above row 11 in the Figure 4.3 spreadsheet. *Important:* Note that the row must be inserted **above** row 11 and not below it, or the cell references in the equations in range B22:F25 of the modified spreadsheet will be incorrect.

After the new row is inserted, the range B12:G12 of the new spreadsheet is copied to the row above (B11:G11), and the label "Base:" is inserted into cell A11.

The next step in the procedure requires deciding how the weights for Cost Increase and Security should be automatically changed as the weight for Productivity Enhancement is varied. An approach that often makes sense is to keep the ratio of the weights for Cost Increase and Security the same as in the base case analysis. The ratio of the base case Security weight to the base case Cost Increase weight is $0.43/0.35 = 1.23$. Therefore, we want to keep these weights in this same ratio as we vary the weight for Productivity Enhancement.

Some algebraic manipulation establishes that the appropriate formulas to do this are as follows: The weight for Cost Increase is given by

$$w_c = (1 - w_p) \times \left(\frac{w_c^o}{w_c^o + w_s^o} \right) \tag{4.3}$$

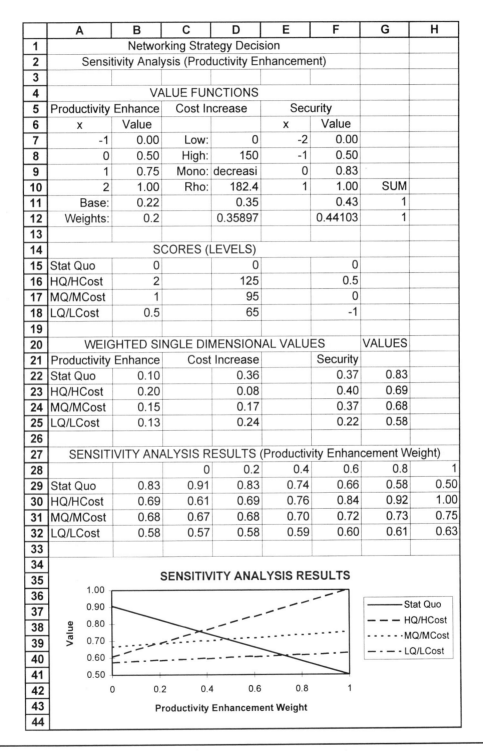

	A	B	C	D	E	F	G	H
1			Networking Strategy Decision					
2		Sensitivity Analysis (Productivity Enhancement)						
3								
4			VALUE FUNCTIONS					
5	Productivity Enhance		Cost Increase		Security			
6	x	Value			x	Value		
7	-1	0.00	Low:	0	-2	0.00		
8	0	0.50	High:	150	-1	0.50		
9	1	0.75	Mono:	decreasi	0	0.83		
10	2	1.00	Rho:	182.4	1	1.00	SUM	
11	Base:	0.22		0.35		0.43	1	
12	Weights:	0.2		0.35897		0.44103	1	
13								
14			SCORES (LEVELS)					
15	Stat Quo	0		0		0		
16	HQ/HCost	2		125		0.5		
17	MQ/MCost	1		95		0		
18	LQ/LCost	0.5		65		-1		
19								
20		WEIGHTED SINGLE DIMENSIONAL VALUES					VALUES	
21	Productivity Enhance		Cost Increase		Security			
22	Stat Quo	0.10		0.36		0.37	0.83	
23	HQ/HCost	0.20		0.08		0.40	0.69	
24	MQ/MCost	0.15		0.17		0.37	0.68	
25	LQ/LCost	0.13		0.24		0.22	0.58	
26								
27		SENSITIVITY ANALYSIS RESULTS (Productivity Enhancement Weight)						
28			0	0.2	0.4	0.6	0.8	1
29	Stat Quo	0.83	0.91	0.83	0.74	0.66	0.58	0.50
30	HQ/HCost	0.69	0.61	0.69	0.76	0.84	0.92	1.00
31	MQ/MCost	0.68	0.67	0.68	0.70	0.72	0.73	0.75
32	LQ/LCost	0.58	0.57	0.58	0.59	0.60	0.61	0.63
33								
34								
35			SENSITIVITY ANALYSIS RESULTS					
36–44								

Figure 4.6 *Sensitivity analysis spreadsheet*

and the weight for Security is given by

$$w_s = (1 - w_p) \times \left(\frac{w_s^o}{w_c^o + w_s^o} \right) \tag{4.4}$$

where w_p is the weight for Productivity Enhancement that is being used in a particular sensitivity analysis case, w_c^o is the base case weight for Cost Increase, and w_s^o is the base case weight for Security. (That is, $w_c^o = 0.35$ and $w_s^o = 0.43$.)

The logic shown in these equations works in general. For each weight except the one that you are varying, the equation is given by the following: Multiply 1 minus the weight that is being varied by the ratio of the base case value for the weight being considered to the sum of the base case weights for all the weights except the one that is being manually varied.

To continue constructing the sensitivity analysis spreadsheet, enter equations 4.3 and 4.4 into cells D12 and F12 of the Figure 4.6 spreadsheet as shown in Figure 4.7. Note that this can be done by entering an equation into cell D12 and copying this to cell F12 if a combination of relative and absolute cell references are used to refer to w_p, as well as the base cases weights w_c^o and w_s^o. The necessary use of relative and absolute references is shown in Figure 4.7 in cells D12 and F12. (Figure 4.7 shows the equations for the Figure 4.6 spreadsheet.)

Once these equations are entered, the spreadsheet will correctly calculate values for any Productivity Enhancement weight w_p that is entered in cell B12. For example, Figure 4.6 shows the results when this weight is set equal to 0.2. From this, we see that the corresponding weights for Cost Increase and Security are approximately 0.359 and 0.441. Note that the ratio of these two weights is 1.23, and this is equal to the ratio of their base case weights. Of course, this has to be true because we have set up the equations to make it true.

In Figure 4.6, the results of a sensitivity analysis over the range $0 \le w_p \le 1$ are shown in the table in range A27:H32. Six different values of w_p were used to construct this table (0, 0.2, 0.4, 0.6, 0.8, and 1), as shown in range C28:H28. This table is most easily constructed using the Excel Data Table command. To use this command, first copy the value calculation equations in range G22:G25 to the range B29:B32. *Important:* Before doing this, change the addressing for the equations in range G22:G25 from relative to absolute, as shown in Figure 4.7. If this is not done, then the copied equations will refer to the wrong cells in the spreadsheet.

After copying the equations to range B29:B32, highlight range B28:H32. Then select the Table entry from the Data menu. In the "Row Input Cell" entry for the dialog that is displayed, enter $B$12, which is the cell for the Productivity Enhancement weight that is to be varied. Then click OK. The table in range C29:H32 will then be automatically filled in. Note also that if you now change one of the values for the Productivity Enhancement weight at the top of the table in range C28:H28, the entries in that column of the table will be automatically updated to correspond to the new value of the Productivity Enhancement weight.

A graph of the sensitivity analysis table in range A28:H32 is shown at the bottom of the Figure 4.6 spreadsheet. While this shows the same information as the table, the graphical representation allows you to quickly determine two important conclusions: (1) Only the Status Quo and High Quality/High Cost alternatives are most preferred, regardless of the weight on Productivity Enhancement, and (2) the value of this weight at which the most preferred alternative changes between these two alternatives is slightly less than 0.4.

Also, note that the curve representing the value of each alternative in the sensitivity analysis graph is a straight line. An examination of the equation derived earlier for the sensitivity analysis over a weight shows that this is a general property of sensitivity analyses for weights.

4.8 Decisions with Continuous Decision Variables

This section extends the procedures presented so far to consider situations where the decision alternatives are specified by continuous decision variables. This material is somewhat more advanced than that in earlier sections of this chapter, and an understanding of this section is not required for the remainder of this book, except for Sections 7.7 and 8.4.

The computer networking decision that has been analyzed in this chapter has four alternatives, and many other practical decisions have a relatively small number of alternatives. In such cases, the procedures presented in earlier sections of this chapter are effective. However, an important class of decisions involves allocating some scarce resource or resources among different activities, where the allocation can be done in an infinite number of different ways. Examples of this include allocation of an annual budget among different activities or allocation of scarce personnel time to different projects.

This section reviews procedures for analyzing such decisions using a specific example based on an application presented by Keefer and Kirkwood (1978). The original work was conducted with the director of a large product engineering department within a major corporation to help him allocate his annual operation budget. This department had engineering design responsibility for several major product lines involving the same general product type. Engineering effort was concentrated in three areas:

1 Cost improvement: Reducing the cost of the product.
2 Quality: Preventing and responding to field incidence of product failures.
3 New features and models: Developing new features for existing product models, periodically revising the major model lines, and responding to requests for special limited edition models.

The annual decision problem was to allocate the department's operating budget among the three areas in order to do "as well as possible" in each. In the actual decision problem, there was uncertainty about the levels of performance that would result from various allocations among the areas. In this section, we will assume that there is no uncertainty and concentrate on considering the tradeoffs among the different performance areas. The analysis will be extended to consider uncertainty in Section 7.7.

	A	B	C	D
1				
2				
3				
4				
5	Prod			Cost Increase
6	x	Value		
7	-1	0	Low:	0
8	0	0.5	High:	150
9	1	0.75	Mono:	decreasing
10	2	1	Rho:	182.4
11	Base:	0.22		0.35
12	Weights:	0.2		=(1-$B12)*(D11/SUM($D11:$F11))
13				
14				
15	Stat Quo	0		0
16	HQ/HCost	2		125
17	MQ/MCost	1		95
18	LQ/LCost	0.5		65
19				
20				
21		Productivity Enhance		Cost Increase
22	Stat Quo	=B$12*ValuePL(B15,A		=D$12*ValueE(D15,D$7,D$8,D$9,D
23	HQ/HCost	=B$12*ValuePL(B16,A		=D$12*ValueE(D16,D$7,D$8,D$9,D
24	MQ/MCost	=B$12*ValuePL(B17,A		=D$12*ValueE(D17,D$7,D$8,D$9,D
25	LQ/LCost	=B$12*ValuePL(B18,A		=D$12*ValueE(D18,D$7,D$8,D$9,D
26				
27				
28			0	0.2
29	Stat Quo	=SUM($B$22:$F$22)	=TABLE(B12,)	=TABLE(B12,)
30	HQ/HCost	=SUM($B$23:$F$23)	=TABLE(B12,)	=TABLE(B12,)
31	MQ/MCost	=SUM($B$24:$F$24)	=TABLE(B12,)	=TABLE(B12,)
32	LQ/LCost	=SUM($B$25:$F$25)	=TABLE(B12,)	=TABLE(B12,)

Figure 4.7a   *Equations for sensitivity analysis spreadsheet (part 1)*

	E	F	G	H
1				
2				
3				
4				
5		Security		
6	x	Value		
7	-2	0		
8	-1	0.5		
9	0	0.83		
10	1	1	SUM	
11		0.43	=SUM(B11:F11)	
12		=(1-$B12)*(F11/SUM($D11:$F11))	=SUM(B12:F12)	
13				
14				
15		0		
16		0.5		
17		0		
18		-1		
19				
20			VALUES	
21		Security		
22		=F$12*ValuePL(F15,E$7:E$10,F$7:	=SUM($B$22:$F$22)	
23		=F$12*ValuePL(F16,E$7:E$10,F$7:	=SUM($B$23:$F$23)	
24		=F$12*ValuePL(F17,E$7:E$10,F$7:	=SUM($B$24:$F$24)	
25		=F$12*ValuePL(F18,E$7:E$10,F$7:	=SUM($B$25:$F$25)	
26				
27				
28	0.4	0.6	0.8	1
29	=TABLE(B12,)	=TABLE(B12,)	=TABLE(B12,)	=TABLE(B12,)
30	=TABLE(B12,)	=TABLE(B12,)	=TABLE(B12,)	=TABLE(B12,)
31	=TABLE(B12,)	=TABLE(B12,)	=TABLE(B12,)	=TABLE(B12,)
32	=TABLE(B12,)	=TABLE(B12,)	=TABLE(B12,)	=TABLE(B12,)

Figure 4.7b *Equations for sensitivity analysis spreadsheet (part 2)*

Evaluation considerations and value function   The following four evaluation measures were used to measure the results of the budget allocation:

1 First Year Cost Improvement: The cost improvement in the first year due to engineering efforts during that year.

2 Carryover Cost Improvement: The cost improvement in future years due to effort in a particular year.

3 Quality: A five-level constructed scale, with levels from 0 to 4 (where higher scores are more preferred).

4 New Features and Models: A five-level constructed scale, with levels from 0 to 4 (where higher scores are more preferred).

The single dimensional value functions over First Year Cost Improvement and Carryover Cost Improvement were determined to be linear, with a range from 0.10 to 0.80 for First Year Cost Improvement and a range of 0.80 to 2.00 for Carryover Cost Improvement. The value function over Quality was piecewise linear with two different line segments joining at a Quality score of 1, which had a single dimensional value of 0.40. The single dimensional value function over New Features and Models was exponential with a midvalue of 3. Applying the usual procedure involving Table 4.2 determined that the exponential constant for the New Features and Models value function was -1.640.

Finally, the weights for the four evaluation measures were determined to be the following:

1 First Year Cost Improvement: 0.10
2 Carryover Cost Improvement: 0.15
3 Quality: 0.55
4 New Features and Models: 0.20

Alternatives   The decision to be made was the allocation of the budget to each of the three areas (cost improvement, quality, and new features and models). Thus, any particular alternative was specified by the fraction of the budget to be allocated to each area. Factors such as organizational pressures exerted by various groups that interface with the product engineering department and the skills available within the department limited the director's flexibility in carrying out this allocation. This was specified in the decision problem by setting lower and upper bounds on the budget fraction that could be allocated to each area as follows:

Budget Area	Lower Bound	Upper Bound
Cost improvement	0.20	0.40
Quality	0.20	0.32
New features and models	0.30	0.60

In addition, of course, the total of the budget fractions allocated to the three areas had to be 1.

Thus, there were an infinite number of possible different alternatives for this decision problem, with each alternative being specified by the budget fraction that was allocated to the three areas. The fact that there are an infinite number of possible alternatives leads to two analysis difficulties. First, how will the scores on the four evaluation measures be determined for each of the (infinite number of) alternatives? Second, how will the value calculations be carried out for this infinite number of different alternatives? Each of these difficulties will now be considered.

Scores   In this decision problem, the scores on the First Year Cost Improvement and Carryover Cost Improvement evaluation measures are determined by the budget allocation to cost improvement. The score on the Quality evaluation measure is determined by the budget allocation to quality, and the score on the New Features and Models evaluation measure is determined by the budget allocation to new features and models.

The difficulty is that there are an infinite number of possible allocations to each of these three areas. Therefore, it was impossible to determine the scores associated with each possible allocation by directly asking the decision maker.

However, it is reasonable to assume that the performance with respect to each evaluation measure will change smoothly as the budget allocation to the associated area is changed. Thus, for example, if the budget is increased to the quality area, the Quality score should increase, and furthermore it should do this in a smooth way—the score with respect to Quality should gradually increase as the budget for quality work is increased with no abrupt jumps.

Given that this behavior is valid, then the score with respect to each evaluation measure can be estimated by a two-step process:

1 Determine the scores for a small number of different levels of the associated budget allocation, and
2 Draw a smooth curve through the assessed points, and use this to estimate scores for other budget allocations.

Figure 4.8 shows the type of curves desired. For each of these curves, the x-axis shows the budget allocation to the area, and the y-axis shows the score for each budget allocation. Curves like this are sometimes referred to as *response functions* since they show the response of the evaluation measure scores as a function of the budget allocation.

The estimated score was determined by the director for the lowest and highest allowed budget allocations in each area, as well as for the allocation that was the average of the lowest and highest allowed allocations. A quadratic (second-order polynomial) curve was then fit to these points, and this was used to estimate the scores for intermediate budget allocations. The resulting curves are shown in Figure 4.8.

Value calculations   Figure 4.9 shows a spreadsheet to carry out the calculations required for the product engineering budget decision, and Figure 4.10 shows the equations for this spreadsheet. Range B4:J10 of this spreadsheet shows the parameters for the value function in the same format that we used previously for the networking decision.

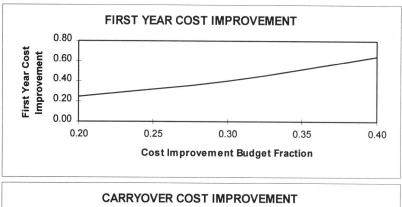

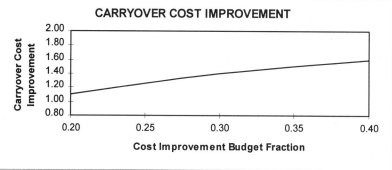

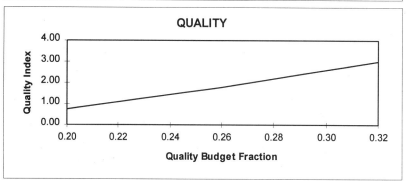

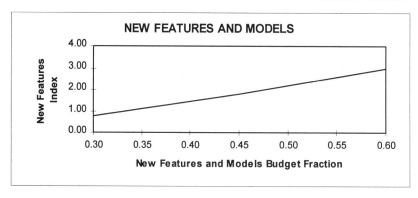

Figure 4.8   *Response functions for product engineering decision*

	A	B	C	D	E	F	G	H	I	J
1			PRODUCT ENGINEERING BUDGET ALLOCATION							
2										
3				VALUE FUNCTIONS						
4		First Year Improve		Carryover Improve		Quality		New Features		
5		x	Value	x	Value	x	Value	x	Value	
6		Low:	0.10	Low:	0.80	0.00	0.00	Low:	0.00	
7		High:	0.80	High:	2.00	1.00	0.40	High:	4.00	
8		Mono:	Increasing	Mono:	Increasing	4.00	1.00	Mono:	Increasing	
9		Rho:	Infinity	Rho:	Infinity			Rho:	-1.64	SUM
10		Weights:	0.10		0.15		0.55		0.20	1.00
11										
12				RESPONSE FUNCTION PARAMETERS						
13		First Year Improve		Carryover Improve		Quality		New Features		
14		b	x	b	x	b	x	b	x	
15	Low:	0.20	0.25	0.20	1.10	0.20	0.75	0.30	0.75	
16	Middle:	0.30	0.40	0.30	1.40	0.26	1.80	0.45	1.80	
17	High:	0.40	0.65	0.40	1.60	0.32	3.00	0.60	3.00	
18										
19				BUDGET ALLOCATION						
20			COST	QUALITY	FEATURES	TOTAL				
21			0.30	0.28	0.42	1.00				
22		Lower Bound:	0.20	0.20	0.30	1.00				
23		Upper Bound:	0.40	0.32	0.60					
24										
25					SCORES					
26		First Year Improve		Carryover Improve		Quality		New Features		
27			0.40		1.40		2.18		1.58	
28										
29				WEIGHTED SINGLE DIMENSIONAL VALUES						
30		First Year Improve		Carryover Improve			Quality		New Features	VALUE
31			0.04		0.08		0.35		0.03	0.50
32										
33					SOLUTION TABLE					
34					Quality					
35			0.50	0.20	0.22	0.24	0.26	0.28	0.30	0.32
36			0.20	0.324	0.376	0.405	0.436	0.470	0.506	0.546
37			0.22	0.324	0.378	0.407	0.439	0.474	0.511	0.551
38			0.24	0.325	0.380	0.411	0.444	0.479	0.517	0.558
39			0.26	0.328	0.384	0.415	0.449	0.485	0.524	0.565
40			0.28	0.332	0.388	0.420	0.455	0.492	0.531	0.573
41		Cost	0.30	0.336	0.394	0.426	0.461	0.499	0.539	0.581
42			0.32	0.342	0.400	0.433	0.469	0.507	0.547	0.589
43			0.34	0.348	0.407	0.441	0.477	0.515	0.556	0.598
44			0.36	0.355	0.414	0.448	0.485	0.524	0.565	0.608
45			0.38	0.362	0.422	0.457	0.494	0.533	0.574	0.617
46			0.40	0.370	0.431	0.466	0.503	0.542	0.584	

Figure 4.9 *Spreadsheet for product engineering decision*

	A	B	C	D	E	F
1						
2						
3						
4					Carryover Improve	
5		x	Value	x	Value	x
6		Low: 0.1		Low: 0.8		0
7		High: 0.8		High: 2		1
8		Mono: Increasing		Mono: Increasing		4
9		Rho: Infinity		Rho: Infinity		
10		Weights: 0.1			0.15	
11						
12						
13					Carryover Improve	
14		b	x	b	x	b
15	Low: 0.2	0.25	0.2	1.1	0.2	
16	Middle: 0.3	0.4	0.3	1.4	0.26	
17	High: 0.4	0.65	0.4	1.6	0.32	
18						
19						
20			COST	QUALITY	FEATURES	TOTAL
21			0.3	0.28	=1-SUM(C21:D21)	=SUM(C21:E21)
22		Lower Bound: 0.2	0.2	0.3	1	
23		Upper Bound: 0.4	0.32	0.6		
24						
25					SCORES	
26				Carryover Improve		
27			=Quad(C21,B15:B17,C15:C17)	=Quad(C21,D15:D17,E15:E17)		
28						
29						
30			First Year Improve	Carryover Improve		
31			=C10*ValueE(C27,C6,C7,C8,C9)	=E10*ValueE(E27,E6,E7,E8,E9)		
32						
33					S	
34						
35			=SUM($C$31:$I$31)	0.2	0.22	0.24
36			0.2	=TABLE(D21,C21)	=TABLE(D21,C21)	=TABLE(D21,C21)
37			0.22	=TABLE(D21,C21)	=TABLE(D21,C21)	=TABLE(D21,C21)
38			0.24	=TABLE(D21,C21)	=TABLE(D21,C21)	=TABLE(D21,C21)
39			0.26	=TABLE(D21,C21)	=TABLE(D21,C21)	=TABLE(D21,C21)
40			0.28	=TABLE(D21,C21)	=TABLE(D21,C21)	=TABLE(D21,C21)
41		Cost	0.3	=TABLE(D21,C21)	=TABLE(D21,C21)	=TABLE(D21,C21)
42			0.32	=TABLE(D21,C21)	=TABLE(D21,C21)	=TABLE(D21,C21)
43			0.34	=TABLE(D21,C21)	=TABLE(D21,C21)	=TABLE(D21,C21)
44			0.36	=TABLE(D21,C21)	=TABLE(D21,C21)	=TABLE(D21,C21)
45			0.38	=TABLE(D21,C21)	=TABLE(D21,C21)	=TABLE(D21,C21)
46			0.4	=TABLE(D21,C21)	=TABLE(D21,C21)	=TABLE(D21,C21)

Figure 4.10a   *Spreadsheet equations for product engineering decision (part 1)*

	G	H	I	J
1				
2				
3				
4			New Features	
5	Value	x	Value	
6	0	Low:	0	
7	0.4	High:	4	
8	1	Mono:	Increasing	
9		Rho:	-1.64	SUM
10	0.55		0.2	=SUM(C10:I10)
11				
12				
13			New Features	
14	x	b	x	
15	0.75	0.3	0.75	
16	1.8	0.45	1.8	
17	3	0.6	3	
18				
19				
20				
21				
22				
23				
24				
25				
26			New Features	
27	=Quad(D21,F15:F17,G15:G17)		=Quad(E21,H15:H17,I15:I17)	
28				
29				
30	Quality		New Features	VALUE
31	=G10*ValuePL(G27,F6:F8,G6:G8)		=I10*ValueE(I27,I6,I7,I8,I9)	=SUM($C$31:$I$31)
32				
33				
34				
35	0.26	0.28	0.3	0.32
36	=TABLE(D21,C21)	=TABLE(D21,C21)	=TABLE(D21,C21)	=TABLE(D21,C21)
37	=TABLE(D21,C21)	=TABLE(D21,C21)	=TABLE(D21,C21)	=TABLE(D21,C21)
38	=TABLE(D21,C21)	=TABLE(D21,C21)	=TABLE(D21,C21)	=TABLE(D21,C21)
39	=TABLE(D21,C21)	=TABLE(D21,C21)	=TABLE(D21,C21)	=TABLE(D21,C21)
40	=TABLE(D21,C21)	=TABLE(D21,C21)	=TABLE(D21,C21)	=TABLE(D21,C21)
41	=TABLE(D21,C21)	=TABLE(D21,C21)	=TABLE(D21,C21)	=TABLE(D21,C21)
42	=TABLE(D21,C21)	=TABLE(D21,C21)	=TABLE(D21,C21)	=TABLE(D21,C21)
43	=TABLE(D21,C21)	=TABLE(D21,C21)	=TABLE(D21,C21)	=TABLE(D21,C21)
44	=TABLE(D21,C21)	=TABLE(D21,C21)	=TABLE(D21,C21)	=TABLE(D21,C21)
45	=TABLE(D21,C21)	=TABLE(D21,C21)	=TABLE(D21,C21)	=TABLE(D21,C21)
46	=TABLE(D21,C21)	=TABLE(D21,C21)	=TABLE(D21,C21)	=TABLE(D21,C21)

Figure 4.10b   *Spreadsheet equations for product engineering decision (part 2)*

Range A12:I17 shows the assessed information for the response functions shown in Figure 4.8. For example, range B15:C17 shows the information for the response of the First Year Cost Improvement score to the cost improvement budget allocation. This shows that for a budget allocation of 0.20 to cost improvement, the First Year Cost Improvement score is 0.25; for a budget allocation of 0.30, the First Year Cost Improvement is 0.40; and for a budget allocation of 0.40, the First Year Cost Improvement is 0.65.

Range B19:F23 shows the budget allocations to the three areas. For the example shown in Figure 4.9, the allocation is 0.30 to cost improvement, 0.28 to quality, and 0.42 to new features and models. The lower and upper bounds on the allowed budget allocations are shown in range C22:E23, and the total budget allocation must be equal to 1, as shown in cell F21.

The portion of the spreadsheet below row 25 is calculated based on the information above that row. We will first discuss how the scores in range B25:I27 are calculated. It is easiest to follow this procedure by examining the equations in this range, as shown in Figure 4.10. Note that these equations use a function called "Quad." This function carries out the quadratic interpolation procedure discussed above to estimate the score for any budget allocation. The Visual Basic code for this function is shown in Figure 4.12, and this will be explained below. For the moment, assume that this function is available.

To take a specific example, consider the equation in cell C27, which calculates the score for the First Year Cost Improvement evaluation measure. The first argument of the Quad function in cell C27 is the cost budget allocation for which the First Year Cost Improvement is desired, and this refers to cell C21, where this number is given. The second argument refers to the range of three cells where the low, middle, and high budget allocation levels are specified, and this argument refers to range B15:B17, where these numbers are given. The third argument refers to the range of three cells where the scores associated with the low, middle, and high budget allocation levels are specified, and this argument refers to the range C15:C17, where these numbers are given. The equations in cells E27, G27, and I27 to calculate the scores for Carryover Cost Improvement, Quality, and New Features and Models are analogous.

Range B29:J31 calculates the weighted single dimensional values associated with the scores in range B25:I27, as well as the overall value for a particular budget allocation. The procedure used for these calculations is identical to that used in the networking strategy decision studied earlier in this chapter, and it is shown by the equations in Figure 4.10.

With the portion of the spreadsheet that has been reviewed so far, it is possible to determine the value for any specified budget allocation. Simply enter the budget allocation in range C21:E21, and read the resulting value in cell J31. The difficulty is that there are an infinite number of different possible budget allocations, and we need to determine which one has the highest value.

For this product engineering decision, we can solve this problem by using Excel's Data Table command. This command was used earlier to conduct a sensitivity analysis in the networking decision. In that analysis, one variable was changed, and the impact of this change was investigated on the values for all four of the decision alternatives. There is a also a "two-input" feature for the Data

Table command, and this can be used to solve the product engineering budget allocation.

To apply this approach, it is first necessary to understand that there are only two independent decision variables in the product engineering decision. This is because the three budget allocations must sum to 1, and hence once two of the budget fractions are set, the third one is automatically determined. In the Figure 4.9 spreadsheet, this fact is taken into account by entering into cell E21 an equation that calculates the budget fraction for new features and models once the budget fractions are set for cost improvement and quality. This equation is shown in Figure 4.10.

To use the two-input data table approach, the table shown in range B33:J46 of Figure 4.9 is determined. To do this, first copy the equation from cell J31 to cell C35. *Important:* Before doing this, change the cell reference in the J31 equation to be absolute, as shown in Figure 4.10. If this is not done, the cell references will be incorrect after the equation is copied.

Then enter the numbers 0.20, 0.22, ..., 0.32 in the range D35:J35. These are different levels for the quality budget allocation, and these cover the range from the lower bound to the upper bound for this budget allocation. Then enter the numbers 0.20, 0.22, ..., 0.40 in the range C36:C46. These are different levels for the cost improvement budget fraction, and these cover the range from the lower bound to the upper bound for this budget allocation.

To complete the table, highlight the range C35:J46 and select the Table command from the Data menu entry. In the dialog that is displayed, enter $D$21 for the "Row Input Cell" and $C$21 for the "Column Input Cell." Then select the OK button, and the table will be completed.

Examination of the solution table in range B33:J46 shows that the most preferred budget allocation is to allocate 0.38 to cost improvement and 0.32 to quality. Therefore, $1 - 0.38 - 0.32 = 0.30$ must be allocated to new features and models.

What about the mysterious black rectangle in cell J46? It appears from the table that the budget allocation corresponding to this cell might have a higher value than the 0.617 shown in cell J45. However, this is not a valid allocation because of the lower bound of 0.30 on the budget allocation to new features and models. The allocation associated with cell J46 would assign $1 - 0.40 - 0.32 = 0.28$ to new features and models. To show that this is not a valid allocation, this cell has been blacked out. (The solution table data are plotted in Figure 4.11.)

Quadratic interpolation formula   The Visual Basic function Quad shown in Figure 4.12 carries out the quadratic interpolation discussed earlier in this section. You do not need to understand the derivation of this function to use it, but for those who are interested, here is a brief discussion. In Figure 4.12, $Xi(1)$ represents the lowest level of the x-variable (that is, the budget fraction in the product engineering decision), and $Yi(1)$ represents the corresponding y-variable level. Similarly, $Xi(2)$ represents the middle level of the x-variable, and $Yi(2)$ represents the corresponding y-variable level. Finally, $Xi(3)$ represents the highest level of the x-variable, and Yi(3) represents the corresponding y-variable level.

A transformation of variables is done to transform the x- and y-variables into corresponding normed variables such that the lowest transformed x-variable level is zero, with an associated transformed y-variable level equal to zero, and the highest transformed x-variable level is 1, with an associated transformed y-variable level equal to 1. With this transformation, it is straightforward to determine the quadratic interpolation equation and use this to calculate the normed y-variable value corresponding to the x-variable input. This normed y-variable value is then transformed back to the original coordinate system and returned as the value of the function.

Generalizations   The analysis of the product engineering decision presented above assumes that there is no uncertainty about the evaluation measure scores that will result from a particular budget allocation. In many realistic resource allocation decisions, there is substantial uncertainty. Analysis of such allocation decisions is considered further in Section 7.7.

The Data Table solution procedure works well as long as there are only two decision variables, which can be set independently. However, this solution procedure breaks down if there are more than two independent variables. The Excel Solver can be used in such situations, and this is considered further in Section 8.4.

4.9 References

D. G. Brooks and C. W. Kirkwood, "Decision Analysis to Select a Microcomputer Networking Strategy: A Procedure and a Case Study," *Journal of the Operational Research Society,* Vol. 39, pp. 23–32 (1988).

J. S. Dyer and R. K. Sarin, "Measurable Multiattribute Value Functions," *Operations Research,* Vol. 27, pp. 810–822 (1979).

W. Edwards, "How to Use Multiattribute Utility Measurement for Social Decisionmaking," *IEEE Transactions on Systems, Man, and Cybernetics,* Vol. SMC–7, pp. 326–340 (1977).

W. Edwards and F. H. Barron, "SMARTS and SMARTER: Improved Simple Methods for Multiattribute Utility Measurement," *Organizational Behavior and Human Decision Processes,* Vol. 60, pp. 306–325 (1994).

W. Edwards and J. R. Newman, "Multiattribute Evaluation," in H. R. Arkes and K. R. Hammond (eds.), *Judgment and Decision Making: An Interdisciplinary Reader,* Cambridge University Press, Cambridge, England, 1986, pp. 13–37.

W. Edwards, D. von Winterfeldt, and D. L. Moody, "Simplicity in Decision Analysis: An Example and a Discussion," in D. E. Bell, H. Raiffa, and A. Tversky (eds.), *Decision Making: Descriptive, Normative, and Prescriptive Interactions,* Cambridge University Press, Cambridge, England, 1988, pp. 443–464.

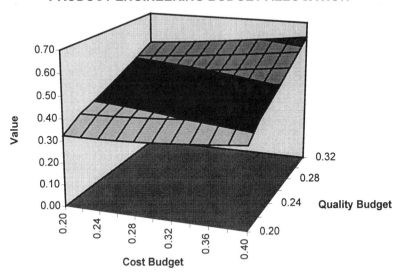

Figure 4.11 *Graph of product engineering solution table*

```
Function Quad(x, Xi, Yi)
  Xnorm = (x - Xi(1)) / (Xi(3) - Xi(1))
  Xm = (Xi(2) - Xi(1)) / (Xi(3) - Xi(1))
  Ym = (Yi(2) - Yi(1)) / (Yi(3) - Yi(1))
  a = (Ym - Xm) / (Xm * Xm - Xm)
  b = 1 - a
  Ynorm = a * Xnorm * Xnorm + b * Xnorm
  Quad = Yi(1) + Ynorm * (Yi(3) - Yi(1))
End Function
```

Figure 4.12 *Equations for quadratic interpolation function*

D. L. Keefer and C. W. Kirkwood, "A Multiobjective Decision Analysis: Budget Planning for Product Engineering," *Journal of the Operational Research Society,* Vol. 29, pp. 435–442 (1978).

C. W. Kirkwood and R. K. Sarin, "Preference Conditions for Multiattribute Value Functions," *Operations Research,* Vol. 28, pp. 225–232 (1980).

D. von Winterfeldt and W. Edwards, *Decision Analysis and Behavioral Research,* Cambridge University Press, Cambridge, England, 1986.

4.10 Review Questions

R4-1   Explain why it is necessary to combine evaluation measures in order to rank alternatives.

R4-2   Describe the purpose(s) of using single dimensional value functions.

R4-3   What are the two types of single dimensional value functions we have studied, and when is each type used?

R4-4   Describe the steps in determining a piecewise linear single dimensional value function.

R4-5   Describe the steps in determining an exponential single dimensional value function.

R4-6   Can a piecewise linear single dimensional value function be used for an evaluation measure with many different levels?

R4-7   Describe the advantages and disadvantages of using a quantitative procedure to combine evaluation measures and rank alternatives.

R4-8   Using a single dimensional value function over profits from $-\$100,000$ to $\$200,000$, a profit of $\$10,000$ has a single dimensional value of 0.5. What does this mean?

R4-9   Describe the steps in finding weights for a value function.

R4-10   Cost ranges from $\$100,000$ to $\$200,000$, and quality of service ranges from 0 to 7 (higher levels are better). Using a value function over these evaluation measures, an alternative with a cost of $\$175,000$ and a quality of service of 3 has a value of 0.4. What does this mean?

R4-11   For the value function in the preceding question, the weight for cost is twice as great as the weight for quality of service. What does this mean?

R4-12   For the value function in the preceding two questions, suppose the range considered for cost is changed to extend from $\$50,000$ to $\$200,000$. What can we say about the weights for cost and quality of service?

R4-13   What does the expression "(swing) weights depend on ranges" mean?

R4-14   Describe advantages and disadvantages of using weights on evaluation measures in a decision analysis.

R4-15   Discuss the role of evaluation measure ranges in (i) the assessment of a value function, and (ii) the interpretation of analysis results when you use the value function.

R4-16   Discuss the following statement: "Combining different evaluation measures using weights seems like 'comparing apples and oranges.' I learned somewhere that you are not supposed to do that."

R4-17   Discuss the following statement about value functions: "All the numbers that you use in a value function are subjective and just a matter of opinion, so you can get any result you want. What good is that?"

R4-18   Discuss the role of sensitivity analysis in decision analysis.

4.11 Exercises

4.1   Upon being introduced to the methods in this chapter, a distinguished senior operations analyst said that this procedure reminded him of a story about how they weigh hogs in Arkansas. "First they find a board and a bunch of rocks. Then they lay the board across the top of the biggest rock and put the hog on one end of the board. After that, they pile rocks on the other end of the board until there are just enough rocks on the board to balance the hog. Finally, they guess the weight of the rocks."

His point with this story is that when you conduct a multiobjective value analysis, you take one subjective task (selecting the most preferred alternative) and replace it with another, possibly more complex, subjective task (specifying evaluation measures and determining a value function). He was questioning whether this was worth doing. Why not just directly make the decision with unaided judgment?

(i) Discuss reasons why it might be preferable to conduct a multiobjective value analysis instead of using unaided judgment to make a decision. In your answer, particularly address the use of value analysis in an organizational decision making setting where there are multiple stakeholders for the decision, and it may be important to have an audit trail to show the basis for the decision.

(ii) The analyst referred to in this exercise spent many years working on air traffic operations. This work addressed such issues as the shortest intervals that should be allowed between planes landing or taking off from an airport. Discuss whether this type of decision making requires consideration of subjective issues such as tradeoffs between cost and safety.

4.2 Suppose that for a decision problem there are two attributes, User Satisfaction and Organizational Impact, where scores for User Satisfaction can range from −1 to 2, and scores for Organizational Impact can range from −1 to 1. Higher scores for User Satisfaction are more preferred, while higher scores for Organizational Impact are less preferred. Furthermore, the single dimensional value function over each of these attributes is linear. (That is, it is a straight line.)

There are two alternatives. Alternative 1 has a User Satisfaction score of −1 and an Organizational Impact score of 0, while Alternative 2 has a User Satisfaction score of 1 and an Organizational Impact score of −1.

(i) Make graphs of the single dimensional value functions for User Satisfaction and Organizational Impact.

(ii) Determine the single dimensional values for Alternative 1 and Alternative 2. (There are four numbers in all.)

4.3 Find the weights for three evaluation measures—Cost, Efficiency, and Effectiveness—over the ranges $\$0 \le \text{Cost} \le \$1,000$, $-1 \le \text{Efficiency} \le 3$, and $-2 \le \text{Effectiveness} \le 2$, given the following assessments:

(i) The value increase is two times as great when Cost is decreased from $\$1,000$ to $\$0$ as when Efficiency is increased from −1 to 3.

(ii) The value increase is three times as great when Effectiveness is increased from −2 to 2 as when Efficiency is increased from −1 to 3.

(Note that higher scores are more preferred for the Efficiency and Effectiveness attributes, while lower scores are more preferred for the Cost attribute.)

4.4 Freeport Power and Light Company must select between two possible sites (Free River and Loco Bend) for the new power plant that it is building. The evaluation measures for this site selection are construction cost, in millions of dollars, and environmental impact, measured using a constructed scale with possible levels 0, 1, 2, 3, and 4. Higher levels on this scale are less preferred. The single dimensional value function over cost is exponential over the range $\$500$ million to $\$1,000$ million, and the midvalue for this range is $\$800$ million.

The value increment for improving from 4 to 3 on the environmental impact evaluation measure scale is the same as the value increment from 3 to 2, and is also the same as the value increment from 1 to 0. The value increment from 2 to 1 is twice as great as the value increment from 4 to 3. The value increment for improving from $\$1,000$ million to $\$500$ million on the cost evaluation measure is the same as the value increment for improving from 4 to 0 on the environmental impact evaluation measure.

Siting the power plant at Free River will result in a cost of $\$600$ million and an environmental impact of 3. Siting the plant at Loco Bend will result in a cost of $\$850$ million and an environmental impact of 1.

(i) Determine an additive value function over cost and environmental impact that is consistent with the information presented above.

(ii) Use the value function found in part (i) to determine the overall value for each of the two sites (Free River and Loco Bend). Which site is more preferred?

(iii) Determine the range over the weight on cost for which the more preferred site found in part (ii) remains the best site.

4.5 Aba Manufacturing Company is considering modifying the batch production process for the Gee-Wizz Widget to one of two new processes. The fixed costs would not change for either of the proposed processes; however, both the marginal cost per production batch and the percent yield per batch would change.

Consider a range for marginal cost from $10 to $20 per batch and a range for percent yield per batch from 25 to 75 percent. Assume that the single dimensional value function over marginal cost is exponential and the single dimensional value function over percent yield is piecewise linear with two linear segments, one from 25 to 40 percent yield and one from 40 to 75. Further assume that the midvalue for the marginal cost range from $10 to $20 is $17, and that the value increase from increasing percent yield from 40 to 75 percent is three times as great as the value increase from increasing percent yield from 25 to 40 percent. In addition, the value increase from increasing the percent yield from 25 to 75 percent is one-half the value increase from decreasing marginal cost per batch from $20 to $10.

The marginal cost per batch and percent yield per batch for the current production process and the two proposed new processes are

> Current Process: $15 and 50%
> Proposed Process 1: $18 and 60%
> Proposed Process 2: $13 and 35%

Answer the following questions:

(i) Determine the multiobjective value function for this decision problem.

(ii) Use this value function to determine the most preferred alternative. If you use a spreadsheet to answer this question, include a copy of the equations for this spreadsheet with your answer.

(iii) Determine to two decimal places the range of values for the weight on marginal cost per batch for which each of the three alternatives is most preferred. Include with your answer any computer output that you use to answer this question. Assume that, while you are varying this weight, the ratio of the other two weights remains constant.

4.6 Zenren ElectroProducts is considering making changes in the way it procures the Z-1254 microprocessor chip. In addition to the Status Quo, which is in-house production at its current California plant, Zenren is considering either purchasing from an offshore manufacturer or making the chip in its new Arizona plant. There are three evaluation measures (attributes) for this decision: cost per batch of 10,000 chips, operational ease, and quality as measured by the number of defective chips per batch of 10,000.

For the following analysis, assume a range from $500 to $1,000 for cost per batch, a range from 1 to 4 for the (constructed) evaluation measure scale for operational ease, and a range from 0 to 100 for the number of defects per batch. Note that preferences are monotonically increasing for operational ease ("more is better"), while preferences are monotonically decreasing for cost and number of defects ("more is worse"). The scores for each of the three alternatives on these three evaluation measures are as follows:

Alternative	Cost	Operational Ease	Number of Defects
Status Quo	750	4	55
Offshore	550	1	40
Arizona	675	2	30

A multiobjective value analysis is being done to evaluate these alternatives. The single dimensional value function over cost per batch is exponential with a midvalue of 750. The single dimensional value function over operational ease is piecewise linear. The value increment going from an operational ease score of 1 to a score of 2 is the same as the value increment going from an operational ease score of 2 to a score of 3. The value increment going from an operational ease score of 3 to a score of 4 is three times as great as the value increment going from an operational ease score of 1 to a score of 2. The single dimensional value function over number of defects per batch is exponential with a midvalue of 30.

The value increment going from an operational ease score of 1 to an operational ease score of 4 is twice as great as the value increment going from a cost per batch of $1,000 to a cost of $500, while the value increment going from a number of defects per batch of 100 to a number of defects per batch of 0 is the same as the value increment going from an operational ease score of 1 to an operational ease score of 4.

(i) Show that the exponential constant ρ for the cost-per-batch single dimensional value function is equal to infinity, while the exponential constant for the number-of-defects-per-batch single dimensional value function is equal to -55.5.

(ii) Show that the single dimensional values for operational ease scores of 2 and 3 are 0.2 and 0.4, respectively.

(iii) Determine the swing weights for all three evaluation measures.

(iv) Create a spreadsheet model for this decision problem, and include a printout of the equations for this spreadsheet with your homework solutions.

(v) Using the model from part (iv), determine the values of the three alternatives, and find which alternative is most preferred.

(vi) Determine to two decimal places the range of values of the swing weight on the operational ease evaluation measure for which the most preferred alternative found in part (v) remains most preferred. Assume that, while you are varying this weight, the ratios of the other two weights remain the same.

4.7 MiTech Foundry plans to purchase a new tester for use in its manufacturing processes for very large scale integrated circuits. Three possible testers are under consideration: the OR9000, the J941, and the Sentry 50. There are four evaluation measures: cost in hundreds of thousands of dollars, accuracy in picoseconds, delivery time in months, and uptime in percent.

For the following analysis, assume a range of one to three hundred thousand dollars for cost, 500 to 1000 picoseconds for accuracy, three to six months for delivery time, and 95 to 99 percent for uptime. Preferences are monotonically increasing ("more is better") for uptime, and monotonically decreasing ("more is worse") for cost, accuracy, and delivery time. The single dimensional value function over cost is linear (that is, this function is piecewise linear with only one straight line segment). The single dimensional value function over accuracy is exponential with a midvalue of 900.

The single dimensional value functions over delivery time and uptime are piecewise linear. The value function for delivery time has two segments, which join at a delivery time of 4 months. The value increment from decreasing delivery time from four months to three months is three times as great as the value increment from decreasing delivery time from six months to four months. The value function for uptime has three segments, one from 95 to 96 percent, one from 96 to 97 percent, and one from 97 to 99 percent. The value increment from increasing uptime from 96 to 97 percent is the same as the value increment from increasing uptime from 97 to 99 percent. The value increment from increasing uptime from 95 to 96 percent is twice as great as the value increment from increasing uptime from 96 to 97 percent.

The value increment obtained by decreasing accuracy from 1000 picoseconds to 500 picoseconds is three times as great as the value increment obtained by decreasing cost from three hundred thousand dollars to one hundred thousand dollars. The value increment obtained by decreasing delivery time from six months to three months is equal to the value increment obtained by increasing accuracy from 95 percent to 99 percent. The value increment obtained by increasing uptime from 95 percent to 99 percent is twice as great as the value increment obtained by decreasing cost from three hundred thousand dollars to one hundred thousand dollars.

The evaluation measure scores (levels) for the three alternatives are given in the following table:

Alternative	Cost	Accuracy	Delivery Time	Uptime
OR9000	1.4	900	3	98
J941	1.0	1000	6	96
Sentry 50	2.8	600	5	95

(i) Determine the parameters for the value function for this decision. This requires finding (1) the exponential constant for any exponential single dimensional value functions, (2) the values for the points at which the segments of piecewise linear single dimensional value functions join, and (3) the evaluation measure weights.

(ii) Create a spreadsheet model for this decision problem, and include a printout of the equations for this model with your homework solution.

(iii) Use your spreadsheet model to calculate the values for the three testers, and determine which alternative is preferred.

(iv) Determine to two decimal points the range of the weight on cost for which the most preferred alternative found in part (iii) remains most preferred. Assume, while you are varying this weight, that the ratios of the other three weights remain the same.

4.8 Project (Part D). This is a continuation of the project from Chapters 1, 2, and 3. In this part of the project, you determine a value function, collect data about your alternatives, and conduct a value analysis for your alternatives.

(i) Assess a value function for your decision. Present the general procedure used for the assessment but not a blow-by-blow description. Include assessed "raw data" used to determine the value function, perhaps in a table. Show the math used to obtain the final value function from the assessed raw data (perhaps in a figure), as well as the parameters for the final value function.

(ii) Present the procedure used to determine the evaluation measure scores (levels) for each alternative, along with the final evaluation measure scores. (The final scores can be presented in a table.) Reference your data sources, including interviews with experts, in standard bibliographic style.

(iii) Present the value calculations for the alternatives and a sensitivity analysis. Briefly describe how computations were done, but you do not have to present the actual computations if you use a spreadsheet program to do these calculations. Include a display of the equations for any spreadsheet model that you use. Conduct and present a systematic sensitivity analysis to investigate how variations in key assumptions impact the analysis results. Such a sensitivity analysis typically addresses at least (1) the impact of variations in evaluation measure weights, and (2) the impact of variations in evaluation measure scores for which there is significant uncertainty.

(iv) Present your conclusions based on the analysis in the preceding parts, including a qualitative discussion of the reasons that the preferred alternative is best. The goal of this discussion is that someone who does not understand the details of decision analysis methods will find your discussion to be a convincing argument for the preferred alternative. That is, the analysis should not be a mysterious procedure, but rather a way of developing insight about the key factors in the decision and how these lead to selection of the preferred alternative.

4.9 The Manufacturing Process Improvement Group at Multibolt Industries is planning the allocation of its budget for next year. The group must divide its entire budget between two activities: throughput improvement and defect reduction. Throughput is measured in parts per hour, and defects are measured in defects per thousand parts produced. Currently, throughput is 1,000 parts per hour, and defects are at 10 per thousand parts produced.

If no budget is allocated to throughput improvement, then throughput will remain at 1,000 parts per hour, while if the entire available budget is allocated to throughput improvement, this will increase to 2,000 parts per hour. Improvement in throughput is linear with increasing allocations of budget to throughput improvement. Improvement in defects is also linear with increasing allocations of budget to defect reduction. If no budget is allocated to defect reduction, then defects will remain at 10 per thousand parts produced, and if the entire available budget is allocated to defect reduction, then defects will improve to 5 per thousand parts produced.

The Manufacturing Process Improvement Group will use an additive value function to evaluate different budget allocations. The two evaluation measures to be included in this value function are throughput and defects. The range for throughput is from 1,000 to 2,000 parts per hour, and the range for defects is from 5 to 10 defects per thousand parts produced. Preferences are increasing with respect to throughput and decreasing with respect to defects. The single dimensional value function over throughput is linear, and the single dimensional value function over defects is exponential with a midvalue of 9 defects per thousand parts produced. The increment in value obtained by improving defects from 10 per thousand parts produced to 5 per thousand parts produced is three times as great as the value increment obtained by improving throughput from 1,000 per hour to 2,000 per hour.

(i) Determine the parameters for an additive value function over throughput and defects using the information given above.

(ii) Develop a spreadsheet or other computer model that can be used to determine the overall value of different allocations of budget to throughput improvement and defect reduction. Include the equations for this model with your solution.

(iii) Use the model you developed in part (ii) to prepare a graph of the overall value of a budget allocation as a function of the budget allocated to throughput improvement. [*Hint:* The entire budget must be allocated to either throughput improvement or defect reduction. Thus, the range of possible budget allocations to throughput improvement is from 0 to 1, as a fraction of the available budget. You may find a data table useful in answering this question.]

(iv) Use the model you developed in part (ii) to determine the most preferred budget allocation. Determine this accurate to two decimal places for the fraction of available budget allocated to each activity.

CHAPTER 5

Thinking about Uncertainty

The discovery of large, relatively safe, companies trying to attract buyers for their bonds by offering higher-than-market yields made lots of news in the mid-1980s. In 1986, Drexel Burnham Lambert predicted that this new "high-yield bond" market would be highly lucrative and was paying yearend bonuses in the millions of dollars to keep some first- and second-year bond traders. By the end of 1989, Drexel Burnham was bankrupt.

In April 1990, all 45,000 seats in the Indianapolis Hoosierdome sold out 90 minutes after going on sale for *Farm Aid 4*. Three previous *Farm Aid* programs raised $12 million for farmers. A decade of rising farmland prices in the 1970s led many farm investors to predict even greater increases during the 1980s. Instead, where a typical acre of farmland brought about $1200 in 1981, by 1986 the price had fallen to $600, and the *Farm Aid* programs began.

Most people can supply similar stories of "sure thing" forecasts that led to unattractive consequences, either for themselves or for others they know. It doesn't seem reasonable to attribute these instances to sudden strikes of collective incompetence. What else might be going on?

This chapter reviews what is known about the mistakes that people make when they think about uncertainty, and then presents procedures for thinking more clearly about uncertainty. In particular, the methods of *probability analysis* are introduced, including approaches for eliciting probabilities. In the next chapter, these approaches are applied to analyzing decisions with uncertainty.

Several decades of research show clearly that people do a poor job of unaided reasoning about situations where there is uncertainty. Therefore, the methods presented in this and succeeding chapters can substantially improve the quality of decision making when there is uncertainty.

5.1 What Is Probability?

Most people have an intuitive notion of probability from everyday life. When we hear that the weather forecast is for an *eighty percent chance* of rain, we usually decide that we should be prepared for rain. On the other hand, if the probability of rain is given as ten percent, we are less worried about the possibility of rain. Informally, we expect that it will actually rain about 80 out of 100 times (that is, four out of five times) when the forecast says there is an eighty percent chance of rain. Similarly, we expect that it will rain about 10 out of 100 times (that is, one out of ten times) when the forecast says there is a ten percent chance of rain.

For these sorts of repetitive events, we think of probability as representing the *relative frequency* of occurrence of the event of interest. Thus, for a fair coin, the chance of a head coming up on any toss is fifty percent. That is, we expect that about 50 times out of 100 (or one time out of two) when the coin is tossed, a head will come up.

From a consideration of such situations, we can see that a probability, when expressed in percentage terms, must be a number between zero and one hundred percent, inclusive. Probabilities are also expressed as *proportions*, in which case the number will be between zero and 1, inclusive. Thus, saying that a probability is 0.25 is equivalent to saying that the probability is 25 percent.

The relative frequency interpretation of probability works all right when considering events that repeat, but what about situations that will happen only once, if at all? For example, suppose that an expert says the probability of the local nuclear power plant exploding due to an accident during its operational lifetime is 0.0000001. This seems to be a meaningful statement, and we are comforted to see all those zeros after the decimal point. We would probably be much more concerned if the expert said that the probability was 0.01 (that is, one in a hundred). However, there is no possibility of repetitive events in this case. If the nuclear power plant explodes, then it is gone—there is no possibility of counting many lifetimes of the plant and observing the proportion of those for which the plant explodes. Either it explodes or it doesn't. While the probability statement seems to have meaning, it cannot be interpreted as a relative frequency.

Actually, when you carefully consider the weather forecast and the coin toss mentioned earlier, it isn't clear that relative frequency is relevant even there. You are interested only in a particular day, or a particular coin toss. The specific day or toss of interest will not repeat, but still the probability statements seem to mean something.

Many management uncertainties are similar to the nuclear power plant case in the sense that they deal with situations where it is not meaningful to think about relative frequencies. If you are interested in whether a potential new product that might be introduced will be a success, then relative frequency isn't of much use. Either the product will be a success, or not. You can't introduce it a hundred times, and count the proportion of times that it is a success.

Philosophers and statisticians have debated the meaning of probability for centuries, and a widely held view today is that probabilities are statements about the judgments of individuals. Specifically, a probability is a numerical specification of an individual's *degree of belief* that an uncertain event will occur. Thus, probabilities are ultimately *subjective* in nature, and it is possible that two different people might have different probabilities for the same event.

For example, if you have not yet looked out the window today, and you hear the weather forecast on the radio is for an eighty percent chance of rain, then you might assign a probability of eighty percent to the likelihood of rain. On the other hand, if your spouse has already looked out the window and discovered that it is, in fact, raining, he or she will probably assign a probability of one hundred percent to rain today. Similarly, two marketing specialists might have different views on the probability that a proposed new product will be a success.

If probabilities are subjective, then they must ultimately be determined by asking someone for them. If the person whose probability we need to determine is familiar with probability concepts, then we may be able to directly ask for the required probability. If not, then we need to use some sort of measuring instrument. The *probability wheel* is one instrument for measuring probabilities. Figure 5.1 shows sketches of two probability wheels. These wheels are shown on their stands, and each wheel is free to spin around its center. Each wheel has two sectors, one lighter and one darker, and the proportion of a wheel that is of each shade can be adjusted. (The wheel on the left has one-half of its disk dark, while the wheel on the right has one-quarter of its disk dark.) There is a fixed pointer above each wheel, and after the wheel is spun, the pointer will be pointing to either the dark or light portion of the disk. Suppose each wheel is perfectly balanced like a fine roulette wheel so that when it is spun, the chances are equally likely that the pointer will point to any spot along the edge of the disk.

Using a probability wheel as a measuring instrument, the probability of an event (for example, that it will rain tomorrow) can be elicited as follows: Consider two different situations. The first is out in the real world where it either rains or it doesn't rain. Suppose that you will receive an attractive prize if it rains (for example, $100.00), and you will not receive this prize if it doesn't rain. The second situation involves a single spin of the probability wheel. If, after the spin, the pointer points to the dark sector of the wheel, you will receive the same attractive prize ($100.00 for our example), and if the pointer points to the light sector of the wheel, you will not receive this prize.

To elicit the probability of rain tomorrow, adjust the portion of the probability wheel that is dark until you are indifferent between selecting a spin of the wheel (and hence receiving $100.00 only if the pointer points to the dark sector) or waiting until tomorrow to see whether it rains (in which case you will receive $100.00) or doesn't rain (in which case you receive nothing). When you have adjusted the portion of the wheel that is dark so that you are indifferent between participating in these two uncertain situations (one involving the probability wheel, and the other involving the weather), then your probability of rain is equal to the portion of the wheel that is dark. Thus, if you are indifferent between betting on the left wheel in Figure 5.1a or betting on it raining tomorrow, then your probability that it will rain tomorrow is one-half (fifty percent or 0.50).[1]

When this approach to specifying probabilities is presented to a group, someone often says, "What do you mean, that is the probability of rain!? The wheel doesn't have any influence on whether it will rain or not!" Of course, this is true, but that is the point! Probabilities are judgments. The probability wheel is a measuring gauge that you can use to assess your probabilities for events. The measuring gauge (that is, the wheel) doesn't affect whether it will rain or not, but it measures your probability for rain.

As this discussion makes clear, probability is judgmental (that is, subjective). The experimental results presented later in this chapter show that most people have a difficult time making probability judgments that are well calibrated to the real world. Therefore, you may question whether it is worth the effort necessary to determine accurate probabilities. The next section discusses reasons for using probabilities in analyzing decisions.

5.2 The Importance of Probability

The widespread use of probabilities in weather forecasts means that the idea of using probabilities to express judgments about uncertainty is not completely novel to many people. However, as the review of relevant research in the next section shows, people have difficulty giving probability statements that accurately reflect the real situation.

One possible reaction to these research results is to reject the use of probabilities based on direct judgment and conclude that probabilities should be used only when these have a good basis in measured data. The difficulty with taking this position is that it rules out using probabilities for many management decisions. Historical data are often only loosely relevant to the current situation, and thus judgment is necessary to obtain probabilities in many practical situations. For example, the economic recovery of the early 1990s in the United States stalled out twice before finally taking off. With both of the nonrecoveries, appeals were made to the historical data about other recoveries to support the view that a recovery was under way. Unfortunately, the historical record did not apply in those cases. Similarly, with many management decisions, historical data are not sufficient to make forecasts—judgment must also be applied.

Some observers conclude from the difficulty of obtaining relevant historical information on which to base probabilities that we should not try to use probabilities and instead should be content to use verbal descriptions such as "very likely," "highly unlikely," etc. Merkhofer (1987) describes an example that shows the difficulty with this approach:

---

[1] There is another subtle condition that should be met for probability measurement using a probability wheel—the spin of the wheel, and the payoff from this, should be visualized as occurring at the same time that the uncertain event of interest would be resolved.

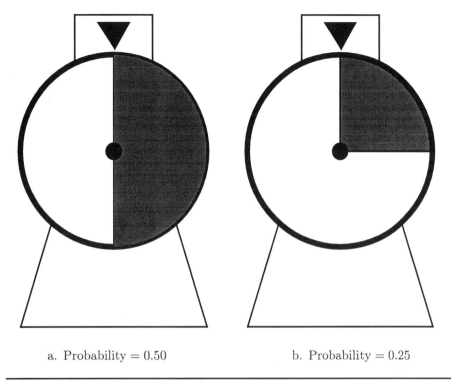

a. Probability = 0.50   b. Probability = 0.25

Figure 5.1   *Two probability wheels*

[In a decision analysis seminar,] participants were individually asked to assign probabilities to common expressions such as "very likely to occur," "almost certain to occur," etc. The fact that different individuals assign very different probabilities to the same expression demonstrates vividly the danger of using words to communicate uncertainty. The seminar leader had just completed the demonstration and was about to move on to another topic when the president of the company said, "Don't remove that slide yet." He turned to one of his vice presidents and said the following: "You mean to tell me that last week when you said the Baker account was almost certain, you meant 60 percent to 80 percent probability? I thought you meant 99 percent!" If I'd known it was so low, I would have done things a lot differently."

Researchers have investigated how people use verbal expressions about uncertainty, and they have found that different people often assign widely differing numerical probabilities to the same verbal expression. Thus, using such expressions to attempt to express judgments about uncertainty poses a serious risk of miscommunication.

Probabilities, on the other hand, are unambiguous. If someone says that the probability of rain tomorrow is fifty percent, then the exact meaning of this is specified in terms of a probability wheel. On the other hand, if that same person says that it is "fairly likely" to rain tomorrow, what does that mean? (To me, it means a fifty to sixty percent chance. What does it mean to you?)

In addition to aiding unambiguous communication about uncertainty, probabilities also allow defensible quantitative analyses of decision problems that have uncertainty. In this day of personal computer spreadsheets and databases, at least some degree of quantitative analysis **will** be conducted by some interested party for most significant management decisions. In the absence of probabilities, the analysis will have to assume no uncertainty or to use informal *ad hoc* methods for considering uncertainty. Analysis methods based on probabilities have been developed over several centuries, and these methods are generally accepted by experts as the correct way to analyze situations with uncertainty.

The difficulty with using a single number, sometimes called a *point estimate*, to represent an uncertain quantity is illustrated by Figure 5.2. Suppose that this figure represents a particular sales representative's judgment about the relative likelihood of various different possible levels of sales for a specified item over the next year in the sales representative's territory. In this figure, the horizontal axis represents the possible different levels of sales, while the vertical axis represents the relative chance that sales will be at any specified level. From this figure, we see that the most likely level of sales is around 200 units, but that there is some chance that sales could be much lower than this (all the way down to almost zero units) or much higher (even as high as 500 units).

If the sales representative is asked to provide a single point estimate forecast of sales, what number will he or she give? The number will probably depend on what the representative sees as the relative benefits and costs of making a forecast that is either under or over the actual number. For example, if he or she thinks the number will be used to set sales targets, and if bonuses are given for exceeding a sales target, then it makes sense to give a low number, perhaps 100. If the sales representative wants to have the best chance that the actual sales are close to forecast, then it makes sense to give a number around 200. On the other hand, if there is an advantage to providing a high forecast (for example, if some sales staff might be laid off because forecasted sales are low), then it makes sense to provide a high number, perhaps 400.

As this example shows, using a single number to represent an uncertain quantity mixes up judgments about uncertainties with assessments of the desirability of various outcomes. In addition, giving a single number does not provide information about how much variation is possible in the actual number.

Of course, it should be kept in mind that judgmentally elicited probabilities are only as good as the people who provide the judgments. When there are relevant historical data, then these should be used. (The field of *statistics* provides methods for drawing defensible conclusions from data when there is uncertainty.) However, it is often necessary to rely on judgment to determine probabilities, and care should be taken in eliciting this judgment to address the various heuristics and biases that humans are prone to when thinking about uncertainty. These are discussed in the next section.

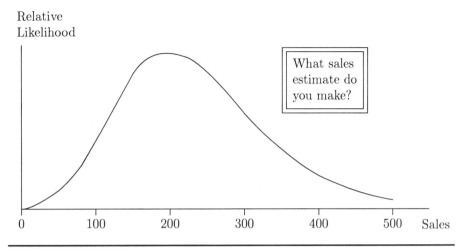

Figure 5.2   *The difficulty with point estimates*

5.3 Heuristics and Biases in Reasoning under Uncertainty

During the academic year 1968-69, Alpert and Raiffa (1982) conducted a probability elicitation experiment involving all 800 students from the first-year class in the MBA program at the Harvard Business School. (Three other subject groups were also included, and the results for the other groups were very similar to those for the MBAs.) The purpose of the experiment was to determine how good the students were at probability elicitation and whether it was possible to train them to be better.

To determine how good someone is at probability elicitation, it is necessary to consider that person's probability distributions for a number of different uncertain quantities. To see why this is true, suppose that someone specifies a probability distribution for a single continuous variable and that the actual value for the variable turns out to be higher than the 0.99 fractile of the specified distribution. While the person thought this outcome was not likely, it is not possible to say from this result that the specified distribution is poor since we expect that one out of 100 actual values will exceed the 0.99 fractile.

On the other hand, suppose someone provides distributions for 1000 variables, and the actual values for all 1000 are higher than the 0.99 fractiles for their respective distributions. This seems to be clear evidence that the person doing the elicitation is not determining distributions that correspond to reality.

In the experiment conducted by Alpert and Raiffa, the 0.01, 0.25, 0.50, 0.75, and 0.99 fractiles were elicited for a number of continuous variables whose actual values could be determined, although the students did not know these actual values when they provided the distributions. Examples of quantities for which distributions were elicited included these:

1 The percentage of respondents expressing an opinion to a March 1968 Gallup Poll surveying a representative sample of adult Americans who felt that public school teachers should be permitted to join unions.

	Group A		Group B	
	Round 1	Round 2	Round 1	Round 2
Actual values falling inside the 0.25 to 0.75 ranges (%)	32.9	40.3	33.4	46.4
Actual values falling outside the 0.01 to 0.99 ranges (%)	38.8	24.9	42.6	22.2

Table 5.1   *The Harvard experiment results*

2 The number of "Physicians and Surgeons" listed in the 1968 Yellow Pages of the phone directory for Boston and vicinity.

3 The total number of students currently enrolled in the Doctoral Program at the Harvard Business School.

4 The total egg production in millions in the United States in 1965.

Two separate questionnaire forms were prepared, each requesting probability distributions for ten variables, and each form was distributed to half of the class. After the students provided their fractiles for each question on the form they had received, the results for each form were summarized and provided to the students. The students were then given the other form and once again provided fractiles. These were also analyzed. The results for each group are summarized in Table 5.1.

If the students' distributions had been well calibrated with the "real world," then 50 percent of the actual values would have fallen within the 0.25 to 0.75 range. Instead, with both groups, approximately 33 percent fell within that range for Round 1, and 40 or 46 percent fell within this range for Round 2, depending on the group. Similarly, if the students' distributions had been well calibrated, 2 percent of the actual values would have fallen outside the 0.01 to 0.99 percent range. Instead, about 39 or 43 percent fell outside that range for Round 1 and 25 or 22 percent for Round 2.

In short, the students were *overconfident* about how well they could predict the values of the variables—their distributions were too narrow. This was particularly dramatic for the extreme tails. The percentage of these "surprises" for Group A during Round 1 was off from the correct result by a factor of about $39/2 = 19.5$. (That's a 1950% error!) Group B was even worse with an error factor of about $43/2 = 21.5 = 2150\%$. Since Alpert and Raiffa's experiment, this result has been replicated innumerable times. The effect holds across all groups (with the limited exceptions discussed below), and the effect is consistently as strong as shown in Table 5.1.

However, the table also shows that even a modest amount of training improves people's ability to specify probability distributions. The distributions for Round 2 were more accurate than for Round 1, although they still displayed substantial overconfidence. In fact, experiments done with people who estimate probabilities for a living (for example, meteorologists providing probabilistic weather forecasts) show that they become much better at assessing these distributions with practice.

Representative Thinking and Availability

In the years since the Harvard experiments, many experiments have been con-ducted investigating specific heuristics and biases in people's judgments under uncertainty that might account for overconfidence. Experiments show that peo-ple tend to draw conclusions about probabilities based on *representative* char-acteristics and ignore other relevant aspects. Dawes (1988, p. 75) quotes an example from *Management Focus*:

> Results of a recent survey of 74 chief executive officers indicate that there may be a link between childhood pet ownership and future career success.
>
> Fully 94% of the CEOs, all of them employed within Fortune 500 com-panies, had possessed a dog, a cat, or both, as youngsters ...
>
> The respondents asserted that pet ownership had helped them to de-velop many of the positive character traits that make them good managers today, including responsibility, empathy, respect for other living beings, generosity, and good communication skills.

However, as Dawes notes, "For all we know, *more* than 94% of children raised in the backgrounds from which chief executives come had pets, in which case the direction of dependency would be negative." To explain this in more detail: The survey shows that the probability of someone having been a childhood pet owner, *given that he or she is a CEO*, is 0.94. However, this is being interpreted to mean that the probability of someone becoming a CEO, *given that he or she was a childhood pet owner*, is 0.94. These are very different statements.

[Here is an analysis of the difficulty for readers who understand conditional probability. The problem with the reasoning in the quote above is that the direction of the conditioning for a conditional probability is misinterpreted. If PET = "Childhood Pet Ownership" and CEO = "Currently a CEO," then the survey shows that $P(\text{PET}|\text{CEO}) = 0.94$. However, it is being interpreted as $P(\text{CEO}|\text{PET}) = 0.94$. The first result implied the second only in very special circumstances.]

Another example illustrates the reasoning difficulty even more clearly. Sup-pose the survey had asked the CEOs whether they had drunk fluids as a child. Presumably all of the respondents would have answered affirmatively, but we would not conclude from this that drinking fluids as a child increases your prob-ability of becoming a CEO relative to other occupations.

After we think about this situation more closely, we realize the error in our thinking. However, when we observe situations in the real world, they do not come with a detailed analysis attached. The strong tendency is to start and finish our analysis within the framework of the representative information that is readily available and to ignore other relevant facts. In the CEO example, the requirements of pet ownership are *representative* of characteristics that we associate with a chief executive officer. Therefore, other relevant information (namely, the large number of childhood pet owners who do not become CEOs) is ignored. Experiments have shown that representative reasoning is common.

Another related difficulty is called the *availability bias*. Dawes (1988, p. 99) illustrates it with the following question that Tversky and Kahneman asked experimental subjects:

How many seven-letter English words have the form

$$\_ \_ \_ \_ \_ n \_$$

Not many? How many seven-letter English words have the form

$$\_ \_ \_ \_ ing$$

More? ... [Experimental subjects] judged seven-letter words ending in *-ing* to be much more frequent than seven-letter words with n in the sixth position ... [However,] there are, of course, more seven-letter words with n in the sixth position ... Every word ending in *-ing* has n in the sixth position, while there are in addition seven-letter words with n in the sixth position that do not end in *-ing*—for example *turbans*.

Experiments show that we judge cases to be more probable if we have more examples of these cases *available* in our thinking processes.

Anchoring and Adjustment

Experiments also show that we often *anchor* our thinking about a situation on (possibly irrelevant) information and fail to sufficiently *adjust* from the anchor. Dawes (1988, p. 121) presents the results of a 1972 experiment by Kahneman and Tversky:

Student subjects [were asked] to estimate the percentage of African countries that were in the United Nations. (The correct answer was 35%.) Prior to making the estimate, however, the subjects were requested to make a simple binary judgment of whether this percentage was greater or less than a number determined by spinning a wheel that contained numbers between 1 and 100. The wheel was in fact rigged so that the number that came up was either 10 or 65. Subjects who first judged whether the percentage was greater or less than 10 made an average estimate of 25%; those who first judged whether it was greater or less than 65 made an average estimate of 45%. Thus, 10 and 65 served as anchors for the estimates *even though the subjects were led to believe that these numbers were generated in a totally arbitrary manner*. What happened was that the final estimates were "insufficiently" adjusted away from these anchors.

This is a startling result. A seemingly arbitrary number, which has *no* relationship to the quantity being estimated, significantly influences estimates of that quantity. This result has been replicated, as have the other reasoning difficulties presented above. Tversky and Kahneman (1974) discuss these reasoning difficulties further.

5.4 General Considerations in Eliciting Probabilities

The primary consideration in eliciting probabilities is to use methods that help to overcome the various reasoning difficulties reviewed above. The discussion of these difficulties in the preceding section shows that (1) we are overly influenced by almost any information we have, even if logic shows that the information is unrelated to the event of interest, and (2) we underestimate the degree of uncertainty in a situation. Thus, the methods we use to elicit probabilities should help to overcome these reasoning difficulties.

Also, in many practical decision situations probabilities will be elicited from experts who are not the primary decision maker. For example, if a decision is to be made about whether or not to produce a new microprocessor, it makes sense to use semiconductor production experts to provide probability judgments about production costs for the microprocessor. When this is done, there may be reasons for the experts to provide inaccurate probabilities. (This problem was discussed for the sales representative example earlier in this chapter.) Thus, a probability elicitation procedure should address this concern.

Merkhofer (1987) and Spetzler and Staël von Holstein (1975) review probability elicitation procedures that are used in practice. Merkhofer's procedure includes five stages: motivating, structuring, conditioning, encoding, and verifying. Each of these stages is discussed below. Appendix C presents an annotated transcript of an actual elicitation of a probability distribution from a business executive using Merkhofer's procedure.

Stage 1: Motivating

The purposes of the motivating stage are (1) to establish good communications with the person whose probabilities you will be eliciting, and (2) to determine whether there is a significant potential for *motivational bias*. Such a bias is a conscious or unconscious adjustment in the person's probabilities based on perceived personal rewards (for example, consider the sales representative example discussed above).

This stage begins with a discussion of the analysis that is being conducted and the role in this analysis of the probabilities that will be elicited. As part of this discussion, the potential is explored for the two primary motivational biases, *management* bias and *expert* bias. Management bias occurs when the person views the uncertainty quantity to be subject to management control. ("If that's the number that management wants, then that's what they'll get!"). Expert bias occurs when the person views himself or herself as the expert on the subject of interest and thinks that experts are not supposed to have uncertainty about their area of expertise.

In some cases, either management or expert bias may be so strong that some other person has to be used as the subject for the elicitation.

Stage 2: Structuring

The purposes of the structuring stage are (1) to specify the uncertain quantity of interest unambiguously, and (2) to establish the underlying assumptions that will be made during the probability elicitation.

An uncertain quantity for which probabilities are to be elicited should pass the *clairvoyance test*. That is, the quantity should be well enough defined that a clairvoyant who could foresee the future would be able to tell you what the actual level will be for the uncertain quantity. Thus, the event "rain tomorrow" does not pass the clairvoyance test (At what location? During what hours? What does it mean to "rain"?) Similarly, the event "the price of wheat in six months" also does not pass the clairvoyance test. (At what location? What terms of sale? What type of wheat?) If an uncertain quantity does not pass the clairvoyance test, then the person from whom you are eliciting probabilities may have a different idea about what the uncertain quantity means than other people participating in the analysis.

Underlying assumptions to be made during the elicitation should be clarified to establish a common understanding about what is not included in the uncertainties being considered. For example, if you are considering the price of oil in six months, then you should reach a common understanding about what assumptions are to be made about OPEC embargoes, unrest in the Middle East, federal taxes on petroleum products, and so forth.

Stage 3: Conditioning

The purpose of the conditioning stage is to make sure that the person is considering all relevant information about the uncertain quantity. The research reviewed earlier in this chapter shows that people have a significant tendency to focus on specific information and insufficiently adjust for the uncertainty about the relevance of that information to the uncertain quantity of interest. For example, suppose you have decided to buy a Volvo after carefully researching the matter, paying particular attention to reliability data about the car. The evening before you are going to the Volvo dealer to buy the car, you meet someone at a party. He says, "You shouldn't buy a Volvo! My third cousin's wife's brother has a Volvo, and I hear that he has had nothing but trouble with it." Almost any of us in such a situation would feel our judgment about Volvos swinging toward the negative, perhaps significantly. But this is only one Volvo out of the many thousands on the road (assuming, of course, that your acquaintance was right about the make of car).

We tend to focus on specific compelling examples like this and not give sufficient weight to other information. If we are considering a new product that we have been working on for the last two years, we can easily think about all of the product's positive features and why it should sell well. We forget the general fact that most new products are failures.

Thus, the purpose of the conditioning stage is to help the elicitation subject expand his or her immediately available information to include the full range of possibilities that are relevant to the uncertain quantity.

Stage 4 (Encoding) and Stage 5 (Verifying)

The encoding and verifying stages are where the probabilities are actually elicited. The specific procedure to be used depends on whether the uncertain quantity is *discrete* or *continuous*. A discrete quantity is one that can take on only a small number of different levels. For example, the event "it rains tomorrow" is discrete. Assuming that this quantity is defined well enough so that it passes the clairvoyance test, then the only possibilities are that it rains or it doesn't. Similarly, events such as "war in the Middle East within the next six months" or "a cure for AIDS within the next five years" are discrete.

A continuous uncertain quantity, on the other hand, can take on a large number of different levels. The sales estimate in Figure 5.2 is continuous. It can take on levels from zero to over five hundred. Similarly, quantities like "the price of oil in six months" and "the number of months until there is a cure for AIDS" are continuous uncertain quantities.

The procedures used for encoding and verifying differ somewhat, depending on whether the uncertain quantity is discrete or continuous. Each of these cases is now considered.

5.5 Encoding for Discrete Uncertain Quantities

If an uncertain event can either occur or not, then basically you need to ask the person what the probability is that it will occur. However, there are two potential difficulties. First, some people do not have much intuitive feel for the meaning of a particular probability number, and, second, you need to guard against the various heuristics and biases that were reviewed earlier in this chapter.

You can help a person who is not comfortable working with probability numbers by using a graphical aid such as a probability wheel. Either you or the elicitation subject can adjust the wheel until the subject judges the outcome on the wheel to be equally likely to the event. Then, if you have a scale on the wheel, the probability can be read from this scale.

When using a probability wheel, you should guard against the anchoring effect. The experimental evidence reviewed earlier in this chapter implies that the final probability provided by the subject will be influenced by the initial wheel setting. Since you have to provide some initial setting for the wheel, how can you address this difficulty?

One approach is to rapidly provide several different settings of the wheel so that the person does not have a specific "anchor" to concentrate on. For example, suppose you anticipate that the final probability will be between twenty and seventy percent. Then you could initially set the wheel to ninety-five percent and ask whether the actual probability is higher or lower than this. Assuming you receive an answer that the probability is lower, then you could set the wheel to five percent and ask the question again. Assuming the answer is that the probability is higher, then you could swing the setting back to a high number, for example eighty-five percent, and ask the question again.

By rapidly moving back and forth between high and low settings, while moving the high and low settings toward each other, you provide offsetting anchors that help to combat the anchoring bias.

5.6 Stage 4 (Encoding) for Continuous Uncertain Quantities

The procedure given in the preceding section does not work well when an uncertain quantity can take on a substantial range of different levels (for example, the sales estimates in Figure 5.2). The probability of any specified number (for example, sales of 321 units in Figure 5.2) is usually small, and it would take a very long time to consider all the different possibilities using the approach in the preceding section.

Instead, consider *ranges* of possible levels and assess the relative likelihood of outcomes within different ranges. The procedure is illustrated by the diagrams in Figure 5.3. Suppose that you are eliciting from a production manager the probability distribution for the marginal production cost in cents for a proposed new novelty toy item. The procedure for doing this is illustrated below.

The difficulty of "overconfidence" should be kept in mind while doing this elicitation. Remember the experiment with Harvard Business School students discussed earlier in this chapter. That experiment, which has been replicated many times, shows that people usually understate their uncertainty. Thus, you should work diligently to help the elicitation subject reduce this bias.

Because of this bias, one obvious starting point for the elicitation is a poor choice. This poor starting point is to first assess the *median* for the probability distribution. The median is the level of the uncertain quantity such that there are equal chances of the actual level being greater or less. Putting this a different way, there is a fifty percent chance the actual level will be below the median, and there is a fifty percent chance the actual level will be above the median.

When people first started eliciting probabilities several decades ago, they usually began by eliciting the median using the procedure discussed below, but the difficulty with this is that the median, once elicited, provides a cognitive "anchor" for the subject. As the experiments discussed earlier in this chapter show, people will have a powerful tendency to "lock on" to this level and underestimate the likelihood that the actual level will be very far from the median.

Thus, the recommended procedure is to first elicit extreme levels, rather than the median. Specifically, ask for levels such that there is (for the high level) a one percent chance that the actual level will exceed the determined level, and (for the low level) a one percent chance that the actual level will be below the determined level. These are hard questions for many people to think about. You can help by setting the probability wheel first to ninety-nine percent and then one percent so that the subject has a concrete picture of what these numbers mean.

Once you have elicited preliminary numbers for these two levels (which are referred to as the "lower extreme" and the "upper extreme"), then you should discuss with the subject what sets of conditions (sometimes called "scenarios") could lead to a result that is above the upper extreme or below the lower extreme.

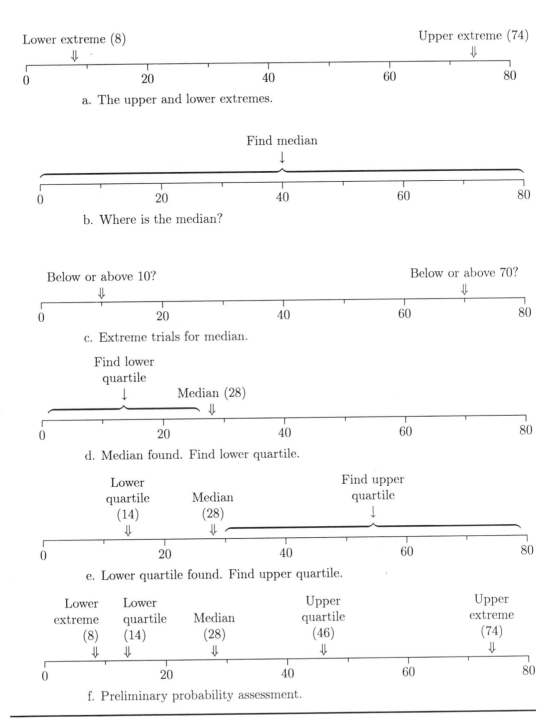

Figure 5.3   *Probability elicitation for marginal production cost*

The subject may realize that there are more cases than he or she had first considered, which could lead to extreme results. The subject may end up moving the two extremes farther apart.

Illustrative lower and upper extremes (8 and 74 for our example) are shown in Figure 5.3a. By eliciting these two widely separated numbers, we have taken steps to avoid anchoring, because there is no single number to anchor on. We can now proceed to assess the median, as illustrated in Figure 5.3b. A method for doing this is to provide a trial level for the median and ask the subject whether it is more likely that the actual level of the uncertain quantity will be above or below this trial level. Two initial possible trial levels (10 and 70) for the median are shown in Figure 5.3c. Note that one of these trial levels is toward the low end of the range of possible levels, and the other is toward the high end. By moving back and forth between the low and high ends of the range, you continue to combat the tendency to anchor on numbers that are provided.

Continue to provide trial levels for the median, alternating between one toward the high end of the scale and one toward the low end. With each trial, move the lower level up some and the higher level down some. Finally, you will reach a point where the elicitation subject finds the chances of the actual level being on either side of the trial level to be equal. This trial level is then the median. In Figure 5.3d, the median is shown to be 28. Thus, the subject judges the chances to be equally likely of the true level being above or below 28. Putting this a different way, if the subject had to bet on the true level being either above 28 or below 28, he or she would be indifferent between picking either of the sides.

Having determined the median, then use a similar procedure to determine the "lower quartile." The lower quartile is the level that splits the interval below the median into two equally likely subintervals. The procedure for determining this is similar to that for the median. Select a trial level that is near one end of the interval—that is, near either the lower extreme or the median. Determine whether the subject assesses the chance to be greater that the actual level will be below the trial level or between the trial level and the median. Then select a trial level near the other extreme of the interval below the median, and repeat the question. Continue to select trial levels near the low end and the median, and move these trial levels toward each other until the subject finds the chances to be equal that the actual level will be below the specified trial level or between the trial level and the median.

This trial level is called the "lower quartile." Some thought shows that the probability must be one-quarter that the actual level will be below the lower quartile and also one-quarter that the actual level will be between the lower quartile and the median. A lower quartile of 14 is shown in Figure 5.3e.

A similar procedure is used to find the "upper quartile," which is the level that splits the interval above the median into two equally likely subintervals. Trial levels are successively selected near the median and the upper extreme. These trial levels are moved toward each other until the subject finds the chances to be equal that the actual level is between the median and the trial level or above the trial level. This trial level is then the upper quartile. The probability is one-quarter that the actual level will be between the median and the upper quartile and also one-quarter that the actual level will be above the upper quartile.

Another way to say this is that the probability is three-quarters that the actual level will be below the upper quartile.

The (still preliminary) probability numbers that were elicited for the production cost example are shown in Figure 5.3f. In this example, the upper quartile was elicited to be 46. When you have obtained a set of numbers like those shown in Figure 5.3f, it is usual to make some checks to determine whether the subject is firmly settled on the numbers.

5.7 Stage 5 (Verifying) for Continuous Uncertain Quantities

A variety of checks can be made of the probability numbers that you have elicited. Figure 5.4 illustrates one checking procedure. If the subject has provided accurate information, then it must be true that he or she judges the chances to be equal that the actual level for the uncertain quantity will be either inside the interval between the lower quartile and the upper quartile, or outside this interval. For the production cost example, these intervals are shown in Figure 5.4a. The interval between the lower and upper quartiles (between 14 and 46 in the example) is called the *fifty-percent credible interval* or the *interquartile range*. Of course, the interval outside the interquartile range includes everything below 14 and above 46 for this example.

Suppose the subject responds that these two intervals are equally likely, which is consistent with the earlier elicitation. Then a second check is shown in Figure 5.4b. The subject is asked whether it is more likely that the actual level will be below the lower quartile (14 in the example) or above the upper quartile (46 in the example). If these are accurate quartiles, then the subject should respond that the chances are equal that the actual level will be in either of these two intervals. Suppose, however, that the subject responds that it is more likely that the actual level will be above the upper quartile (that is, above 46).

This response means that either or both of the quartiles must be changed until equality is achieved. However, changing one or more of the quartiles often leads an elicitation subject to change the median also (and perhaps also the upper and lower extremes). Often when you identify a difficulty with a quartile, the best thing to do is to re-elicit the median and the two quartiles. With the background that has been built up during the initial elicitation, the re-elicitation is likely to be much quicker than the initial elicitation. A possible result of this re-elicitation is shown in Figure 5.4c. Note that the lower quartile, median, and upper quartile have all changed slightly from the preliminary numbers shown in Figure 5.3f.

With this final set of numbers, it should be true that the subject finds it to be equally likely that the actual level will lie below the lower quartile (below 16 for the Figure 5.4c example), between the lower quartile and the median (between 16 and 30), between the median and the upper quartile (between 30 and 50), or above the upper quartile (above 50). That is, the probability is one-quarter that the actual level will lie in each of these four intervals.

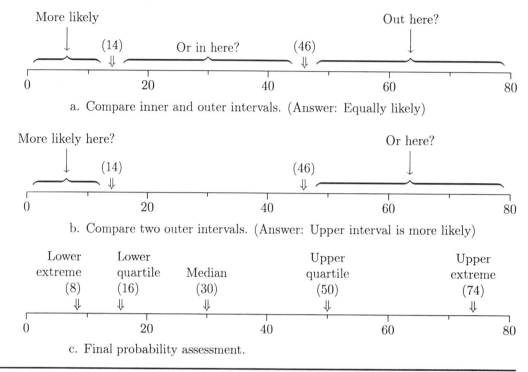

a. Compare inner and outer intervals. (Answer: Equally likely)

b. Compare two outer intervals. (Answer: Upper interval is more likely)

c. Final probability assessment.

Figure 5.4   *Consistency checks*

5.8 Cumulative Distribution

The five elicited probability numbers can be used to specify a probability curve that can be used to determine probabilities for any interval of the uncertain quantity. Such a curve is called a *cumulative distribution* or a *cumulative probability distribution*, and an example based on the information in Figure 5.4c is shown in Figure 5.5. In this curve, the horizontal axis is the uncertain quantity of interest, and the vertical axis is the probability that the actual level of the uncertain quantity will be less than or equal to a specified level. For example, this curve shows that there is about a 0.64 probability (64 percent chance) that the marginal production cost will be 40 cents or below.

Once a cumulative distribution has been drawn, some additional consistency checks can be made, if desired. For example, a specific level can be picked for the uncertainty quantity, and the elicitation subject can be asked to give the probability that the actual level will be below the specified level. (A probability wheel can be used to aid in the assessment of this probability.) The assessed probability can then be compared with that determined from the cumulative distribution curve. For example, in the Figure 5.5 example, the subject might be asked to determine the probability that the marginal production cost will be below 25 cents. The Figure 5.5 cumulative distribution curve shows that this

Probability of c or less

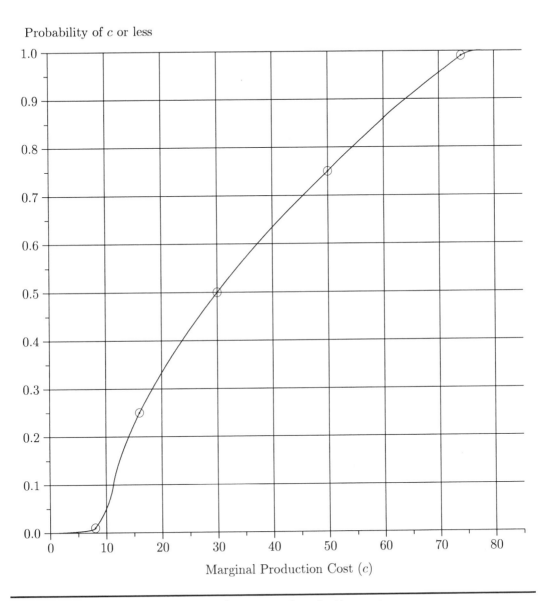

Marginal Production Cost (c)

Figure 5.5 *Cumulative distribution for marginal production cost*

probability is about 0.42. If the directly assessed probability is different from this, then further elicitation might be appropriate.

Merkhofer (1987) comments that a probability elicitation can take anywhere from a half hour to a half day, with two hours being typical. My own experience is that the elicitation of a single probability distribution with someone who knows nothing about probability can take that long, but that with someone who has some experience with either probability or statistics, the process is faster. Furthermore, it is often true that you will need to elicit more than one probability distribution from an individual, and each succeeding elicitation usually takes less time.

5.9 References

M. Alpert and H. Raiffa, "A Progress Report on the Training of Probability Assessors," In D. Kahneman, P. Slovic, and A. Tversky (Editors), *Judgment Under Uncertainty: Heuristics and Biases*, Cambridge University Press, Cambridge, England, 1982.

R. M. Dawes, *Rational Choice in an Uncertain World*, Harcourt Brace Jovanovich, San Diego, 1988.

M. W. Merkhofer, "Quantifying Judgmental Uncertainty: Methodology, Experiences, and Insights," *IEEE Transactions on Systems, Man, and Cybernetics*, Vol. SMC-17, pp. 741–752, 1987.

C. S. Spetzler and C.-A. S. Staël von Holstein, "Probability Encoding in Decision Analysis," *Management Science*, Vol. 22, pp. 340–358, 1975.

A. Tversky and D. Kahneman, "Judgment Under Uncertainty: Heuristics and Biases," *Science*, Vol. 185, pp. 1124–1131 (1974).

5.10 Review Questions

R5-1 Define and discuss the significance of the following cognitive biases: overconfidence, framing, representative thinking, availability, and anchoring and adjustment.

R5-2 Define probability.

R5-3 Discuss how the standard probability elicitation procedures address cognitive biases.

5.11 Exercises

5.1 Define the availability and representative cognitive biases. Discuss the differences between these biases. Use examples of these biases as part of your answer.

5.2 For a continuous uncertain quantity of your choice, assess a cumulative probability distribution for someone outside of this class using the procedure reviewed in this chapter. Discuss briefly the assessment procedure you used as well as any difficulties you encountered. Present the raw data that resulted from your assessment, and draw a graph of the cumulative distribution.

CHAPTER 6

Decisions with Uncertainty

When there is no uncertainty about the outcome of a decision alternative, the primary complexity in evaluating alternatives comes from the need to consider *tradeoffs* among the evaluation measures. When there is only one evaluation measure in such a situation, then there is generally little difficulty in evaluating alternatives: If more of the evaluation measure is preferred to less, then pick the alternative that gives the largest amount of the evaluation measure; on the other hand, if less of the evaluation measure is preferred to more, then select the alternative that gives the smallest amount of the evaluation measure.

However, when there is uncertainty about what outcome will result from selecting an alternative, then even with a single evaluation measure, the evaluation can be difficult. To illustrate the difficulty, consider the following example: Suppose that someone offers to sell you for $7,000 a lottery ticket that gives you the right to flip a *fair* coin (that is, one that has equal chances of coming up heads or tails) and receive $10,000 if a head comes up or $5,000 if a tail comes up. Should you buy the lottery ticket?

The answer to this is not immediately clear. If a head comes up, then you will end up with $10,000 - $7,000 = $3,000$, which is desirable, but if a tail comes up, then you will end up with $5,000 - $7,000 = -$2,000$, which is undesirable. What would you do? Some people who have not studied how to think about situations with uncertainty throw up their hands and say that it is not possible to formally analyze such a situation. However, this is not correct. Much is known about how to analyze such situations. The analysis approach uses methods based on probabilities.

6.1 Probabilities

The meaning of probabilities was reviewed in Chapter 5, and procedures for determining probabilities were also presented there. Some properties of probabilities are as follows:

1 A probability is a number between zero and 1 (including zero and 1).
2 The relative sizes of the probabilities for different possible outcomes show the relative likelihood of the outcomes occurring.
3 The total of the probabilities for all possible distinct outcomes is 1.

For the coin flip discussed above, heads and tails are equally likely; therefore, they must have equal probabilities. Since there are only two possible outcomes, the probabilities of heads and tails must sum to 1. Hence, the probability of a head (and also of a tail) must be 0.5.

The most useful way to work with probabilities is often to organize the (uncertain) evaluation measure levels into an *event space*. An event space is a list of possible outcomes on the evaluation measure, which has the following properties:

1 Every possible evaluation measure level is included in at least one item of the list, and
2 No possible evaluation measure level is included in more than one item of the list.

The first property of an event space means that the list collectively includes all possible evaluation measure levels. This is sometimes stated by saying the list of items in an event space is *collectively exhaustive*. The second property of an event space means that each possible evaluation measure level is included in no more than one of the list items. This is sometimes stated by saying that the items in an event space are *mutually exclusive*.

To take a slightly more complex example, assume that rather than flipping a fair coin, the lottery involves throwing a fair die (that is, one that has equal chances of each of the six faces coming up). Suppose that in this lottery, you will win only if a six comes up. Are your chances greater or less of winning than they are with the coin flip lottery?

This can be answered by once again going back to the three properties given above for probabilities. There are six possible distinct outcomes for the die. Since each face is equally likely to come up, each face must have the same probability of occurring. Since there are six faces, and the sum of the probabilities must be 1, the probability of any face coming up is 1/6. Therefore, your chance of winning in the lottery is only 1/6, which is lower than with the coin flip lottery.

The list of possible outcomes for the die toss (1, 2, 3, 4, 5, and 6) is an event space since each possible outcome of a die toss is included, and each outcome is included only once. There are other possible event spaces for the die toss. For example, the two outcomes "odd outcome" and "even outcome" are an event space because these two outcomes include between them all possible results of a die toss, and each possible result is included in only one of the list items.

The reason that we are interested in event spaces is that they have a special property with respect to working with probabilities. Specifically, if the probabilities are listed for all the items in an event space, then the probability for some *combination* of event space items is equal to the sum of the probabilities for the event space items in the combination. To illustrate this, consider the die toss again. We showed above that the probability of each possible number from one to six is one-sixth, and that this list of outcomes is an event space. Suppose we are interested in the probability of obtaining an odd outcome from a toss of the die. This outcome occurs if either a 1, 3, or 5 comes up. Thus, the result "odd" is a combination of the three items 1, 3, and 5. Since the original list of outcomes is an event space, then the probability of an odd outcome can be determined by adding the probabilities for the outcomes 1, 3, and 5. That is, the probability of an odd outcome is $1/6 + 1/6 + 1/6 = 1/2$.

6.2 Expected Value

It is intuitively clear that the coin flip lottery should be worth more than the die throw lottery since the chances of winning with the coin flip are higher. However, it is still not clear whether it is reasonable to pay \$7,000 for the coin flip lottery ticket. To begin to answer this question, we investigate a quantity called the *expected value*.

The expected value of an uncertain situation is calculated by multiplying each possible outcome by its associated probability and adding the results. Thus, for the coin flip lottery,

$$\begin{aligned}
\text{Expected Value} &= 0.5 \times (\$10,000 - \$7,000) + 0.5 \times (\$5,000 - \$7,000) \\
&= 0.5 \times \$3,000 + 0.5 \times (-\$2,000) \\
&= \$500
\end{aligned}$$

while for the die throw lottery,

$$\begin{aligned}
\text{Expected Value} &= (1/6) \times (\$10,000 - \$7,000) + (5/6) \times (\$5,000 - \$7,000) \\
&= (1/6) \times \$3,000 + (5/6) \times (-\$2,000) \\
&= -\$1,166.67
\end{aligned}$$

The importance of the concept of expected value comes from a mathematical result called the "Weak Law of Large Numbers," which is presented in most probability or statistics textbooks (Olkin, Gleser, and Derman 1994; Ghahramani 1996). This law proves that under very general conditions the probability is very high that the average outcome for a large number of independent decisions will be very close to the average of the expected values for the decisions. Thus, using the expected value calculations given above, if you pay \$7,000 for the coin flip lottery and repeat this process many times (paying \$7,000 each time), then you are very likely to end up making an average of about \$500 for each time you play. On the other hand, if you repeat the die throw lottery many times, you are very likely to lose an average of about \$1,166.67 for each time you play.

This result seems to argue for placing a value on an alternative that is equal to its expected value. This is because if you pay more than the expected value for alternatives, then from the Weak Law of Large Numbers, you are very likely to lose money over many decisions because the probability is high that on average the alternatives will return only about their expected values. Similarly, if you are not willing to pay as much as the expected value for alternatives, you will lose the opportunity to make additional money that you could have made over the long run because the probability is very high that alternatives will, on average, return more than you are willing to pay for them.

In fact, it is often appropriate to use expected value as a criterion for choosing among alternatives in business decisions. However, there are also situations where this is not appropriate. We will consider such situations after discussing how to calculate expected values for continuous uncertain quantities, and how probabilities can be determined.

6.3 Expected Value for Continuous Uncertain Quantities

The procedure in Section 6.2 is straightforward to apply to an uncertain quantity that can take on only a discrete number of different levels, but what about a continuous uncertain quantity that can take on a range of different levels? Such uncertain quantities often arise in practice, for example, from elicitations of judgmental (subjective) probability distributions. Figure 6.1 shows an example cumulative distribution for a continuous uncertain quantity. In this figure, the horizontal axis shows the uncertain quantity levels, while the vertical axis shows the probability that the level of the uncertain quantity will be less than or equal to any specified number for the uncertain quantity. (For example, the probability is 0.5 that the uncertain quantity will be less than or equal to 30.)

Since there are an infinite number of possible different levels for a continuous uncertain quantity, it is not possible to directly apply the procedure in the preceding section to determine the expected value. It turns out (Olkin, Gleser, and Derman 1994; Ghahramani 1996) that the theoretically correct way to calculate expected value involves using the derivative

$$f(x) = \frac{dF(x)}{dx}$$

for the cumulative distribution $F(x)$. (This derivative is called the *probability density function*.) The expected value of the uncertain quantity is then calculated by integrating $f(x)$:

$$E(x) = \int_{-\infty}^{\infty} x f(x) dx$$

There are two difficulties with applying this procedure: (1) Many people who have decisions to make do not know how to differentiate or integrate, and (2) often the cumulative distribution is available only in graphical form so that it is not straightforward to determine the derivative even if you know how to do this.

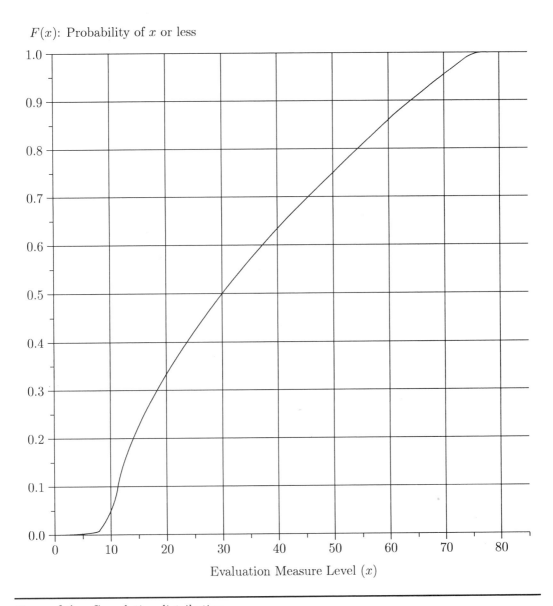

$F(x)$: Probability of x or less

Evaluation Measure Level (x)

Figure 6.1 *Cumulative distribution*

An alternate procedure that is generally adequate for practical situations is to approximate the continuous uncertain quantity with a discrete uncertain quantity, and then use the method of the preceding section to calculate the expected value for this (approximate) discrete uncertain quantity. The problem of finding a good discrete approximation for a continuous uncertain quantity has been studied (Keefer and Bodily 1983), and the two procedures given below usually give accurate results in practice.

To understand these procedures, you need to know the definition of the term *fractile*. Examine Figure 6.1, which shows a graph of the probability $F(x)$ that the uncertain quantity will be less than or equal to x for any x. If a particular probability p is specified, then the p *fractile* is the level x_p of the uncertain quantity such that $F(x_p) = p$. To take a specific example, suppose we are interested in the 0.75 fractile for the uncertain quantity shown in Figure 6.1. To find this, determine the level $x_{0.75}$ such that $F(x_{0.75}) = 0.75$. Figure 6.1 shows that this level is 50. Hence, 50 is the 0.75 fractile for this uncertain quantity. Thus, there is a 0.75 probability that the uncertain quantity will be less than or equal to 50. That is, $F(50) = 0.75$.

Here are the two discrete approximations that can be used to determine the expected value for a continuous uncertain quantity:

Extended Pearson-Tukey approximation

With this approximation, the continuous uncertain quantity is approximated with a discrete uncertain quantity having three different possible levels. The three levels are set equal to the 0.05, 0.50, and 0.95 fractiles of the continuous uncertain quantity. The 0.05 and 0.95 fractile levels are assigned probabilities of 0.185 each, and the 0.50 fractile level is assigned a probability of 0.630.

Extended Swanson-Megill approximation

With this approximation, the continuous uncertain quantity is also approximated with a discrete uncertain quantity having three different possible levels. The three levels are set equal to the 0.10, 0.50, and 0.90 fractiles of the continuous uncertain quantity. The 0.10 and 0.90 fractile levels are assigned probabilities of 0.3 each, and the 0.50 fractile level is assigned a probability of 0.4.

For the example in Figure 6.1, the extended Pearson-Tukey approximation is developed as follows: On the vertical axis of Figure 6.1, determine the 0.05 level. Move horizontally across the figure until you intersect the cumulative distribution curve $F(x)$. Then determine the evaluation measure level x that goes with this point on the $F(x)$ curve. When this procedure is followed in Figure 6.1, the resulting x level is 10, and this is the 0.05 fractile of the uncertain quantity. Follow an analogous procedure to determine the 0.50 fractile (which is 30) and the 0.95 fractile (which is 70). Thus, the approximate discrete probability distribution resulting from the extended Pearson-Tukey approximation has a probability of 0.185 for a level of 10, a probability of 0.630 for a level of 30, and a probability of 0.185 for a level of 70. Using these levels to calculate the expected value results in $0.185 \times 10 + 0.630 \times 30 + 0.185 \times 70 = 33.70$.

Using an analogous procedure for the extended Swanson-Megill approximation yields a 0.10 fractile level of 12, a 0.50 fractile level of 30, and a 0.90 fractile level of 64. Thus, the approximate discrete probability distribution has a probability of 0.3 for a level of 12, a probability of 0.4 for a level of 30, and a probability of 0.3 for a level of 64. Using these levels to calculate the expected value results in $0.3 \times 12 + 0.4 \times 30 + 0.3 \times 64 = 34.80$, which is close to the result determined using the extended Pearson-Tukey approximation.

Research results (Keefer and Bodily 1983) have shown that the extended Pearson-Tukey approximation is usually somewhat more accurate than the extended Swanson-Megill approximation, but both are generally accurate enough for practical applications unless the continuous uncertain quantity has an unusually shaped cumulative distribution.

6.4 Two Caveats about Using Expected Value

We have reviewed some reasons for using expected value as a criterion for making decisions. However, there are two caveats that need to be made about the use of expected value as a criterion for decision making:

1 What if your probabilities are bad?

2 What if you can't play the long-run averages?

Bad probabilities   The phrasing of the Weak Law of Large Numbers is somewhat subtle. It states that the *probability is very high that* the average outcome for a large number of independent decisions will be very close to the average of the expected values for the decisions. It is easy to misinterpret this and leave out the italicized phrase ("the probability is very high that") so that the phrase reads "the average outcome for a large number of independent decisions will be very close to the average of the expected values for the decisions." However, this (incorrect phrasing) may not be true if the probabilities that you use to calculate the expected values are not relevant to the real world.

This can be illustrated with an extreme example. Suppose that your friend Fred insists that the only possible outcome from a throw of a fair die is a six. Then (for Fred) the probability of a six coming up on a throw of the die is 1, and the expected value for any throw of the die is $1 \times 6 = 6$. Thus, applying the Weak Law of Large Numbers results in the statement that "the probability is very high that the average outcome for a large number of independent throws of the die will be very close to six."

Now, in fact, for a fair die the chances are equal that any of the numbers one through six will come up. Thus, the "real world" expected value for a roll of the die is $(1/6) \times (1 + 2 + 3 + 4 + 5 + 6) = 3.5$. Thus, out there in the real world (rather than in Fred's mind), the average outcome for a large number of independent throws of the die is very likely to be very close to 3.5, and not to 6. If Fred insists on paying up to six dollars to participate in lotteries that return the amount of the die number that comes up, then he is highly likely to lose money over the long run.

The point of this is that we must remember that probabilities are subjective and hence a matter of judgment or opinion, but that most outcomes in the real world are not a matter of opinion. If your opinions do not agree with the facts, then the facts usually ultimately win out. Thus, it is important to work diligently to determine probabilities that are in accord with the real world. For the example of the die toss, it would be better to go to an expert on die tosses to obtain probabilities, rather than asking Fred.

Long-run averages   Notice that the Weak Law of Large Numbers applies to "a large number of independent decisions." There is nothing in the Weak Law of Large Numbers that guarantees that you will obtain the expected value for any particular decision. In fact, for the coin flip example above, it is impossible to obtain the expected value for any single flip, since you will either make $10,000 - $7,000 = $3,000 or lose $7,000 - $5,000 = $2,000, and neither of these amounts is equal to the expected value of $500. The Weak Law of Large Numbers guarantees that over many repetitions of this lottery the probability is high that you will average out to a net of $500 for each time you play. However, it may take a while for this to happen. In the meantime, you may go broke and not be able to continue to play the lottery!

This can be illustrated by considering a lottery with lower stakes where there are equal chances of winning $10.00 or losing $5.00. The expected value for this lottery is $0.5 \times \$10.00 + 0.5 \times (-\$5.00) = \$2.50$, and most of us would probably be willing to pay up to this amount to play the lottery based on the Weak Law of Large Numbers. However, suppose that the winning and losing amounts are now multiplied by a factor of 10,000 to $100,000 and −$50,000, respectively. It is easy to show that the expected value for this lottery is $25,000. However, most of us would not be willing to pay this much to participate in this lottery, even though the Weak Law of Large Numbers says that we will win about this much *on the average over many plays of the lottery.* Why wouldn't we pay this much? Because the stakes are high enough that we may not be able to play the lottery many times—a few $50,000 losses would leave most of us without the resources to continue. We cannot "play the averages" over a series of decisions when the stakes are this high, and thus considerations of long-run average returns are less relevant to our decision making. (A vice president of a Fortune 500 company once commented to me, "Most of the decisions we analyze are for a few million dollars. It is adequate to use expected value for these." Whether or not this is true depends on your asset position!)

6.5 Attitude Toward Risk Taking and Expected Utility

Often, organizations and individuals take a more conservative attitude toward significant risks than expected value would suggest. (A *risk* is something bad that might, or might not, actually come to pass.) That is, many organizations or individuals will sell an uncertain alternative for less than its expected value to get rid of the risk of an undesirable outcome. This attitude toward risk taking is called *risk averse*. While it is less common, some organizations or individuals will sell alternatives only for more than their expected values. This attitude toward risk taking is called *risk seeking*. (Finally, to complete the definitions, if an

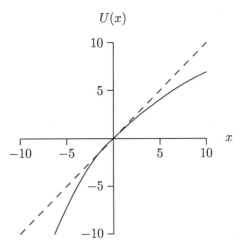

Figure 6.2   *Example utility function*

organization or individual will sell alternatives for exactly their expected values, the organization or individual is said to be *risk neutral*. Many organizations or individuals are risk neutral when the risks are small relative to their assets.)

The presence of risk aversion can be taken into account in an evaluation by replacing the use of expected value by the use of *expected utility*. This is done by assigning to each possible outcome of an uncertain decision a corresponding *utility* number that takes into account the risk aversion of the decision maker, and then using these utility numbers in the expected value calculations. For risk averse decision makers, these utility numbers are determined in such a way that desirable outcomes do not get as much weight as when using expected value, and undesirable outcomes get more weight. Thus, the result of using utilities rather than actual outcomes is that risky alternatives are evaluated as less desirable than would be true using expected value.

The basic idea of the utility concept is illustrated by Figure 6.2. In this figure, the solid line represents a utility function, while the dashed line illustrates what happens when expected value is used to rank alternatives. The solid line shows an example utility function $U(x)$ for which the utility of a positive level of the evaluation measure x is not as high as (is less desirable than) the actual level of x, which is represented by the dashed line, while the utility of a negative level of x is more negative (that is, less desirable) than the actual level of x as represented by the dashed line. Suppose this utility function is used to evaluate alternatives. Then two alternatives with the same expected value may be evaluated differently. Specifically, the alternative with more uncertainty will be ranked lower than the alternative with less uncertainty, even though both alternatives have the same expected value. This is because of the curvature of the utility function. For more details about utility see Clemen 1996, Keeney and Raiffa 1976, and Pratt, Raiffa, and Schlaifer 1964.

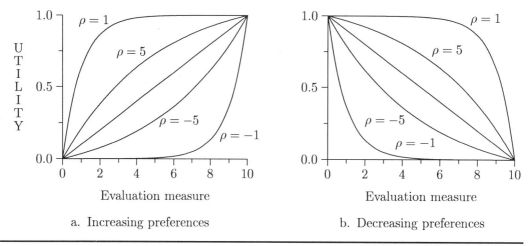

Figure 6.3 *Example exponential single dimensional utility functions*

Further analysis shows that a risk averse decision maker always has a utility function with the "hill-like" shape shown by the solid curve in Figure 6.2, although the specific details of the shape depend on the particular decision maker. On the other hand, a risk seeking decision maker always has a "bowl-like" utility function shape, which curves upward above the dashed straight line in Figure 6.2. Finally, a risk neutral decision maker has a utility function that is a straight line.

6.6 Determining a Utility Function

In practice, a utility function can often be approximately determined in a straightforward manner. Pratt (1964) showed conditions under which a utility function has an *exponential* shape. These conditions are reviewed in Section 9.3, Theorem 9.28, and these conditions are often approximately met in practice. In addition, Kirkwood (1992) notes that even when the conditions are not met, the exponential can often provide an adequate approximation for the utility function in a decision analysis.

Examples of exponential utility functions are shown in Figure 6.3. This figure illustrates that the specific shape of the exponential utility function depends on a constant ρ ("rho"), which is called the *risk tolerance*. In practice, it is generally possible to use a calculator or spreadsheet to make necessary calculations using the exponential utility function. If preferences are *monotonically increasing* over an evaluation measure x (that is, larger amounts of x are preferred to smaller amounts), then the exponential utility function $u(x)$ can be written as

$$u(x) = \begin{cases} \dfrac{1 - \exp\left[-(x - \text{Low})/\rho\right]}{1 - \exp\left[-(\text{High} - \text{Low})/\rho\right]}, & \rho \neq \text{Infinity} \\[2ex] \dfrac{x - \text{Low}}{\text{High} - \text{Low}}, & \text{otherwise} \end{cases} \tag{6.1}$$

and if preferences are *monotonically decreasing* over x, then

$$u(x) = \begin{cases} \dfrac{1 - \exp\left[-(\text{High} - x)/\rho\right]}{1 - \exp\left[-(\text{High} - \text{Low})/\rho\right]}, & \rho \neq \text{Infinity} \\[3ex] \dfrac{\text{High} - x}{\text{High} - \text{Low}}, & \text{otherwise} \end{cases} \tag{6.2}$$

where "Low" is the lowest rating level of x that is of interest, "High" is the highest level of interest, and ρ (rho) is the risk tolerance for the value function. The utility function is scaled so that it varies between 0 and 1 over the range from $x = \text{Low}$ to $x = \text{High}$. That is, for monotonically increasing preferences, $u(\text{Low}) = 0$ and $u(\text{High}) = 1$, while for monotonically decreasing preferences $u(\text{Low}) = 1$ and $u(\text{High}) = 0$.

Based on the discussion in the preceding section, we see from Figure 6.3 that risk averse decision makers have exponential utility functions with risk tolerances greater than zero ("hill-like"), while risk seeking decision makers have risk tolerances less than zero ("bowl-like"). While it is not marked on the figures, the diagonal straight lines in the center of each of the two diagrams in Figure 6.3 correspond to the case where the risk tolerance is infinity, and risk neutral decision makers have this utility function. Note that the discussion in this paragraph holds whether preferences are monotonically increasing ("more is better") or monotonically decreasing ("less is better").

Methods for determining the risk tolerance depend on using the concept of the *certainty equivalent* (also called the *certain equivalent*) for an uncertain alternative. The certainty equivalent for an (uncertain) alternative is the certain level of the evaluation measure that is equally preferred to the alternative. In situations where expected value is used to rank alternatives, the certainty equivalent is equal to the expected value. From the discussion of risk aversion above, we see that when we are dealing with an evaluation measure where "more is better," a risk averse decision maker will have certainty equivalents that are less than the expected values, while a risk seeking decision maker will have certainty equivalents that are greater than the expected values. Conversely, when "less is better" for the evaluation measure, a risk averse decision maker will have certainty equivalents that are greater than the expected values, while a risk seeking decision maker will have certainty equivalents that are less than the expected values.

The basic approach to determining the risk tolerance for a particular decision maker is to have that person specify the certainty equivalent for a particularly simple alternative, and use this certainty equivalent to determine the risk tolerance. Once the risk tolerance is determined, the resulting exponential utility function can be used to calculate expected utilities for more complicated alternatives, and we can rank the alternatives based on the results of these calculations.

Approximately Determining the Risk Tolerance

We will shortly present a general procedure for determining the value of the risk tolerance based on an assessed certainty equivalent. However, there are two special cases of practical interest where it is straightforward to approximately determine the risk tolerance from the certainty equivalent. For the first case, the decision maker is asked to consider a hypothetical alternative that has equal chances of yielding a positive amount R, where R is not specified, or a negative amount $R/2$. The decision maker is asked to specify the value of R for which he or she is indifferent between receiving or not receiving the alternative. Putting this another way, the decision maker is asked to adjust R until the alternative has a certainty equivalent of zero.

When the decision maker has selected a value for R, this value is approximately equal to the risk tolerance ρ. Note that the expected value for this uncertain alternative is equal to $0.5 \times R - 0.5 \times (R/2) = 0.25 \times R$. Hence, the greater the value of R that the decision maker selects, the greater the difference between the certainty equivalent of zero and the expected value of $0.25 \times R$. (The accuracy of this approximation, as well as the one discussed in the next paragraph, is considered further in Section 6.8.)

In the second approach to determining the risk tolerance, the decision maker is asked to consider an alternative that has a 0.75 probability of yielding a positive amount R and a 0.25 probability of yielding the same negative amount R. The decision maker is asked to adjust R until he or she is indifferent between receiving or not receiving the uncertain alternative. (That is, find the value of R for which the uncertain alternative has a certainty equivalent of zero.) As with the first approach to finding the risk tolerance, the assessed value of R is approximately equal to the risk tolerance ρ. (Note that the expected value for this uncertain alternative is $0.75 \times R - 0.25 \times R = 0.50 \times R$.)

Exact Procedure for Determining the Risk Tolerance

The approximate procedures just presented for estimating the risk tolerance are often adequate in practical applications, but it is sometimes necessary to use other approaches. For example, both of the approximate procedures require that a decision maker consider uncertain situations with both negative and positive outcomes. In some practical situations, the actual decision alternatives do not have the possibility of both negative and positive outcomes. In such situations, it may be difficult for a decision maker to think about the types of uncertain alternatives used in the approximations.

A more general approach can be developed by considering the equations for the exponential utility function presented in equations 6.1 and 6.2.

A comparison of these equations with equations 4.1 and 4.2 in Chapter 4 for the exponential single dimensional value function shows that the two sets of equations are identical. Thus, it is possible to use a procedure to determine the risk tolerance that is somewhat analogous to that used to determine the exponential constant for an exponential single dimensional value function. This procedure makes use of Table 4.2 in Chapter 4, which is also used in the procedure to determine the exponential constant for an exponential single dimensional value function.

To determine the risk tolerance, first find the certainty equivalent for an alternative with equal probabilities of an outcome at the lowest possible level in the range of outcomes that are of interest or an outcome at the highest possible level. (These two outcomes are labeled "Low" and "High" in equations 6.1 and 6.2.) Equations 6.1 and 6.2 show that the utility for the least preferred end of the range of outcomes is zero, and the utility for the most preferred end of the range of outcomes is 1. Therefore, an alternative with equal probabilities of yielding either end of the range has an expected utility of $0.5 \times 0 + 0.5 \times 1 = 0.5$. Designate the certainty equivalent for this alternative by $x_{0.5}$. Since this is equally preferred to the uncertain alternative, which has an expected utility of 0.5, it must be true that $u(x_{0.5}) = 0.5$.

If you compare the development in the preceding paragraph with that in Section 4.3, you will see that the equation that must be solved to determine the risk tolerance is identical to the equation that must be solved to determine the exponential constant for an exponential single dimensional value function. Therefore, the procedure reviewed in Section 4.3 using Table 4.2 to determine the exponential constant can also be applied to determine the risk tolerance. Specifically, use the following procedure:

1 Calculate the *normalized certainty equivalent* by taking the difference between the certainty equivalent $x_{0.5}$ and the *less preferred* of the two ends of the range of interest and dividing this by the difference between the highest and lowest scores in the range. When doing this, take each of the two differences in such a way that the result has a positive sign.

2 Look up the normalized certainty equivalent in Table 4.2 under the column marked $z_{0.5}$, and find the *normalized risk tolerance R* that corresponds to this.

3 The value of the risk tolerance ρ that corresponds to this value of R is found by multiplying R by the length of the range between the highest and lowest scores in the range.

As an example, suppose that outcomes are measured in terms of net profit, so that preferences are monotonically increasing. Consider an alternative with equal probabilities of yielding a net profit of ten million dollars or twenty million dollars. Suppose that the certainty equivalent for this alternative is thirteen million dollars. Then the normalized certainty equivalent is $z_{0.5} = (13-10)/(20-10) = 0.3$. From Table 4.2, the corresponding value of R is 0.555; therefore, $\rho = 0.555 \times (20 - 10) = 5.55$ million dollars.

As another example, suppose that outcomes are measured in terms of cost so that preferences are monotonically decreasing. Suppose that the certainty equivalent for an alternative with equal probabilities of yielding a cost of five million dollars or ten million dollars is eight million dollars. Then $z_{0.5} = (10 - 8)/(10 - 5) = 0.4$. Therefore, from Table 4.2, $R = 1.216$, and $\rho = 1.216 \times (10 - 5) = 6.08$ million dollars.

6.7 Realistic Values for the Risk Tolerance

The procedures for finding the risk tolerance given in the preceding section demonstrate that the utility function shape is ultimately a matter of personal judgment or preference, since the shape depends on the risk tolerance, and this is determined by judgment. However, some values of the risk tolerance are more realistic than others. Howard (1988) provides some rules of thumb about the value of the risk tolerance based on his practical experience. He notes that he has found in business practice that the risk tolerance for a monetary evaluation measure is about (1) six percent of total sales for an organization, (2) one to one-and-a-half times net income, (3) one-sixth of equity, or (4) one-fifth of market value. These rules of thumb were developed based on experience in the oil and chemical industry, and they assume a particular relationship among total sales, net income, equity, and market value. McNamee and Celona (1990) comment that the equity or market value rules of thumb appear to translate most accurately to other industries.

The rules of thumb provide insight into an observation that is sometimes heard from top executives of businesses. They sometimes comment that managers lower down in their organizations are not willing to take as many risks as they (the top executives) would like. Howard's rules of thumb show that the value of the risk tolerance is linked to the size of the organization. Larger organizations (at least within a particular industry) tend to have higher sales, net income, equity, and market value. From the perspective of the top executives of a business, the relevant numbers for sales, income, equity, or market value are those for the entire organization. However, managers lower down in the organization will be most concerned with the sales, income, equity, or market value for their particular business unit. Since this business unit will be smaller than the entire business, the rules of thumb indicate that the lower-level managers will tend to have smaller values for their risk tolerance than the top executives.

The two approximate procedures reviewed in the preceding section for determining the risk tolerance show that a utility function with a smaller risk tolerance has greater risk aversion. That is, risky alternatives will have smaller certainty equivalents (and hence be viewed as less desirable) for smaller values of the risk tolerance. Thus, it follows from Howard's rules of thumb that lower-level managers will be more averse to taking risks. The stakes at risk are a larger portion of the size of the business unit (as specified by sales, income, equity, and market value) that is relevant for the lower-level managers than for the top executives.

Insight can also be gained into the rules of thumb by translating them into personal terms. The personal equivalent of a business firm's equity is your total assets. Thus, if you have assets of $250,000, Howard's rule of thumb yields a risk tolerance of $(1/6) \times \$250,000 = \$41,667$. Using the approximation discussed in the preceding section, this risk tolerance means that you would be willing to participate in an uncertain situation with equal chances of winning $41,667 or losing $(1/2) \times \$41,667 = \$20,834$. Similarly, you would be willing to participate in an uncertain situation with a 0.75 probability of winning $41,667 and a 0.25 probability of losing this same amount.

One final comment on realistic values for the risk tolerance: As Figure 6.3 shows, the smaller the risk tolerance, the more curved the shape of the exponential utility function. If the risk tolerance is smaller than one-tenth the range of evaluation measure levels of interest, then the shape of the utility function is so bowed that there is probably some very unusual situation that requires special study. On the other hand, if the risk tolerance is greater than ten times the range of evaluation measure levels being considered, then the utility function is almost a straight line, and thus you can usually use expected value as a criterion for decision making.

6.8 Spreadsheet Calculation Examples

It is straightforward to use either a calculator or a spreadsheet program to do calculations involving exponential utility functions. As discussed earlier, the certainty equivalents for uncertain alternatives can be used to rank the alternatives. It is straightforward to develop an equation that can be used to determine the certainty equivalent for an alternative when you are using an exponential utility function. From the definition presented earlier, the certainty equivalent CE for an uncertain alternative a is determined by the equation

$$u(\text{CE}) = \text{E}[u(x)|a] \tag{6.3}$$

where $u(x)$ is the utility function, and $\text{E}[u(x)|a]$ is the expected utility for the alternative a.

When the appropriate one of the equations 6.1 or 6.2 is substituted into 6.3, the result can be solved for the certainty equivalent CE. When the substitution is done, it turns out that many common factors drop out. In particular, all references to High and Low drop out, and when equation 6.1 holds (that is, preferences are monotonically increasing), the certainty equivalent CE for an uncertain alternative is

$$\text{CE} = \begin{cases} -\rho \ln \text{E}[\exp(-x/\rho)], & \rho \neq \text{Infinity} \\ \text{E}(x), & \text{otherwise} \end{cases} \tag{6.4}$$

and when equation (6.2) holds (that is, preferences are monotonically decreasing),

$$\text{CE} = \begin{cases} \rho \ln \text{E}[\exp(x/\rho)], & \rho \neq \text{Infinity} \\ \text{E}(x), & \text{otherwise} \end{cases} \tag{6.5}$$

These are the equations that you use in a calculator or spreadsheet analysis involving exponential utility functions. These analysis procedures are illustrated below with two specific examples.

Example 1

Earlier in this chapter, two approaches were presented for approximating a decision maker's risk tolerance based on determining the certainty equivalents for simple uncertain alternatives that have only two possible outcomes.

The first of the simple alternatives has equal chances of yielding either $+R$ or $-R/2$, and the second has a 0.75 probability of yielding $+R$ and a 0.25 probability of yielding $-R$. In both cases, you adjust the value of R until the decision maker is indifferent between either accepting the uncertain alternative or not. That is, you adjust the value of R until the uncertain alternative has a certainty equivalent of zero. The approximation states that the risk tolerance is approximately equal to R.

Assume we have determined that $R = \$100,000$ for a particular decision maker. Use a spreadsheet to show that the values for the risk tolerance ρ found using the two approximations are in error by about four percent and nine percent, respectively. This result holds in general. (Such an error in the risk tolerance will often not change the ranking of alternatives.)

Solution for Example 1   Since preferences are monotonically increasing for this situation, equation 6.4 is used to determine the certainty equivalents for the two simple alternatives. For both the alternatives, the approximation says that $\rho = R$, and hence $\rho = \$100,000$. For the first alternative, it follows from equation 6.4 that

$$
\begin{aligned}
CE &= -\$100,000 \times \ln[0.5 \times \exp(-\$100,000/\$100,000) \\
&\quad + 0.5 \times \exp(\$50,000/\$100,000)] \\
&= -\$100,000 \times \ln[0.5 \times \exp(-1) + 0.5 \times \exp(0.5)] \\
&= -\$826.61
\end{aligned}
$$

[This calculation can easily be done using any calculator that has exponential (exp) and natural logarithm (ln) functions.] From this we see that the calculated certainty equivalent of $826.61 is slightly less than zero, which is the actual certainty equivalent for the alternative. Thus, the actual risk tolerance ρ must be larger than $100,000. That is, the decision maker is actually slightly less risk averse than what the approximation says.

The example question statement says that the approximation is in error by about four percent, and thus the actual risk tolerance must be four percent higher than $100,000, which is $104,000. Checking this, we find

$$
\begin{aligned}
CE &= -\$104,000 \times \ln[0.5 \times \exp(-\$100,000/\$104,000) \\
&\quad + 0.5 \times \exp(\$50,000/\$104,000)] \\
&= \$19.64
\end{aligned}
$$

For this situation, which involves sums of $100,000 and $50,000, the error of $19.64 is very close to zero, and thus the four percent error is verified.

While this calculation can easily be done by calculator, it is worthwhile understanding how a spreadsheet can be used for the calculation. This is because the spreadsheet procedure can also be used for more complex examples or situations where the calculation has to be done repeatedly. In such situations, a calculator approach is tedious and error prone. Figure 6.4a shows a spreadsheet for this calculation. The assumed risk tolerance is in cell B1, and the two possible outcomes for the uncertain alternative are in cells A3 and A4. The probabilities for these are in cells B3 and B4, respectively. For this simple example, it would be possible just to enter equation 6.4 into another cell to determine the certainty equivalent for the alternative. However, this becomes more complex as the alternative has more possible outcomes, and hence a procedure is presented that is more generally applicable.

Specifically, the calculation is split up into several parts. The equations for this procedure are shown in Figure 6.4b. In this figure, row 2 contains labels that indicate what operations are done in each column. In column C, the quantity $\exp(-x/\rho)$ is calculated for each x. Note that in this column, the absolute reference $B$1 is used for the cell that contains the risk tolerance. When this is done, it is necessary only to enter the expression EXP(-A3/$B$1) once (in cell C3), and then this can be copied down to the cell below. While the saving over re-entering the expression in cell C4 is slight in this example, it is more substantial when there are more possible outcomes.

In column D, the quantity $p(x) \times \exp(-x/\rho)$ is calculated by multiplying the quantities in columns B and C. In cell D5, which is below the two rows that represent each of the two possible outcomes for the alternative, the column D entries are summed. Note that using the SUM(D3:D4) expression means that it is straightforward to include more possible outcomes for the alternative without needing to add additional terms to the expression as would be true if the expression D3+D4 had been used. Finally, in cell D6, the certainty equivalent is determined by taking the natural logarithm of the entry in cell D5 and multiplying it by $-\rho$. Once again, an absolute reference to cell B1 for the risk tolerance is used so that the equation will remain correct even if additional rows are added to represent additional possible outcomes for the alternative.

Once this spreadsheet is set up, the answers to all parts of this example can be quickly determined. Figure 6.4c shows the result when the risk tolerance (cell B1) is changed to $104,000. Figure 6.4d shows the calculation for the second approximation (with 0.75 and 0.25 probabilities) for a risk tolerance of $100,000, and Figure 6.4e shows the calculation for the second approximation for a risk tolerance of $91,000. From Figure 6.4d, we see that a risk tolerance of $100,000 is too low because the certainty equivalent of $4,554.14 is greater than the desired certainty equivalent of zero. When the risk tolerance is reduced by nine percent to $91,000, the risk tolerance is −$13.14, which is extremely close to the desired value of zero.

Example 2

A manager is considering which of two uncertain alternatives to select. An assessment of the net profit from each alternative has been done, and the result is displayed in cumulative probability distributions, one for each alternative. For the first alternative, the 0.05 fractile of the cumulative probability distribution is −10 (in thousands of dollars), the 0.50 fractile is +100, and the 0.95 fractile is +150. For the second alternative, the 0.05 fractile is −30, the 0.50 fractile is +110, and the 0.95 fractile is +170.

a Confirm, using the extended Pearson-Tukey approximation, that the expected net profit for the first alternative is 88.90, and the expected net profit for the second alternative is 95.20. Hence, using expected value as the criterion for making decisions, the second alternative is preferred.

b Suppose that the manager has a certainty equivalent of zero for an alternative with equal chances of yielding +50 or −25. Use the approximation presented in this chapter to estimate the corresponding risk tolerance. With this risk tolerance, use a spreadsheet to determine that the certainty equivalents for the two alternatives in this example are 56.90 and 44.21, respectively, and hence that the first alternative is preferred. Explain why it makes sense that the certainty equivalent for a risk averse manager might be higher for the first alternative, even though the second one has a higher expected value.

c Show that even with a ten percent increase in the risk tolerance found in part b, the first alternative is still preferred to the second, and that the certainty equivalents for the two alternatives with this increased risk tolerance are 59.94 and 48.65, respectively. However, also show that the certainty equivalents for the two alternatives are closer together with this modified risk tolerance than they were in part b. Explain why it makes sense that the certainty equivalents are closer in this case than in part b.

Solution for Example 2   The answer for part a requires applying the extended Pearson-Tukey approximation to replace the continuous probability distributions with discrete distributions that have three possible outcomes. The first outcome is the 0.05 fractile of the distribution, and it is assigned a probability of 0.185; the second outcome is the 0.50 fractile of the distribution, and it is assigned a probability of 0.63; and the third outcome is the 0.95 fractile of the distribution, and it is assigned a probability of 0.185. Once this approximation is made, it is straightforward to calculate the expected net profit for the discrete distribution using the method discussed earlier in this chapter.

Specifically, the expected net profit for the first alternative is

$$0.185 \times (-10) + 0.63 \times 100 + 0.185 \times 150 = 88.90$$

and the expected net profit for the second alternative is

$$0.185 \times (-30) + 0.63 \times 110 + 0.185 \times 170 = 95.20$$

Thus, using expected net profit as the decision criterion, the second alternative is preferred.

a. Equal chance approximation

	A	B	C	D
1	rho=	100000		
2	x	p(x)	exp(-x/rho)	product
3	100000	0.5	0.37	0.18
4	-50000	0.5	1.65	0.82
5			sum=	1.01
6			CE=	-826.61

b. Equations for equal chance approximation

	A	B	C	D
1	rho=	100000		
2	x	p(x)	exp(-x/rho)	product
3	100000	0.5	=EXP(-A3/$B$1)	=B3*C3
4	-50000	0.5	=EXP(-A4/$B$1)	=B4*C4
5			sum=	=SUM(D3:D4)
6			CE=	=-$B$1*LN(D5)

c. Four percent increase in the risk tolerance

	A	B	C	D
1	rho=	104000		
2	x	p(x)	exp(-x/rho)	product
3	100000	0.5	0.38	0.19
4	-50000	0.5	1.62	0.81
5			sum=	1.00
6			CE=	19.64

d. Seventy-five/ twenty-five percent approximation

	A	B	C	D
1	rho=	100000		
2	x	p(x)	exp(-x/rho)	product
3	100000	0.75	0.37	0.28
4	-100000	0.25	2.72	0.68
5			sum=	0.96
6			CE=	4554.14

e. Nine percent decrease in the risk tolerance

	A	B	C	D
1	rho=	91000		
2	x	p(x)	exp(-x/rho)	product
3	100000	0.75	0.33	0.25
4	-100000	0.25	3.00	0.75
5			sum=	1.00
6			CE=	-13.14

Figure 6.4 *Spreadsheet calculations for Example 1*

To answer part b, it is first necessary to determine the risk tolerance (ρ). The alternative presented in part b, which has a certainty equivalent of zero, has outcomes of R and $-R/2$, where $R = 50$, and each outcome has an equal chance of occurring. Thus, using the approximation presented earlier, the risk tolerance is approximately equal to 50. Once the risk tolerance is known, equation 6.4 can be used to calculate the certainty equivalents for the two alternatives.

A spreadsheet approach to doing the calculations is shown in Figure 6.5. Figure 6.5a shows the calculations for part b of Example 2, and Figure 6.5b shows the calculations for part c of Example 2. The spreadsheet equations are shown in Figure 6.5c. If you developed a spreadsheet like the one shown in Figure 6.4 to solve Example 1, then you can quickly modify this to solve parts b and c of Example 2. Here are the steps:

1   For the spreadsheet shown in Figure 6.4b, insert a new row between the existing rows 3 and 4.
2   Copy row 3 to (the new blank) row 4.
3   Fill in the risk tolerance value of 50 in cell B1. Fill in the values -10, 100, and 150 in rows 3, 4, and 5, respectively, of column A.
4   Fill in the probabilities 0.185, 0.630, and 0.185 in rows 3, 4, and 5, respectively, of column B.
5   The result of Steps 1 through 4 is the range A1:D7 in the spreadsheet of Figure 6.5a. Cell D7 shows that the certainty equivalent for the first alternative is 56.90.
6   In the remaining steps, you copy the part of the spreadsheet that calculates the certainty equivalent for the first alternative, and then change the numbers to calculate the certainty equivalent for the second alternative. First, copy the range A2:D7 to cell A9.
7   Fill in the values -30, 110, and 170 in rows 10, 11, and 12 of row A.
8   The result of Steps 6 and 7 is the spreadsheet shown in Figure 6.5a. Cell D14 shows that the certainty equivalent for the second alternative is 44.21.
9   Change the risk tolerance value in cell B1 to 55. The result is the spreadsheet shown in Figure 6.5b. This shows that the certainty equivalents for the two alternatives change to 59.94 and 48.65, respectively, when the risk tolerance is changed from 50 to 55.

One point that sometimes trips up spreadsheet users: Be sure that in Step 1 you insert the new row between rows 3 and 4, rather than before row 3 or after row 4. If you insert the new row before row 3 or after row 4, then the range for the SUM function in cell D6 of Figure 6.5c will not expand to include the new row, and hence you will obtain an incorrect answer for the certainty equivalent. Also, it is important that you use the absolute reference $B$1 for the cell that contains the risk tolerance, or the formulas in range C10:C12 and cell D14 will be incorrect after you copy these in Step 6.

Now complete the solution for parts b and c. When you examine the original specification of the two alternatives, you see that the second one can result in a worse loss (-30) than the first one (-10). Thus, it is not surprising that the preference order for the two alternatives switches when risk aversion is considered. Because both alternatives are risky, the certainty equivalents for both are less than their expected values. (With a risk tolerance of 50, the first alternative

	A	B	C	D
1	rho=	50		
2	x	p(x)	exp(-x/rho)	product
3	-10	0.185	1.22	0.23
4	100	0.630	0.14	0.09
5	150	0.185	0.05	0.01
6			sum=	0.32
7			CE=	56.90
8				
9	x	p(x)	exp(-x/rho)	product
10	-30	0.185	1.82	0.34
11	110	0.630	0.11	0.07
12	170	0.185	0.03	0.01
13			sum=	0.41
14			CE=	44.21

a. Solution for part b

	A	B	C	D
1	rho=	55		
2	x	p(x)	exp(-x/rho)	product
3	-10	0.185	1.20	0.22
4	100	0.630	0.16	0.10
5	150	0.185	0.07	0.01
6			sum=	0.34
7			CE=	59.94
8				
9	x	p(x)	exp(-x/rho)	product
10	-30	0.185	1.73	0.32
11	110	0.630	0.14	0.09
12	170	0.185	0.05	0.01
13			sum=	0.41
14			CE=	48.65

b. Solution for part c

	A	B	C	D
1	rho=	50		
2	x	p(x)	exp(-x/rho)	product
3	-10	0.185	=EXP(-A3/$B$1)	=B3*C3
4	100	0.63	=EXP(-A4/$B$1)	=B4*C4
5	150	0.185	=EXP(-A5/$B$1)	=B5*C5
6			sum=	=SUM(D3:D5)
7			CE=	=-$B$1*LN(D6)
8				
9	x	p(x)	exp(-x/rho)	product
10	-30	0.185	=EXP(-A10/$B$1)	=B10*C10
11	110	0.63	=EXP(-A11/$B$1)	=B11*C11
12	170	0.185	=EXP(-A12/$B$1)	=B12*C12
13			sum=	=SUM(D10:D12)
14			CE=	=-$B$1*LN(D13)

c. Spreadsheet equations

Figure 6.5 *Spreadsheet calculations for Example 2*

has a certainty equivalent of 56.90 versus its expected value of 88.90, and the second alternative has a certainty equivalent of 44.21 versus its expected value of 95.20.) However, the greater riskiness of the second alternative results in a greater difference between the certainty equivalent and expected value for the second alternative than the first one. This leads to a reversal of the preference order for the two alternatives.

In part c, the risk tolerance is increased by ten percent. This represents a less risk averse attitude than in part b, and hence the certainty equivalents for both alternatives will increase from those in part b. Specifically, the certainty equivalent for the first alternative increases from 56.90 in part b to 59.94 in part c, and the certainty equivalent for the second alternative increases from 44.21 to 48.65.

In addition, the certainty equivalents become closer together in part c. In part b, the difference between the two certainty equivalents is $56.90 - 44.21 = 12.69$, while in part c the difference reduces to $59.94 - 48.65 = 11.29$. This makes sense because when the risk tolerance increases, there is less concern about the possible negative outcome. We know from part a that when the risk tolerance goes to infinity (that is, there is risk neutrality), then the second alternative is preferred to the first. Therefore, as the risk tolerance increases, we expect the relative preference for the first alternative to get smaller. You may wish to experiment and see how large the risk tolerance has to be before the second alternative becomes preferred. Once you have the spreadsheet in Figure 6.5 set up, it is easy to do this experiment. (*Answer:* The second alternative is preferred to the first when the risk tolerance is greater than approximately 143.)

6.9 References

R. T. Clemen, *Making Hard Decisions: An Introduction to Decision Analysis*, Second Edition, Duxbury Press, Belmont, California, 1996

S. Ghahramani, *Fundamentals of Probability*, Prentice Hall, Upper Saddle River, New Jersey, 1996.

R. A. Howard, "Decision Analysis: Practice and Promise," *Management Science*, Vol. 34, pp. 679–695 (1988).

D. L. Keefer and S. E. Bodily, "Three-Point Approximations for Continuous Random Variables," *Management Science*, Vol. 29, pp. 595–609, 1983.

R. L. Keeney and H. Raiffa, *Decisions with Multiple Objectives: Preferences and Value Tradeoffs*, Wiley, New York, 1976.

C. W. Kirkwood, "An Overview of Methods for Applied Decision Analysis," *Interfaces*, Vol. 22, No. 6, pp. 28–39 (November–December 1992).

P. McNamee and J. Celona, *Decision Analysis with Supertree*, Second Edition, The Scientific Press, South San Francisco, California, 1990.

I. Olkin, L. J. Gleser, and C. Derman, *Probability Models and Applications*, Macmillan, New York, 1994.

J. W. Pratt, "Risk Aversion in the Small and in the Large," *Econometrica*, Vol. 32, pp. 122–136 (1964).

J. W. Pratt, H. Raiffa, and R. O. Schlaifer, "The Foundations of Decision Under Uncertainty: An Elementary Exposition," *Journal of the American Statistical Association*, Vol. 59, pp. 353–375 (1964).

6.10 Review Questions

R6-1   Give an example that illustrates why uncertainty adds complexity to decision making.

R6-2   Define an *event space*, and discuss why this concept is important in probability analysis.

R6-3   Discuss the *Weak Law of Large Numbers* and its significance for decision analysis.

R6-4   Explain the use of *cumulative distributions* in decision analysis.

R6-5   Review two limitations on using expected value as a criterion for making decisions when there is uncertainty.

R6-6   Describe the extended Pearson-Tukey and extended Swanson-Megill procedures, and discuss the reasons for using these in decision analysis.

R6-7   Define the certainty equivalent and discuss its significance for decision analysis.

R6-8   Formally define the terms risk averse, risk neutral, and risk seeking. Explain why the formal definitions make sense intuitively.

R6-9   Explain the significance of the exponential utility function in decision analysis, and discuss the role of the risk tolerance (ρ) in the exponential utility function.

R6-10   Summarize Howard's rules of thumb for determining the risk tolerance, and explain the implications of these rules of thumb for the relative risk aversion of different levels within an organization.

R6-11   Describe two approximations for determining the risk tolerance from the certainty equivalent of a simple uncertain alternative that has only two possible outcomes.

R6-12   Explain the relationship between the expected utilities of uncertain alternatives and the certainty equivalents of those alternatives.

R6-13   Describe the use and interpretation of expected utilities and certainty equivalents for uncertain alternatives.

R6-14   Describe qualitatively how the certainty equivalent for an uncertain alternative changes as the risk tolerance varies.

6.11 Exercises

6.1 Many U.S. states now sponsor lotteries, and many people routinely buy lottery tickets. Since the purpose of a lottery is to make money for the government, it must be true that the expected value of buying a lottery ticket is negative. That is, on average over many lottery ticket purchases, the probability is very high that the purchaser will lose money. Discuss whether a person who claims to be risk averse, but who purchases lottery tickets, is expressing his or her true attitude toward risk taking. In your answer, consider both the situation where the only evaluation measure for the ticket purchaser is profit, and also the situation where there may be other evaluation measures.

6.2 Suppose that you have a friend who insists that he is lucky with new coins and that for him the probability is 1 (100%) of a head coming up on the first flip by a neutral person of any brand new coin fresh from the U.S. Mint. Therefore, he is willing to pay $10 to play a game where he will win $15 if a head comes up when the coin is flipped or lose $100 if a tail comes up. His argument is as follows: His expected value for the coin flip is $15. Therefore, the Weak Law of Large Numbers assures that he will win $15 on the coin flip, and hence he will make $15 − $10 = $5 on the game. Determine whether he has misinterpreted the Weak Law of Large Numbers, and also whether he is likely to make money with this opinion regarding flips of new coins.

6.3 Show that the following probability distribution obeys the properties for an event space: $P(X = 1) = 0.10$, $P(X = 2) = 0.15$, $P(X = 3) = 0.25$, $P(X = 4) = 0.20$, $P(X = 5) = 0.15$, $P(X = 6) = 0.10$, $P(X = 7) = 0.05$. Plot the cumulative probability distribution, and calculate the expected value.

6.4 Show that the following probability distribution obeys the properties for an event space: $P(X = 1) = 0.15$, $P(X = 2.5) = 0.10$, $P(X = 3) = 0.25$, $P(X = 4) = 0.20$, $P(X = 5.3) = 0.20$, $P(X = 6) = 0.10$. Plot the cumulative probability distribution, and calculate the expected value.

6.5 Small Retailer has 100 Zupper Zips in stock, and cannot receive any more for at least one week. The probability distribution for the number of Zupper Zips that customers will want to purchase from Small Retailer over the next week is as follows: $P(X = 90) = P(X = 100) = P(X = 110) = 1/3$. (If customers want to purchase more Zupper Zips than Small Retailer has in stock, then the customers purchase the number in stock, and they go elsewhere to obtain the rest.)
  (i) Verify that this is a valid probability distribution.
  (ii) Find the cumulative probability distribution and the expected value for the number of Zupper Zips that customers will want to purchase over the next week.

(iii) Find the probability distribution, the cumulative probability distribution, and the expected value for the number of Zupper Zips that Small Retailer will actually sell during the next week.

6.6 Erhart's Retail Pots has 85 Zirconium Pots in stock, and cannot receive any more for at least one month. The probability distribution for the number of potential customers who will want to purchase a Zirconium Pot from Erhart's over the next month is as follows: $P(X = 70) = 0.50$, $P(X = 80) = P(X = 100) = 0.25$. (No customer will purchase more than one Zirconium Pot, and if Erhart's doesn't have the pot in stock, the customer will purchase it elsewhere.)

   (i) Verify that this is a valid probability distribution.
   (ii) Find the cumulative probability distribution and the expected value for the number of Zirconium Pots that potential customers will want to purchase from Erhart's over the next month.
   (iii) Find the probability distribution, the cumulative probability distribution, and the expected value for the number of Zirconium Pots that Erhart's will actually sell during the next month.

6.7 The Johnson Steel Company must renegotiate labor contracts next year. Management knows that if there is no strike, then the company's profits for the year will be $4.5 million. However, if there is a strike, the production delays and the higher priced labor will result in company profits of $3 million. The president of Johnson Steel tells you that, because of the possible strike, the expected value for profits next year is $4.2 million. Determine what probability of a strike is implied by this statement.

6.8 For the probability distribution you elicited in Exercise 5.2, determine the extended Pearson-Tukey approximation and the extended Swanson-Megill approximation, and use both of these approximations to calculate the expected value of the uncertain quantity.

6.9 Suppose that a decision maker claims
   a. To be indifferent between accepting or not accepting an uncertain alternative with equal chances of winning $100,000 or losing $50,000. (That is, the decision maker's certainty equivalent for this uncertain alternative is $0.)
   b. To be indifferent between accepting or not accepting an uncertain alternative with a 0.75 probability of winning $100,000 and a 0.25 probability of losing $100,000.

Determine whether the approximate risk tolerances implied by claims a and b are consistent. As part of your answer, determine what risk tolerance is implied by each answer.

6.10 Cornflowers Ltd. has a total equity of $10 million and total annual sales of $30 million with an annual net income of $1 million. Using Howard's rules of thumb, determine the value for Cornflowers' risk tolerance that is implied by each of these numbers.

6.11 Suppose that a decision maker has a certainty equivalent profit of 25 for an uncertain alternative with equal chances of yielding profits of either 100 or 0. Determine the decision maker's risk tolerance. Suppose instead that the decision maker has a certainty equivalent cost of 25 for an uncertain alternative with equal chance of yielding *costs* of 100 or 0. Determine the decision maker's risk tolerance.

6.12 Suppose that a decision maker has a certainty equivalent *cost* of 45 for an uncertain alternative with a 0.4 probability of yielding a cost of 90 and a 0.6 probability of yielding a cost of 0. Determine the decision maker's risk tolerance to two decimal places of accuracy.

6.13 Suppose that an uncertain alternative has the following probability distribution for profit: $P(X = 1.2) = 0.05$, $P(X = 2) = 0.15$, $P(X = 3) = 0.25$, $P(X = 4.3) = 0.20$, $P(X = 5) = 0.10$, $P(X = 6.1) = 0.10$, $P(X = 7) = 0.15$. Suppose further that the decision maker's exponential utility function has a risk tolerance of 10. Determine the certainty equivalent for the uncertain alternative.

CHAPTER 7

Multiple Objectives and Uncertainty

This chapter extends the approaches presented in preceding chapters to decisions where there are both multiple evaluation concerns and uncertainty. As we will see, the procedure presented earlier can be extended in a natural way to consider both uncertainty and multiple evaluation concerns. Essentially, this requires extending the value function approach of Chapter 4 to incorporate the methods for analyzing uncertainty that were presented in Chapters 5 and 6.

7.1 Uncertainty with Multiple Evaluation Measures

This section extends the networking strategy decision from Chapter 4 to illustrate a procedure for analyzing decisions with both multiple objectives and uncertainty. Chapter 4 developed the following value function to analyze this decision when there is no uncertainty:

$$v(X_p, X_c, X_s) = 0.22v_p(X_p) + 0.35v_c(X_c) + 0.43v_s(X_s)$$

where X_p represents Productivity Enhancement, X_c represents Cost Increase, and X_s represents Security. The single dimensional value functions developed in Chapter 4 for the three evaluation measures are shown in Figure 7.1.

Suppose that there is uncertainty about the evaluation measure levels that will result from selecting a particular alternative in the networking decision. The Weak Law of Large Numbers, which we studied in Chapter 6, applies to a situation with multiple evaluation measures in the same way that it does to a situation with a single evaluation measure. Thus, an expected value calculation can be done by using the value function given above to determine what the average value is that we expect to receive over the long run from each repeated "play" of an uncertain alternative with multiple evaluation measures.

To illustrate the procedure, consider a modified version of the networking strategy decision from Chapter 4, where three of the four alternatives have uncertainty about what level will occur for one or more of the evaluation measures.

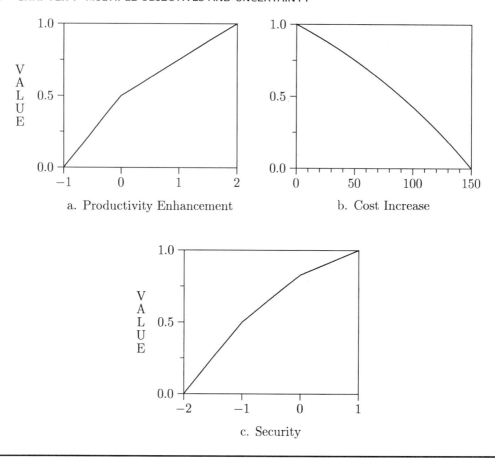

Figure 7.1   *Single dimensional value functions for networking decision*

Specifically, assume that the alternatives have the probability distributions given in Table 7.1. In this table, there are three columns for each of the three evaluation measures; the first gives a probability, and the second gives the score (level) that has this probability. (The third column for each evaluation measure will be discussed shortly.) Thus, the Medium Quality/Medium Cost alternative has a 0.33 probability of having a score of 0.5 for Security and a 0.67 probability of having a score of 0 for that evaluation measure.

Determining the expected value for each of the four alternatives requires calculating the expected value for the value function given above. To do this, it is necessary to first determine the single dimensional values for each of the various possible evaluation measure scores shown in Table 7.1. This can be done using the procedure discussed in Section 4.5, and the resulting single dimensional values are shown in Table 7.1 in the third column (labeled "Value") under each evaluation measure.

The expected value for each alternative can be calculated using the value numbers in Table 7.1. These calculations are simplified by using two properties from elementary probability:

| | Evaluation Measure | | | | | | | | |
| Alternative | Productivity Enhancement ($w_p = 0.22$) | | | Cost Increase ($w_c = 0.35$) | | | Security ($w_s = 0.43$) | | |
	Prob.	Score	Value	Prob.	Score	Value	Prob.	Score	Value
Status Quo	1.00	0	0.50	1.00	0	1.00	1.00	0	0.83
High Quality/	0.75	2	1.00	1.00	125	0.23	0.60	1	1.00
High Cost	0.25	1	0.75				0.40	0.5	0.92
Medium Quality/	1.00	1	0.75	1.00	95	0.46	0.33	0.5	0.92
Medium Cost							0.67	0	0.83
Low Quality/	0.50	1	0.75	1.00	65	0.66	0.40	0	0.83
Low Cost	0.50	0	0.50				0.40	-1	0.50
							0.20	-2	0.00

Table 7.1 *Probabilities and single dimensional values for networking alternatives*

- **Property 1:** The expected value for a sum of uncertain variables is equal to the sum of the expected values for the uncertain variables.
- **Property 2:** The expected value for a constant multiple of an uncertain variable is equal to the expected value of the uncertain variable multiplied by the constant.

The first property is illustrated by the following example: Suppose there are two uncertain variables X and Y, where X is equally likely to be 1 or 2, and Y has a 0.3 probability of being 4 and a 0.7 probability of being 6. Then the expected value of X is $0.5 \times 1 + 0.5 \times 2 = 1.5$, and the expected value of Y is $0.3 \times 4 + 0.7 \times 6 = 5.4$. From Property 1, it follows that the expected value of $X + Y$ is equal to $1.5 + 5.4 = 6.9$. Similarly, from Property 2, it follows that the expected value of 4 times X is equal to 4 times the expected value of X, which is $4 \times 1.5 = 6$.

These two properties can be applied to determine the expected value for the networking decision. For this example,

$$v(X_p, X_c, X_s) = 0.22v_p(X_p) + 0.35v_c(X_c) + 0.43v_s(X_s)$$

Thus, from Property 1, we know that the expected value of this is given by

$$\begin{aligned} EV &= E[0.22v_p(X_p) + 0.35v_c(X_c) + 0.43v_s(X_s)] \\ &= E[0.22v_p(X_p)] + E[0.35v_c(X_c)] + E[0.43v_s(X_s)] \end{aligned}$$

Applying Property 2 to this leads to

$$EV = 0.22 \times E[v_p(X_p)] + 0.35 \times E[v_c(X_c)] + 0.43 \times E[v_s(X_s)]$$

Thus, the expected value for any alternative can be determined by first calculating the expected value for each single dimensional value function, next multiplying the expected single dimensional values by the corresponding weights, and finally summing the results. This procedure works in general.

Thus, using the data in Table 7.1, the expected values for the various networking alternatives are as follows:

EV(Status Quo)

$$= 0.22 \times 1.00 \times 0.50 + 0.35 \times 1.00 \times 1.00 + 0.43 \times 1.00 \times 0.83$$
$$= 0.82$$

EV(High Qual./High Cost)

$$= 0.22 \times (0.75 \times 1.00 + 0.25 \times 0.75) + 0.35 \times 1.00 \times 0.23$$
$$+ 0.43 \times (0.60 \times 1.00 + 0.40 \times 0.92)$$
$$= 0.70$$

EV(Med. Qual./Med. Cost)

$$= 0.22 \times 1.00 \times 0.75 + 0.35 \times 1.00 \times 0.46$$
$$+ 0.43 \times (0.33 \times 0.92 + 0.67 \times 0.83)$$
$$= 0.70$$

EV(Low Qual./Low Cost)

$$= 0.22 \times (0.50 \times 0.75 + 0.50 \times 0.50) + 0.35 \times 1.00 \times 0.66$$
$$+ 0.43 \times (0.40 \times 0.83 + 0.40 \times 0.50 + 0.20 \times 0.00)$$
$$= 0.60$$

Hence, if expected value is used as the criterion for ranking alternatives, the Status Quo alternative is highest ranked with an expected value of 0.82. The High Quality/High Cost and Medium Quality/Medium Cost alternatives are tied for second place with expected values of 0.70, and the Low Quality/Low Cost alternative is ranked last with an expected value of 0.60.

7.2 Multiattribute Risk Aversion

From the discussion in Chapter 6, we know that expected value is not appropriate as a criterion for making decisions when a decision maker is risk averse. In such situations, it is necessary to determine a utility function and conduct an expected utility analysis. The concept of risk aversion that we studied in that chapter for situations with a single evaluation measure also applies to a situation with multiple evaluation measures.

To illustrate the meaning of risk aversion in the multiple evaluation measure situation, consider two hypothetical alternatives for the networking example. (Neither of these is an actual alternative for that decision as given in Table 7.1. They are presented only to illustrate the meaning of risk aversion in a situation with multiple evaluation measures.) Suppose that the first alternative, which we will call "Uncertain Alternative 1," has uncertainty about its Cost Increase and Security. Specifically, Uncertain Alternative 1 has a 0.5 probability of yielding an outcome with $X_c = \$0$ and $X_s = 1$, and also a 0.5 probability of yielding an outcome with $X_c = \$150$ and $X_s = -2$. Regardless of which of these outcomes occurs, the alternative will have $X_p = 1$.

With these outcomes, the expected value for Uncertain Alternative 1 is

$$
\begin{aligned}
\text{Expected Value} &= 0.5 \times v(1,0,1) + 0.5 \times v(1,150,-2) \\
&= 0.5 \times [0.22v_p(1) + 0.35v_c(0) + 0.43v_s(1)] \\
&\quad + 0.5 \times [0.22v_p(1) + 0.35v_c(150) + 0.43v_s(-2)] \\
&= 0.5 \times [0.22 \times 0.75 + 0.35 \times 1 + 0.43 \times 1] \\
&\quad + 0.5 \times [0.22 \times 0.75 + 0.35 \times 0 + 0.43 \times 0] \\
&= 0.56
\end{aligned}
$$

Thus, from the Weak Law of Large Numbers, if we could repeatedly select this alternative many times, we expect that we would receive an average of 0.56 each time.

Now consider a second alternative, which we will call "Uncertain Alternative 2." This alternative has the same score for X_p as Uncertain Alternative 1, but it has different uncertainty about X_c and X_s. Specifically, it has a 0.5 probability of yielding an outcome with $X_c = \$0$ and $X_s = -2$, and a 0.5 probability of yielding an outcome with $X_c = 150$ and $X_s = 1$. With these outcomes, the expected value for Uncertain Alternative 2 is

$$
\begin{aligned}
\text{Expected Value} &= 0.5 \times v(1,0,-2) + 0.5 \times v(1,150,1) \\
&= 0.5 \times [0.22v_p(1) + 0.35v_c(0) + 0.43v_s(-2)] \\
&\quad + 0.5 \times [0.22v_p(1) + 0.35v_c(150) + 0.43v_s(1)] \\
&= 0.5 \times [0.22 \times 0.75 + 0.35 \times 1 + 0.43 \times 0] \\
&\quad + 0.5 \times [0.22 \times 0.75 + 0.35 \times 0 + 0.43 \times 1] \\
&= 0.56
\end{aligned}
$$

Thus, we see that Uncertain Alternative 2 has the same expected value as Uncertain Alternative 1. Hence, using expected value, the two alternatives are rated as equally desirable. Is this reasonable?

Perhaps the two alternatives should be rated as equally preferable, but they certainly are different. Uncertain Alternative 1 either yields the best scores for both X_c and X_s together, or it yields the worst scores for these two evaluation measures together. This is an "all or nothing" alternative with respect to these two evaluation measures. Uncertain Alternative 2, on the other hand, will yield either the best score for X_c combined with the worst score for X_s, or the worst score for X_c combined with the best score for X_s. Thus, with this alternative, you are assured of obtaining good performance with respect to either Cost Increase or Security, but, in contrast to Uncertain Alternative 1, you cannot obtain good performance with respect to both of these evaluation measures at the same time.

Most likely, you would prefer not to have to select either of these alternatives, but, if forced to select, many people would select Uncertain Alternative 2 because with that alternative you eliminate the possibility of receiving the highly undesirable outcome where both Cost Increase and Security are at their worst levels.

A decision maker who prefers Uncertain Alternative 2 to Uncertain Alternative 1 is called *multiattribute risk averse*, while a decision maker who prefers Uncertain Alternative 1 to Uncertain Alternative 2 is called *multiattribute risk seeking*. Finally, a decision maker who finds the two alternatives equally preferable is called *multiattribute risk neutral*. It is common for decision makers to be multiattribute risk neutral if the range of possible variation for each evaluation measure is small in a value sense, and to be multiattribute risk averse if the range of variation for each evaluation measure is large in a value sense.

For readers with additional probability background: Note that Uncertain Alternative 1 and Uncertain Alternative 2 both have the same marginal probability distributions for each of the single dimensional value functions. Thus, Property 1 says that the two alternatives must have the same expected values since the expected value for a sum of uncertain variables is equal to the sum of the expected values for the variables. We can see from this that if two uncertain alternatives have the same marginal probability distributions for each evaluation measure, the alternatives will have the same expected value when a weighted-additive value function is used.

7.3 The Power-additive Utility Function

The methods presented in Chapter 6 for analyzing decisions with a single evaluation measure and a risk averse decision maker can be extended to decisions with multiple evaluation measures. Specifically, if a decision maker is multiattribute risk averse, then it is necessary to determine a utility function to convert values calculated using a multiattribute value function into utilities. This utility function can then be used to rank alternatives that have uncertainty about their outcomes.

Both theory (Keeney and Raiffa 1976, Theorem 6.11) and practical applications show that in many realistic decision making situations, it is appropriate to use an *exponential* form for this utility function. This form is similar to the exponential single dimensional value function that was considered in Chapter 4 and to the exponential single dimensional utility function considered in Chapter 6. (The theory for this utility function is presented in Section 9.3, but it is not necessary to understand that section to make practical use of this utility function.)

Examples of exponential utility functions are shown in Figure 7.2. Spreadsheet methods for using this utility function will be presented below, and when these

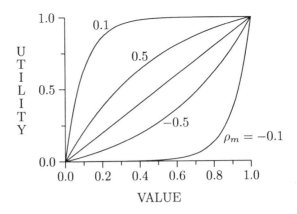

Figure 7.2 *Exponential utility function*

methods are used, it is not necessary to work directly with the equation for this function. However, for interested readers, the equation is as follows:

$$u(x_1, x_2, \ldots, x_n) = \begin{cases} \dfrac{1 - \exp[-v(x_1, x_2, \ldots, x_n)/\rho_m]}{1 - \exp(-1/\rho_m)}, & \rho_m \neq \text{Infinity} \\ v(x_1, x_2, \ldots, x_n), & \text{otherwise} \end{cases} \quad (7.1)$$

where $x_1, x_2, \ldots, x_n$ are the evaluation measures, and

$$v(x_1, x_2, \ldots, x_n) = w_1 v_1(x_i) + w_2 v_2(x_2) + \cdots + w_n v_n(x_n) \quad (7.2)$$

where the w_i are *weights* greater than zero that sum to 1, the $v_i(x_i)$ are *single dimensional value functions* scaled to lie between zero and 1 over the evaluation measure range of interest, and ρ_m is the *multiattribute risk tolerance*. Thus, the value function $v(x_1, x_2, \ldots, x_n)$ in equation 7.2 is the same one that was studied in Chapter 4.

As Figure 7.2 shows, the specific shape of the exponential utility function in equation 7.1 depends on a single constant, called the *multiattribute risk tolerance*, which is represented by the symbol ρ_m (pronounced "rho sub m"). In Figure 7.2, the numbers next to the various curves are the values of ρ_m for each curve. The curves shown in the figure represent a wider range of cases than is usually seen in practice. In most applications, the value of ρ_m is positive and greater than 0.2. Furthermore, if ρ_m is greater than ten, the curve is essentially a straight line, and hence expected value can be used to evaluate alternatives. (The straight line in the center of Figure 7.2 represents the case where ρ_m is equal to infinity.)

The combination of an exponential curve to convert value into utility (as given by equation 7.1 and shown in Figure 7.2) with the weighted sum form of value function (as given in equation 7.2) is called a *power-additive* utility function.

A detailed study of the equations for the power-additive utility function in equations 7.1 and 7.2 shows that a multiattribute risk averse utility function has a positive multiattribute risk tolerance, a multiattribute risk seeking utility function has a negative multiattribute risk tolerance, and for the multiattribute risk neutral case the value function can be used as a utility function. (That is, for the multiattribute risk neutral case, expected values can be used to rank alternatives.) The multiattribute risk seeking case is not often seen in practice.

Determining the Multiattribute Risk Tolerance

As was discussed in Chapter 6, a utility function puts into mathematical form a decision maker's attitude toward risk taking so that this can be taken into account in the evaluation of alternatives. To use the power-additive utility function, it is necessary to specify a numerical value for the multiattribute risk tolerance ρ_m. Because the value of ρ_m shows the decision maker's degree of aversion to taking risks, it is not surprising to learn that the decision maker must consider an uncertain situation that has risk in order to determine the value of ρ_m. *For many people, this is the single most difficult question that must be considered in the assessment of a value function and utility function.* However, the good news is that once this question is answered, the exponential utility function is completely determined. After this point, determining the expected utilities for alternatives is just arithmetic.

The other good news is that the exact value for ρ_m often does not impact the ranking of alternatives in a decision problem. Howard (1988) notes that he has found risk aversion to be of practical concern in only five to ten percent of business decision analyses. That is, in ninety to ninety-five percent of business decisions, you would determine the same alternative to be most preferred using expected value as if you used expected utility. It is important to note that this does *not* say that you will get the same evaluation results when you consider uncertainty that you would get if you ignored the uncertainty in a decision problem and simply conducted a multiobjective value analysis. Analyzing uncertainty using probabilities and expected values can often change the results from a naive analysis that does not consider uncertainty. However, it is often true that going beyond an expected value analysis to consider risk aversion will not change the evaluation results.

Because of this, it is often not necessary to determine an exact number for ρ_m in an evaluation. If you are using a spreadsheet to do calculations, then you can easily conduct a *sensitivity analysis* to determine whether or not the specific value of ρ_m affects the results of the analysis. If this sensitivity analysis shows that the preferred alternative does not change for a wide range of different values of ρ_m, then you generally do not have to consider risk aversion any further. Specifically, as noted earlier, ρ_m is almost always positive and greater than 0.2 in applications. Therefore, if a sensitivity analysis for ρ_m that covers the range from 0.2 up to infinity shows that the preferred alternative does not change over this range, then you should not need to consider the multiattribute risk tolerance any further, and hence you do not have to assess the specific value of ρ_m.

In situations where this sensitivity analysis shows that the preferred alternative does depend on the specific value of ρ_m, then you need to determine a value for ρ_m in order to conduct an expected utility analysis. Here is one procedure for doing this: First consider a hypothetical uncertain alternative that has equal chances of yielding an outcome with all the evaluation measures at their most preferred levels or all the evaluation measures at their least preferred levels. As discussed in Chapter 4, a value function is specified in such a way that a situation with all the evaluation measures at their most preferred levels has a value of 1, and a situation with all the evaluation measures at their least preferred levels has a value of zero. Figure 7.2 shows that a situation with a value of 1 also has a power-additive utility of 1, and a situation with a value of zero has a utility of zero. Thus, the expected utility for this hypothetical alternative is $0.5 \times 1 + 0.5 \times 0 = 0.5$.

Now determine a hypothetical certain alternative that is equally preferred to the uncertain alternative. For the remainder of the procedure, it does not matter exactly how you find this hypothetical certain alternative, but a systematic way to do this is to consider alternatives that have most of the evaluation measures at their least preferred levels but one or more evaluation measures at their most preferred levels. Keep varying the set of evaluation measures that are set to their most preferred levels until you find a hypothetical certain alternative that is equally preferred to the uncertain alternative with equal chances of all evaluation measures being at either their most preferred or least preferred levels. From the discussion in the preceding paragraph, this alternative must have an expected utility of 0.5 since it is equally preferred to an alternative that has this expected utility.

In some cases, it may not be possible to find a certain alternative that is equally preferred to the uncertain alternative for a certain alternative that has all of its evaluation measures at either their most or least preferred levels. It may then be necessary to consider varying one of the evaluation measure scores to find an intermediate level that yields the desired equal preferences. However, it is worth remembering that variations in the value of ρ_m often do not have much effect on the evaluation results. Thus, it may not be worth spending too much time fine tuning the value of ρ_m.

Once the hypothetical certain alternative that is equally preferred to the uncertain alternative has been found, then Table 4.2 in Chapter 4 can be used to determine the multiattribute risk tolerance. While it is not necessary to understand why this table should be used in order to make practical use of the procedure given below, here is an explanation for interested readers: The mathematical form of the power-additive utility function (which is given in equation 7.1) is a special case of the form for the exponential single dimensional value function given in equation 4.1. Specifically, if Low is set equal to zero and High is set equal to 1 in equation 4.1, then the resulting equation has the same form as equation 7.1. Thus, determining ρ_m is mathematically analogous to determining ρ in equation 4.1. Table 4.2 was used to find ρ, and hence it can also be applied to determine ρ_m.

The procedure to determine ρ_m from Table 4.2 is as follows: Look up, in the $z_{0.5}$ column of Table 4.2, the value for the hypothetical certain alternative that has been determined. Then the value of R in this table corresponding to that $z_{0.5}$ is the needed value of ρ_m.

To illustrate the process of finding the multiattribute risk tolerance, consider the computer networking decision discussed earlier. The Security evaluation measure has the largest weight ($w_s = 0.43$), followed by the Cost Increase evaluation measure ($w_c = 0.35$), and the Productivity Enhancement evaluation measure ($w_p = 0.22$). For this example, the hypothetical uncertain alternative to be considered has equal chances of yielding the best levels of all three evaluation measures (that is, $X_p = 2$, $X_c = 0$, and $X_s = 1$) or the worst levels of all three evaluation measures (that is, $X_p = -1$, $X_c = 150$, and $X_s = -2$). Now consider various hypothetical certain alternatives that might be equally preferred to this uncertain alternative. For the first hypothetical certain alternative, consider one with the Security evaluation measure (which has the highest weight) at its most preferred level and the other two evaluation measures at their least preferred levels (that is, $X_p = -1$, $X_c = 150$, and $X_s = 1$).

Suppose this certain alternative is more preferred than the uncertain alternative. Then reduce the preferability of the certain alternative by replacing Security as the evaluation measure at the most preferred level with Cost Increase (which has the second highest weight). That is, consider a certain alternative with $X_p = -1$, $X_c = 0$, and $X_s = -2$. Suppose now that this new certain alternative is less preferred than the uncertain alternative. We have now found two certain alternatives that "bracket" the uncertain alternative in a value sense. (That is, one of the alternatives is more preferred than the uncertain alternative, while the other is less preferred.) We could continue until we found a certain alternative that is equally preferred to the uncertain alternative, but it may be sufficient to determine the ρ_m value that is implied by each of the two certain alternatives, and conduct a sensitivity analysis to see if the evaluation results change for values of ρ_m over that range.

The value of the first certain alternative is equal to w_s, which is 0.43. From Table 4.2, the corresponding value for ρ_m is 1.762. The value of the second certain alternative is equal to w_c which is 0.35. From Table 4.2, the corresponding value for ρ_m is 0.782. Therefore, the value of ρ_m is somewhere between 0.782 and 1.762.

7.4 Certainty Equivalent Calculations

This section develops methods to carry out certainty equivalent calculations for situations with multiple evaluation measures. These methods are tedious to do by hand, and hence the calculations are usually done using a spreadsheet or specialized software. The next section demonstrates spreadsheet-based calculation procedures.

Carrying out certainty equivalent calculations by hand requires using equations 7.1 and 7.2 for the power-additive utility function. This utility function has a form similar to the exponential single dimensional utility function in equation 6.1 of Section 6.6. A development analogous to that used to derive equation 6.4 in Section 6.8 shows that the *certainty equivalent value* for an uncertain alternative when using a power-additive utility function has the following form:

$$v_{ce} = \begin{cases} -\rho_m \times \ln \mathrm{E}\{\exp[-v(x_1, x_2, \ldots, x_n)/\rho_m)]\}, & \rho \neq \text{Infinity} \\ \mathrm{E}[v(x_1, x_2, \ldots, x_n)], & \text{otherwise} \end{cases} \quad (7.3)$$

where $v(x_1, x_2, \ldots, x_n)$ is given by equation 7.2.

This equation can be used to calculate the certainty equivalent value for any uncertain alternative. This certainty equivalent value has exactly the same meaning as the values that we calculated in Chapter 4 for alternatives with no uncertainty. For example, if equation 7.3 is used to determine that the certainty equivalent value for an uncertain alternative is 0.4, then this means that the uncertain alternative is 0.4 of the distance, in a value sense, between a hypothetical alternative that has all its evaluation measures at the least preferred levels being considered and another hypothetical alternative that has all its evaluation measures at the most preferred levels being considered.

Certainty Equivalents with Probabilistic Independence

In many cases of practical interest, the evaluation measures for each alternative are *probabilistically independent*. That is, once the alternative is specified, the probability distribution for any evaluation measure does not change for different levels of the other evaluation measures. When this is true, then a third property for expected values can be shown in addition to the two properties given earlier:

- **Property 3:** When two uncertain quantities X and Y are probabilistically independent, then $E(XY) = E(X) \times E(Y)$. That is, the expected value of a product of two uncertain quantities is equal to the product of the expected values of the uncertain quantities when the uncertain variables are probabilistically independent.

This result can be applied to equation 7.3 to derive that

$$v_{ce} = CE_1 + CE_2 + \cdots + CE_n \tag{7.4}$$

where

$$CE_i = \begin{cases} -\rho_m \times \ln E\{\exp[-w_i \times v_i(x_i)/\rho_m)]\}, & \rho_m \neq \text{Infinity} \\ w_i \times E[v_i(x_i)], & \text{otherwise} \end{cases} \tag{7.5}$$

(Here is the derivation process: The value function $v(x_1, x_2, \ldots, x_n)$ in equation 7.3 has the additive form of equation 7.2. Since an exponential of a sum is equal to the product of the exponentials of each term in the sum, the exponential in equation 7.2 can be written as a product of exponentials. Then Property 2 can be applied to this product to yield the result shown in equations 7.4 and 7.5.)

Equations 7.4 and 7.5 show that when the evaluation measures are probabilistically independent, we obtain the intuitively appealing result that the overall certainty equivalent value v_{ce} for an alternative is equal to the sum of the certainty equivalent values for that alternative with respect to each evaluation measure. (Decisions where the evaluation measures are not probabilistically independent are considered in Section 7.6.)

To apply equations 7.4 and 7.5 to the networking strategy decision, now assume that the evaluation measures for that decision are probabilistically independent. In order to calculate the expected values on the right-hand side of equation 7.5, it is necessary to have a value for the multiattribute risk tolerance ρ_m. The discussion in Section 7.3 showed that ρ_m for the networking example is

Alternative	Certainty Equiv.			Overall Value
	CE_p	CE_c	CE_s	
Status Quo	0.11	0.35	0.36	0.82
High Qual./High Cost	0.21	0.08	0.42	0.70
Med. Qual./Med. Cost	0.17	0.16	0.37	0.70
Low Qual./Low Cost	0.14	0.23	0.22	0.59

Table 7.2   *Certainty equivalent values for networking alternatives*

between 0.782 and 1.762. For this calculation example, we will use the average of these values, which is $(0.782 + 1.762)/2 = 1.272$.

The arithmetic to calculate the expected values on the right-hand side of equation 7.5 is tedious, and therefore the spreadsheet method presented in Section 7.5 is recommended. However, we will illustrate the required calculation procedure. As an example, consider the probabilities and single dimensional values in Table 7.1 for the Medium Quality/Medium Cost alternative. Then for the Security evaluation measure, it follows from equation 7.5 that

$$
\begin{aligned}
CE_s &= -\rho_m \times \ln E[\exp(-0.43 v_s / \rho_m)] \\
&= -1.272 \times \ln[0.33 \times \exp(-0.43 \times 0.92/1.272) \\
&\qquad\qquad + 0.67 \times [0.67 \exp(-0.43 \times 0.83/1.272)] \\
&= 0.37
\end{aligned}
$$

The single dimensional certainty equivalent values for all the alternatives can be calculated in a similar manner, and these are shown in Table 7.2. This table also shows the overall certainty equivalent value for each alternative, which is the sum of the single dimensional certainty equivalent values for the alternative, as shown by equation 7.4.

This table shows that the Status Quo alternative is most preferred with a certainty equivalent value of 0.82, the High Quality/High Cost and Medium Quality/Medium Cost alternatives are tied for next most preferred with values of 0.70, and the Low Quality/Low Cost alternative is least preferred with a value of 0.59. Note that the ranking of the alternatives using this risk averse utility function is the same as the ranking determined earlier using expected values. In fact, with this moderately risk averse utility function there is virtually no difference between the certainty equivalent values and the expected values for the alternatives, which were calculated in Section 7.1.

7.5 Spreadsheet Calculation Procedures

This section reviews spreadsheet methods for doing certainty equivalent calculations. As with the spreadsheet calculation procedures presented in Chapter 4, the approach presented in this section uses Microsoft Excel, Version 5.0. Some use is made of Visual Basic, Applications Edition, which is included in Excel starting with Version 5.0. The required Visual Basic code is presented so that you do not need to know Visual Basic programming to apply the approach presented here.

This section focuses on extending the spreadsheet methods in Section 4.7 to handle decisions with uncertainty. The types of graphical analysis and sensitivity analysis methods presented in that earlier section can also be applied to decisions with uncertainty.

Figure 7.3 shows a spreadsheet to do certainty equivalent calculations for the networking strategy example, and Figure 7.4 shows the equations for this spreadsheet. There are four logical parts to this spreadsheet, and these are similar to the parts of the value calculation spreadsheet discussed in Section 4.7 and shown in Figure 4.3. The first two parts include the input data for the decision analysis:

1 The range A4:H13 specifies the utility function for the decision.
2 The range A15:G24 specifies the probabilities and scores (levels) for each alternative.

The last two parts of the spreadsheet show the analysis results, and these are the parts that the spreadsheet calculates:

3 The range A26:G34 shows the single dimensional values for each of the scores in the A15:G24 input table.
4 The range A36:H40 shows the single dimensional certainty equivalent values and the overall certainty equivalent value for each alternative.

Each of these parts of the spreadsheet will now be discussed in further detail.

Utility function data   Cell E4 contains the multiattribute risk tolerance for the utility function, which is 1.272. Range B6:H13 is identical to the corresponding range in the Figure 4.3 spreadsheet (A4:G11), and Section 4.7 discusses the meaning of each entry in this range.

Probabilities and scores for alternatives   Range A15:G24 contains the probability and score information from Table 7.1. For example, range F20:G21 shows that the Medium Quality/Medium Cost alternative has two possible scores with respect to the Security evaluation measure, 0.5 and 0, and that the probabilities for these are 0.33 and 0.67, respectively.

Single dimensional values   Range A26:G34 contains a table that the spreadsheet calculates by applying the appropriate value function to the scores in the probability and score table (range A15:G24) to determine the single dimensional value for each score. For example, the entry in cell C29 shows that the single dimensional value for a Productivity Enhancement score of 1 (which is in cell C19) is 0.75.

Certainty equivalent values   The entries in range C37:G40 are calculated by applying equation 7.5 to the probabilities in the probability and score table (range A15:G24) and the single dimensional value in the single dimensional value table (range A26:G34). Finally, the overall certainty equivalent values in range H37:H40 are calculated by applying equation 7.4 to the single dimensional values in range C37:G40.

Equations to Calculate Certainty Equivalent Values

Figure 7.4 shows the equations to calculate the Figure 7.3 spreadsheet. The single dimensional value table in range A26:G34 is calculated by using the functions ValuePL and ValueE, which were discussed in Section 4.7. These functions are coded in Visual Basic, as shown in Figure 4.5, and they must be available in your spreadsheet for the equations shown in range A26:G34 to work. Section 4.7 discusses how to add them to a spreadsheet.

If the equations in the range A26:G34 are entered into the spreadsheet with the combination of relative and absolute references shown in Figure 7.4, then you need to enter only three different equations into the table. The remaining equations can be created by copying these three equations. Specifically, note that in each of the equations the cell reference for the first argument of the function ValuePL or ValueE is relative, while for the remaining arguments the row references are absolute.

The first argument of each of these functions is the cell containing the evaluation measure score (level) for which we wish to determine a single dimensional value, while the remaining arguments refer to the parameters for the value function that is to be used. With a relative reference in the first argument and absolute row references in the remaining arguments, the appropriate equation can be entered in the top cell of each relevant column in the single dimensional value table and then filled down into the remaining rows of the table. Then any extra entries can be deleted.

For example, consider the equation in the upper left cell of this table (that is, cell C27). When this equation is entered as shown in Figure 7.4a, then it can be filled down into the range C28:C33. Finally, the equation in cell C31 can be deleted. For this procedure to work, the cell reference in the first argument of the ValuePL function, which is entered into cell C27, must be relative, while the row references in the remaining arguments must be absolute so that these arguments continue to refer to the value function definition in range B9:C12 when the equation is filled down from cell C27 into C28:C33.

A similar approach works with the equations in ranges E27:E32 and G27:G34. The correct equation can be entered into the topmost cell of those ranges (E27 and G27, respectively) and then filled down into the remainder of the range provided that relative cell references are used for the first argument in the function ValuePL or ValueE, and absolute row references are used for the remaining arguments.

	A	B	C	D	E	F	G	H
1				Networking Strategy Decision				
2				Certainty Equivalent Calculations				
3								
4			multiattribute risk tolerance:		1.272			
5								
6					VALUE FUNCTIONS			
7		Productivity Enhance		Cost Increase			Security	
8		x	Value			x	Value	
9		-1	0.00	Low:	0	-2	0.00	
10		0	0.50	High:	150	-1	0.50	
11		1	0.75	Mono:	decreasing	0	0.83	
12		2	1.00	Rho:	182.4	1	1.00	SUM
13		Weights:	0.22		0.35		0.43	1
14								
15				PROBABILITIES AND SCORES (LEVELS)				
16		Prob.	Score	Prob.	Score	Prob.	Score	
17	Stat Quo	1	0	1	0	1	0	
18	HQ/HCost	0.75	2	1	125	0.6	1	
19		0.25	1			0.4	0.5	
20	MQ/MCost	1	1	1	95	0.33	0.5	
21						0.67	0	
22	LQ/LCost	0.5	1	1	65	0.4	0	
23		0.5	0			0.4	-1	
24						0.2	-2	
25								
26				SINGLE DIMENSIONAL VALUES				
27	Stat Quo		0.50		1.00		0.83	
28	HQ/HCost		1.00		0.23		1.00	
29			0.75				0.92	
30	MQ/MCost		0.75		0.46		0.92	
31							0.83	
32	LQ/LCost		0.75		0.66		0.83	
33			0.50				0.50	
34							0.00	
35								
36			SINGLE DIMENSIONAL CERTAINTY EQUIVALENT VALUES					CE Value
37	Stat Quo		0.11		0.35		0.36	0.82
38	HQ/HCost		0.21		0.08		0.42	0.70
39	MQ/MCost		0.17		0.16		0.37	0.70
40	LQ/LCost		0.14		0.23		0.22	0.59

Figure 7.3 *Spreadsheet to calculate certainty equivalents*

	A	B	C	D
1				
2				
3				
4			multiattribute risk tolerance:	
5				
6				
7			Productivity Enhance	
8		x	Value	
9		-1	0.00	Low:
10		0	0.50	High:
11		1	0.75	Mono:
12		2	1.00	Rho:
13	Weights:		0.22	
14				
15				P
16	Prob.		Score	Prob.
17	Stat Quo	1	0	1
18	HQ/HCost	1	2	1
19		0	1	
20	MQ/MCost	1	1	1
21				
22	LQ/LCost	1	1	1
23		1	0	
24				
25				
26				
27	Stat Quo		=ValuePL(C17,B$9:B$12,C$9:C$12)	
28	HQ/HCost		=ValuePL(C18,B$9:B$12,C$9:C$12)	
29			=ValuePL(C19,B$9:B$12,C$9:C$12)	
30	MQ/MCost		=ValuePL(C20,B$9:B$12,C$9:C$12)	
31				
32	LQ/LCost		=ValuePL(C22,B$9:B$12,C$9:C$12)	
33			=ValuePL(C23,B$9:B$12,C$9:C$12)	
34				
35				
36			SINGLE DI	
37	Stat Quo		=CEValue($E$4,C$13,B17,C27)	
38	HQ/HCost		=CEValue($E$4,C$13,B18:B19,C28:C29)	
39	MQ/MCost		=CEValue($E$4,C$13,B20:B21,C30:C31)	
40	LQ/LCost		=CEValue($E$4,C$13,B22:B24,C32:C34)	

Figure 7.4a   *Equations for Figure 7.3 spreadsheet (part 1)*

	E	F	G	H
1	Networking Strategy Decision			
2	Certainty Equivalent Calculations			
3				
4	1.272			
5				
6	VALUE FUNCTIONS			
7	Cost Increase		Security	
8		x	Value	
9	0	-2	0.00	
10	150	-1	0.50	
11	decreasing	0	0.83	
12	182.4	1	1.00	SUM
13	0.35		0.43	=SUM(C13:G13)
14				
15	OBABILITIES AND SCORES (LEVELS)			
16	Score	ob.	Score	
17	0	1	0	
18	125	1	1	
19		0	0.5	
20	95	0	0.5	
21		1	0	
22	65	0	0	
23		0	-1	
24		0	-2	
25				
26	SINGLE DIMENSIONAL VALUES			
27	=ValueE(E17,E$9,E$10,E$11,E$12)		=ValuePL(G17,F$9:F$12,G$9:G$12)	
28	=ValueE(E18,E$9,E$10,E$11,E$12)		=ValuePL(G18,F$9:F$12,G$9:G$12)	
29			=ValuePL(G19,F$9:F$12,G$9:G$12)	
30	=ValueE(E20,E$9,E$10,E$11,E$12)		=ValuePL(G20,F$9:F$12,G$9:G$12)	
31			=ValuePL(G21,F$9:F$12,G$9:G$12)	
32	=ValueE(E22,E$9,E$10,E$11,E$12)		=ValuePL(G22,F$9:F$12,G$9:G$12)	
33			=ValuePL(G23,F$9:F$12,G$9:G$12)	
34			=ValuePL(G24,F$9:F$12,G$9:G$12)	
35				
36	ENSIONAL CERTAINTY EQUIVALENT VALUES			CE Value
37	=CEValue($E$4,E13,D17,E27)		=CEValue($E$4,G$13,F17,G27)	=SUM(B37:G37)
38	=CEValue($E$4,E13,D18:D19,E28:E29)		=CEValue($E$4,G$13,F18:F19,G28:G29)	=SUM(B38:G38)
39	=CEValue($E$4,E13,D20:D21,E30:E31)		=CEValue($E$4,G$13,F20:F21,G30:G31)	=SUM(B39:G39)
40	=CEValue($E$4,E13,D22:D24,E32:E34)		=CEValue($E$4,G$13,F22:F24,G32:G34)	=SUM(B40:G40)

Figure 7.4b   *Equations for Figure 7.3 spreadsheet (part 2)*

To calculate the single dimensional certainty equivalents in range A37:G40 of Figure 7.3, it is necessary to apply equation 7.5 to the information given above this range in the spreadsheet. To do this calculation, a new function CEValue is used. The definition of this function in Visual Basic is given in Figure 7.5, and the logic for the function is discussed below. For the moment, assume that such a function is available. This function has four arguments as follows:

$$\text{CEValue} = (\rho_m, \text{Weight}, P\text{-list}, V\text{-list})$$

where ρ_m is the multiattribute risk tolerance, Weight is the weight for the evaluation measure, P-list is the relevant list of probabilities, and V-list is the list of corresponding single dimensional values.

To illustrate how this function is used, consider the equation in cell G38 of Figure 7.4b. This equation calculates the single dimensional certainty equivalent for the High Quality/High Cost alternative with respect to the Security evaluation measure. The first argument, $E$4, is the cell that contains the multiattribute risk tolerance ρ_m. The second argument, G$13, is the cell that contains the weight for Security. The third argument, F18:F19, refers to the cells that contain the appropriate probabilities to use in this calculation (which are 0.6 and 0.4). The fourth argument, G28:G29, refers to the cells that contain the corresponding single dimensional values (which are 1.00 and 0.92).

In general, it is necessary to enter a different equation into each cell of the single dimensional certainty equivalent value table. However, by making appropriate use of relative and absolute cell references, you can save some effort. Specifically, if you use both absolute column and absolute row references in the first function argument, and if you use a relative column and absolute row reference in the second argument, then these references will be the same throughout the table. Thus, you can enter the correct equation into cell C37, and if you copy it into the remainder of the table, then the first two arguments of CEValue will be correct. You will still need to modify the remaining two arguments in the copied equations.

The final portion of the Figure 7.3 spreadsheet is the table in range H37:H40, which calculates the overall certainty equivalent values for each alternative. This is done using equation 7.4, which requires simply summing the single dimensional certainty equivalent values to the left of each entry in the range H37:H40. Figure 7.4b shows the required equations. The required equation can be entered into cell H37 and then filled down into range H38:H40. (The equation shown in Figure 7.4b uses the shortcut of including all cells in a row of the single dimensional certainty equivalent value table as the argument for the SUM function. This shortcut makes use of the fact that the SUM function assumes that a blank cell contains a zero.)

```
Function CEValue(RhoM, Weight, Pi, Vi)
  If UCase(RhoM) = "INFINITY" Then
    CEValue = Weight * Application.SumProduct(Pi, Vi)
  Else
    EV = 0
    For i = 1 To Application.Count(Pi)
      EV = EV + Pi(i) * Exp(-Weight * Vi(i) / RhoM)
    Next i
    CEValue = -RhoM * Log(EV)
  End If
End Function
```

Figure 7.5   *Equations for certainty equivalent function*

Implementing CEValue

Figure 7.5 shows the Visual Basic code to implement the function CEValue, which is used in the spreadsheet analysis presented above. Section 4.7 reviewed how to enter a Visual Basic function into an Excel spreadsheet. You can add the definition for CEValue to the Visual Basic module where you have placed the definitions for ValuePL and ValueE. It is not necessary to understand the Visual Basic code for CEValue to use it, but this is explained here for interested readers.

The function CEValue implements equation 7.5. There are two cases for this equation, one when the multiattribute risk tolerance is infinity and one when it is not infinity. Function CEValue handles both of these cases. You simply use the term "infinity" (without the quotes) as the first argument of CEValue when you want to use a multiattribute risk tolerance of infinity. Otherwise, you use the numerical value of the multiattribute risk tolerance.

The function CEValue shown in Figure 7.5 first checks to see whether the multiattribute risk tolerance is equal to infinity. If it is, then the standard Excel function SUMPRODUCT is used to calculate the expected value, and this is then multiplied by the evaluation measure weight.

If the multiattribute risk tolerance is not equal to infinity, then the appropriate procedure is carried out to calculate the certainty equivalent value. (One potentially confusing point is that the Visual Basic function to calculate a natural logarithm has the name "Log," while the Excel spreadsheet function to do this calculation has the name "LN.")

To shorten the length of the definition for CEValue, it contains no error-checking code. Therefore, you should be particularly careful that you use this function correctly.

7.6 Decisions with Interdependent Uncertainties

This section considers decisions in which the uncertainties for the various evaluation measures are *interdependent*. That is, the probability distribution for an evaluation measure changes for different levels of the other evaluation measures. To follow the development in this section, you need to understand the elements of conditional probabilities and decision trees. The necessary background is reviewed in Appendix D. The material in this section is useful for certain applications, but an understanding of this material is not necessary to read any other part of the book.

Property 1 for expected values presented in Section 7.1 states that the expected value for a sum of uncertain variables is equal to the sum of the expected values for the uncertain variables whether or not the variables are probabilistically independent. Therefore, the expected value calculation procedure presented in Section 7.1 works even when there are interdependencies among the uncertain evaluation measures.

However, the procedure presented in Section 7.4 for calculating certainty equivalents may not give the correct answer when the uncertain evaluation measures for a decision problem are interdependent. The reason for this is that when there are interdependencies, then Property 3 presented in Section 7.4, which leads to equations 7.4 and 7.5, does not hold. Therefore, these equations are not necessarily valid when the uncertain evaluation measures are interdependent. That is, the certainty equivalent value for an alternative is not necessarily equal to the sum of the single dimensional certainty equivalent values for that alternative when the variables are probabilistically dependent. Equations 7.4 and 7.5 play an important role in the calculation procedure presented in Section 7.4, as well as the spreadsheet calculation procedure in Section 7.5, and hence these procedures do not yield the correct answer when there are interdependencies among evaluation measures.

Whether or not there is probabilistic interdependency among the evaluation measures, the standard decision analysis rollback procedure (reviewed in Appendix D) can be used to calculate the expected utilities for decision problem alternatives. This is illustrated in Figure 7.6a. This presents in tree form the probability information from Table 7.1 for the Low Quality/Low Cost alternative in the networking example. The values shown at each endpoint of the tree are determined by using the weighted-additive value function and the value information in Table 7.1. For example, the value for the topmost path through the tree is calculated as follows:

$$\text{Value} = 0.22v_p(1) + 0.35v_c(65) + 0.43v_s(0)$$
$$= 0.22 \times 0.75 + 0.35 \times 0.66 + 0.43 \times 0.83$$
$$= 0.753$$

The utilities shown for each endpoint of the tree are calculated by applying the exponential form of equation 7.1 to the value for that endpoint. Thus, for the topmost path through the tree,

$$\text{Utility} = \frac{1 - \exp(-0.753/1.272)}{1 - \exp(-1/1.272)} = 0.821$$

The expected utilities for the intermediate nodes in the tree (designated by "EU" in Figure 7.6) are calculated using the standard rollback procedure. For example, the expected utility of 0.707 shown in Figure 7.6a is calculated by

$$EU = 0.4 \times 0.821 + 0.4 \times 0.701 + 0.2 \times 0.491 = 0.707$$

Thus, Figure 7.6a shows that the expected utility for the Low Quality/Low Cost alternative is 0.682. To compare this result with the one determined in Section 7.4, we must convert this expected utility to a certainty equivalent value. From the definition of the certainty equivalent value v_{ce}, this must satisfy the equation

$$\frac{1 - \exp(-v_{ce}/\rho_m)}{1 - \exp(-1/\rho_m)} = EU$$

where EU is the expected utility for the alternative. This equation can be solved to yield the following formula for the certainty equivalent value:

$$v_{ce} = -\rho_m \times \ln\{1 - EU \times [1 - \exp(-1/\rho_m)]\} \tag{7.6}$$

Thus, for the Figure 7.6a alternative

$$v_{ce} = -1.272 \times \ln\{1 - 0.682 \times [1 - \exp(-1/1.272)]\} = 0.590$$

which agrees with the 0.59 number calculated at the end of Section 7.4 and using the spreadsheet in Section 7.5.

To illustrate the impact of interdependencies on expected utility calculations, examine the tree in Figure 7.6b, which represents a modified version of the Low Quality/Low Cost alternative. In this tree, the probability distribution for the Security evaluation measure differs depending on the level of the Productivity Enhancement evaluation measure. Note, however, that the unconditional (that is, marginal) probability distribution for Security is the same in Figure 7.6b as in Figure 7.6a. For example, the unconditional probability that Security will have a level of 0 in Figure 7.6b is equal to $0.5 \times 0.8 = 0.4$, which is the same as in Figure 7.6a, where this probability is calculated as $0.5 \times 0.4 + 0.5 \times 0.4 = 0.4$.

The endpoint value and utility numbers in Figure 7.6b are determined using the same procedure as in Figure 7.6a, and the expected utility determined using the standard rollback procedure for the alternative is 0.680, as shown in the figure. This differs slightly from the 0.682 determined in Figure 7.6a, but the difference is so small that it is of little practical significance, and in fact you might wonder if the difference isn't just due to roundoff error.

Applying equation 7.6 to the Figure 7.6b alternative yields a certainty equivalent value of 0.588, which is very close to the 0.590 certainty equivalent value calculated for the Figure 7.6a alternative. Thus, for this moderately risk averse utility function, there is very little difference between the results assuming probabilistic independence or probabilistic interdependency.

To illustrate the impact of greater risk aversion, consider the two trees shown in Figure 7.7. These are identical to the corresponding trees in Figure 7.6 except that the multiattribute risk tolerance ρ_m has been changed from 1.272 to 0.2.

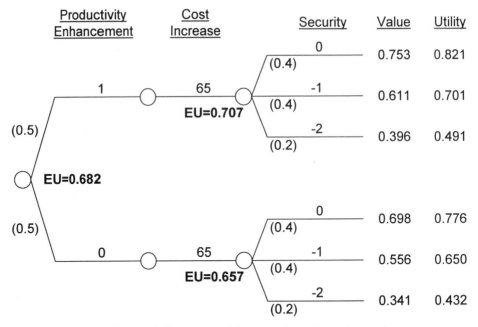

a. Original decision problem (no interdependencies)

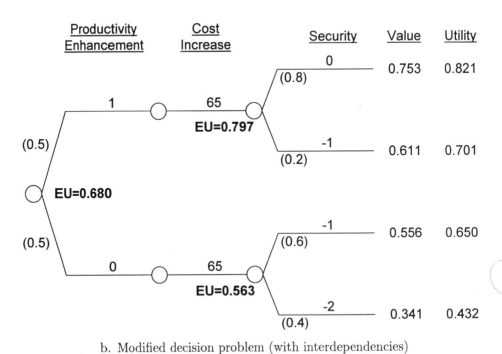

b. Modified decision problem (with interdependencies)

Figure 7.6   *Low Quality/Low Cost alternative* $(\rho_m = 1.272)$

That is, Figure 7.7 illustrates a situation where there is substantial multiattribute risk aversion. (In fact, as noted earlier, the 0.2 value for the multiattribute risk tolerance is at the lower limit of what is observed in practice.) Comparing the expected utilities for parts a and b of Figure 7.7, which are 0.942 and 0.937, we see that with this substantial multiattribute risk aversion there is still only slightly more difference between the expected utilities for the alternatives than in Figure 7.6.

For the Figure 7.7a and Figure 7.7b alternatives, the certainty equivalent values calculated using equation 7.6 are 0.549 and 0.534, respectively. As with the expected utilities, there is still not much difference between the certainty equivalent values for the alternatives calculated assuming probabilistic independence or considering probabilistic interdependencies. However, the difference is somewhat larger than with the less risk averse utility function.

The examples in Figure 7.6 and Figure 7.7 illustrate that when the multiattribute risk tolerance is relatively large (that is, there is only slight risk aversion), then the probabilistic interdependencies between evaluation measures usually do not have much direct impact on the analysis results. This is because when the multiattribute risk tolerance is large, there is not much risk aversion, and hence certainty equivalent values are very close to expected values. We know from the earlier discussion that probabilistic interdependencies do not have any direct impact on expected value calculations. Thus, it makes sense that these interdependencies will not have much impact when there is very little risk aversion.

When the multiattribute risk tolerance is small (that is, there is substantial risk aversion), then the impact of probabilistic interdependencies can be greater since certainty equivalents can differ substantially from expected values in this case.

However, the discussion in the preceding paragraphs does not necessarily mean that you can ignore probabilistic interdependencies when there is a large multiattribute risk tolerance. This is because you may need to consider the interdependencies in order to determine the unconditional probability distributions for the evaluation measures. It is often easier to determine conditional probabilities for a decision problem, and then calculate the unconditional probabilities from these than it is to directly determine unconditional probabilities.

The Figure 7.6a and Figure 7.7a examples illustrate that the tree rollback procedure works even when there are no probabilistic interdependencies. Thus, you may wonder whether you should not just always use this rollback procedure and forget about the more specialized procedures presented in Sections 7.4 and 7.5. To illustrate the difficulty with doing this, consider a decision with ten evaluation measures, each of which has a probability distribution with three different possible levels. Then there are $3^{10} = 59,049$ endpoints in a tree representing each alternative in this decision. This situation, which is representative of some seen in practice, is clearly beyond what can be analyzed with manual tree analysis methods. While modern decision analysis software (ADA Decision Systems 1992) can handle problems of this size, you would need to obtain and learn how to use this software to apply a decision tree approach. On the other hand, the procedures presented in Sections 7.4 and 7.5 can be used with only a spreadsheet program.

Note also that the probabilistic independence conditions required for the procedures in Section 7.4 and 7.5 are met in many actual applications. These conditions are not as restrictive as they might appear at first. In particular, note that these conditions do not require that the probability distributions for the evaluation measures be the same for different alternatives. As the networking example shows, the probability distributions can differ from one alternative to another. To use the procedures, it is necessary only that the probability distributions for the various evaluation measures do not depend on each other *for a particular alternative.* This condition is met in a substantial number of applications (Corner and Kirkwood 1991).

Spreadsheet Calculations with Interdependencies

This section reviews spreadsheet procedures that can be used to carry out the calculations required to determine expected utilities and certainty equivalent values when there are interdependent uncertainties. Specialized software that is specifically designed to analyze such situations is available (ADA Decision Systems 1995), and that software simplifies the analysis. However, to use such software, you have to first obtain it and then learn how to use it. Spreadsheets have the advantage over such software of being more widely available. Thus, if you only occasionally need to analyze decisions with interdependent uncertainties, it may be more efficient to use a spreadsheet.

The spreadsheet analysis procedure will be illustrated using the situation that was previously analyzed and that is shown in Figure 7.7b. A spreadsheet to analyze this situation is shown in Figure 7.8, and the equations for this spreadsheet are shown in Figure 7.9.

The top portion of the Figure 7.8 spreadsheet, in range A1:G10, contains the usual information that defines a power-additive utility function. Range A12:H17 calculates the utilities required for the decision tree rollback procedure, and range A18:H26 does the actual rollback.

The calculation of utilities in range A12:H16 is done in three steps. First, the single dimensional values are calculated for each evaluation measure score. This is done in range A13:F16 using the ValueE and ValuePL functions in the way we have previously studied. In range G13:G16, the overall values are calculated for each endpoint in the decision tree by multiplying each single dimensional value by its corresponding weight and summing the results. In range H13:H16, the utility corresponding to each of these values is calculated using equation 7.1.

The utilities in range H13:H16 are transferred to the endpoints of the decision tree by the equations in range H20:H26. The first stage of the rollback is done by the equations in cells F21 and F25. These equations take the expected values of the appropriate endpoint utilities. The second stage of the rollback is done by the equations in cells D21 and D25 in an analogous manner. The final rollback step is done in a similar manner by the equation in cell B23. Finally, the certainty equivalent corresponding to the utility in cell B23 is calculated in cell A23. This is done by applying equation 7.6, or, in the special case where the multiattribute

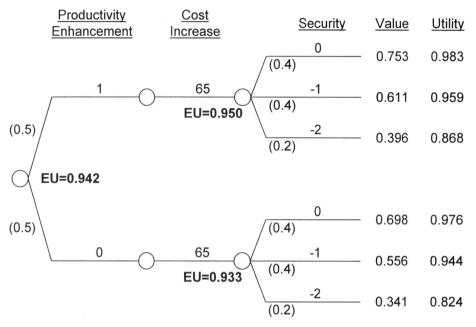

a. Original decision problem (no interdependencies)

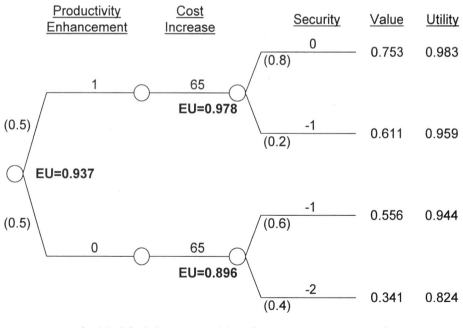

b. Modified decision problem (with interdependencies)

Figure 7.7 *Low Quality/Low Cost alternative ($\rho_m = 0.20$)*

	A	B	C	D	E	F	G	H
1	Decision with Interdependent Uncertainties							
2	multiattribute risk tolerance:			0.2				
3	VALUE FUNCTIONS							
4	Productivity Enhance		Cost Increase		Security			
5	x	Value			x	Value		
6	-1	0.00	Low:	0	-2	0.00		
7	0	0.50	High:	150	-1	0.50		
8	1	0.75	Mono:	decreasing	0	0.83		
9	2	1.00	Rho:	182.4	1	1.00	SUM	
10	Weights:	0.22		0.35		0.43	1	
11								
12	SCORES (LEVELS) AND SINGLE DIMENSIONAL VALUES							
13	Score	SD Value	Score	SD Value	Score	SD Value	VALUES	UTILITIES
14	1	0.750	65	0.664	0	0.830	0.754	0.984
15					-1	0.500	0.613	0.960
16	0	0.500	65	0.664	-1	0.500	0.558	0.945
17					-2	0.000	0.343	0.825
18	DECISION TREE							
19	CE	EU	Prob	EU	Prob	EU	Prob	Utility
20							0.8	0.984
21			0.5	0.979	1	0.979		
22							0.2	0.960
23	0.536	0.938						
24							0.6	0.945
25			0.5	0.897	1	0.897		
26							0.4	0.825

Figure 7.8   *Spreadsheet for calculations with interdependent uncertainties*

risk tolerance is equal to infinity, by simply setting the certainty equivalent equal to the expected utility.

When you compare the results in Figure 7.8 with those shown in Figure 7.7b, you will note that there are slight differences between some numbers. This is because of the rounding error in the hand calculations used to determine the results in Figure 7.7b.

The spreadsheet calculation procedure has advantages over doing the calculations by hand if you need to do more than one similar calculation. Once you have set up the equations, it is easy to consider variations—for example, changes to the evaluation measure weights or the multiattribute risk tolerance. You simply make the desired changes, and the spreadsheet calculates the revised expected utilities and certainty equivalents.

	A	B	C
1		Decision with Interdependent Uncertain	
2			ance:
3		VALUE FUNCTIONS	
4	Productivity Enhance		
5	x	Value	
6	-1	0	Low:
7	0	0.5	High:
8	1	0.75	ono:
9	2	1	Rho:
10	Weights:	0.22	
11			
12		SCORES (LEVELS) AND SINGLE DIMENSION	
13	Score	SD Value	cor
14	1	=ValuePL(A14,A$6:A$9,B$6:B$9)	65
15			
16	0	=ValuePL(A16,A$6:A$9,B$6:B$9)	65
17			
18			
19	CE	EU	Prob
20			
21			0.5
22			
23	=IF(UPPER(D2)="INFINITY",B23,-D2*LN(1-B23*(1-EXP(-1/D2))))	=C21*D21+C25*D25	
24			
25			0.5
26			

a. Left portion of spreadsheet

	D	E	F	G
1				
2	0.2			
3				
4				
5		x	Value	
6	0	-2	0	
7	150	-1	0.5	
8	decreasing	0	0.83	
9	182.4	1	1	SUM
10	0.35		0.43	=SUM(B10:F10)
11				
12				VALUES
13	SD Value	co	SD Value	
14	=ValueE(C14,D$6,D$7,D$8,D$9)	0	=ValuePL(E14,E$6:E$9,F$6:F$9)	=$B$10*$B$14+$D$10*$D$14+$F$10*F14
15		-1	=ValuePL(E15,E$6:E$9,F$6:F$9)	=$B$10*$B$14+$D$10*$D$14+$F$10*F15
16	=ValueE(C16,D$6,D$7,D$8,D$9)	-1	=ValuePL(E16,E$6:E$9,F$6:F$9)	=$B$10*$B$16+$D$10*$D$16+$F$10*F16
17		-2	=ValuePL(E17,E$6:E$9,F$6:F$9)	=$B$10*$B$16+$D$10*$D$16+$F$10*F17
18				
19	EU	Pro	EU	Prob
20				0.8
21	=E21*F21	1	=G20*H20+G22*H22	
22				0.2
23				
24				0.6
25	=E25*F25	1	=G24*H24+G26*H26	
26				0.4

b. Center portion of spreadsheet

Figure 7.9a *Equations for Figure 7.8 spreadsheet (part 1)*

	H
1	
2	
3	
4	
5	
6	
7	
8	
9	
10	
11	
12	UTILITIES
13	
14	=IF(UPPER($D$2)="INFINITY",G14,(1-EXP(-G14/$D$2))/(1-EXP(-1/$D$2)))
15	=IF(UPPER($D$2)="INFINITY",G15,(1-EXP(-G15/$D$2))/(1-EXP(-1/$D$2)))
16	=IF(UPPER($D$2)="INFINITY",G16,(1-EXP(-G16/$D$2))/(1-EXP(-1/$D$2)))
17	=IF(UPPER($D$2)="INFINITY",G17,(1-EXP(-G17/$D$2))/(1-EXP(-1/$D$2)))
18	
19	Utility
20	=H14
21	
22	=H15
23	
24	=H16
25	
26	=H17

c. Right portion of spreadsheet

Figure 7.9b   *Equations for Figure 7.8 spreadsheet (part 2)*

7.7 Continuous Decision Variables and Uncertainty

This section extends the procedure presented in Section 4.8 for analyzing decisions with continuous decision variables to situations where there is uncertainty about the evaluation measure scores (levels) that will result from a particular decision. This material is somewhat more advanced than earlier sections of this chapter, and an understanding of this section is not required for the remainder of this book, except for Section 8.4.

The analysis procedure for this type of decision will be illustrated using an expanded version of the product engineering decision considered in Section 4.8. The analysis of that decision, which was presented in Section 4.8, assumed that the score (level) for each evaluation measure was known with no uncertainty once the budget fraction allocated to the associated area of activity was set. However, in most realistic decisions, there would be uncertainty about exactly what score would be achieved even after the budget was allocated. The analysis method presented in Section 4.8 can be extended to include consideration of this uncertainty in a straightforward manner.

In Section 4.8, a *response function* was specified to give the performance with respect to each evaluation measure as a function of the relevant budget allocation. In order to include an analysis of uncertainty, it is necessary to determine the *probability distribution* for each evaluation measure as a function of the relevant budget allocation, rather than just a specific score. This is difficult in practice because there are an infinite number of different possible budget allocations. Thus, directly determining the probability distributions for all possible budget allocations is generally not feasible.

The quadratic interpolation approach presented in Section 4.8 can be extended to handle this situation. Figure 7.10 shows cumulative probability distributions assessed from the director of the product engineering group for each of the evaluation measures. Specifically, probability distributions were assessed for the lowest allowed budget fraction for the relevant area, the highest allowed budget fraction, and the average of the lowest and highest allowed fractions.

The distributions shown in Figure 7.10 are continuous, and Section 6.3 presents methods for determining expected values for continuous distributions. For example, the expected value can be determined in terms of the 0.05, 0.50, and 0.95 fractiles of the distributions using the extended Pearson-Tukey approximation. Thus, it is possible to use this approximation to determine expected values for each probability distribution shown in Figure 7.10.

The difficulty remains that it is necessary to determine expected values (or certainty equivalents if risk attitude is important) for *any possible* budget allocation in order to determine the preferred allocations. Determining these expected values seems to require that we have probability distributions for all possible different budget allocations.

This difficulty can be overcome by realizing that in order to use the extended Pearson-Tukey approximation to calculate expected values, it is necessary to have only the 0.05, 0.50, and 0.95 fractiles for a probability distribution, and not the entire distribution. It is possible to estimate these fractiles for all budget allocations by interpolating the relevant fractiles for the appropriate three distributions in Figure 7.10. We will do this interpolation using the quadratic interpolation procedure presented in Section 4.8.

The graphs in Figure 7.10 are small to conserve space, but in the actual application these distributions were graphed in more detail, and it was possible to read the 0.05, 0.50, and 0.95 fractiles from the graphs. The actual numbers for these fractiles will be presented shortly. Figure 7.11 shows quadratic interpolation graphs for these fractiles for each evaluation measure. Using these interpolation graphs, it is possible to calculate the expected value for any budget allocation.

Figure 7.12 shows the spreadsheet required to determine the budget allocation with the highest expected value, and the equations for this spreadsheet are shown in Figure 7.13. The value function for this decision is given in range A5:I12 of this spreadsheet, and this is identical to the value function used in the Section 4.8 analysis. Cell F3 shows that the multiattribute risk tolerance is infinity, and therefore expected value is being used to evaluate the budget allocations.

Range A15:I19 gives the 0.05 fractiles for each of the cumulative probability distributions in Figure 7.10, range A20:I23 gives the 0.50 fractiles, and range

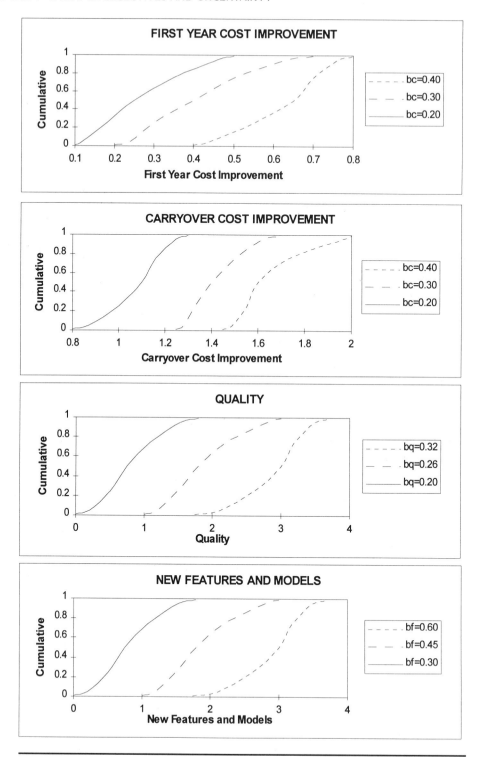

Figure 7.10   *Probability distributions for product engineering decision*

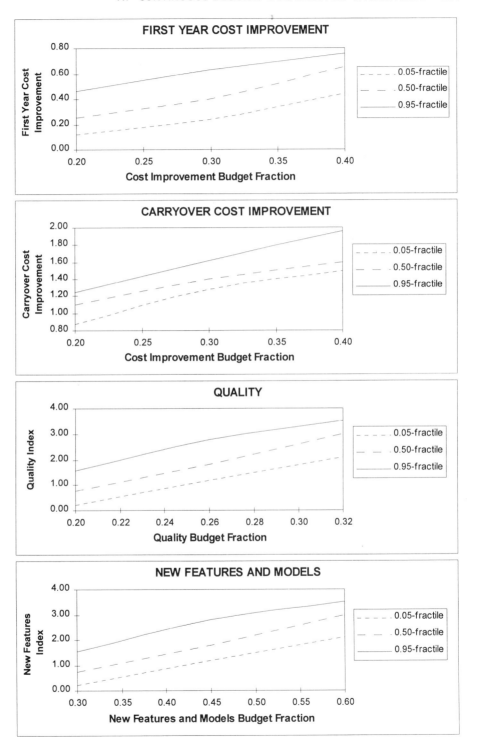

Figure 7.11   *Response functions for product engineering decision*

	A	B	C	D	E	F	G	H	I	J
1				PRODUCT ENGINEERING BUDGET ALLOCATION						
2										
3				multiattribute risk tolerance:	Infinity					
4										
5				VALUE FUNCTIONS						
6		First Year Improve		Carryover Improve		Quality		New Features		
7		x	Value	x	Value	x	Value	x	Value	
8		Low:	0.10	Low:	0.80	0.00	0.00	Low:	0.00	
9		High:	0.80	High:	2.00	1.00	0.40	High:	4.00	
10		Mono:	Increasing	Mono:	Increasing	4.00	1.00	Mono:	Increasing	
11		Rho:	Infinity	Rho:	Infinity			Rho:	-1.64	SUM
12		Weights:	0.10		0.15		0.55		0.20	1.00
13										
14				RESPONSE FUNCTION PARAMETERS						
15		First Year Improve		Carryover Improve		Quality		New Features		
16	0.05-FRACTILES:	b	x	b	x	b	x	b	x	
17	Low:	0.20	0.12	0.20	0.87	0.20	0.20	0.30	0.20	
18	Middle:	0.30	0.24	0.30	1.28	0.26	1.20	0.45	1.20	
19	High:	0.40	0.44	0.40	1.49	0.32	2.10	0.60	2.10	
20	0.50-FRACTILES:									
21	Low:	0.20	0.25	0.20	1.10	0.20	0.75	0.30	0.75	
22	Middle:	0.30	0.40	0.30	1.40	0.26	1.80	0.45	1.80	
23	High:	0.40	0.65	0.40	1.60	0.32	3.00	0.60	3.00	
24	0.95-FRACTILES:									
25	Low:	0.20	0.46	0.20	1.25	0.20	1.55	0.30	1.55	
26	Middle:	0.30	0.63	0.30	1.62	0.26	2.80	0.45	2.80	
27	High:	0.40	0.76	0.40	1.96	0.32	3.50	0.60	3.50	
28										
29				BUDGET ALLOCATION						
30			COST	QUALITY	FEATURE	TOTAL				
31			0.38	0.32	0.30	1.00				
32		Lower Bound:	0.20	0.20	0.30	1.00				
33		Upper Bound:	0.40	0.32	0.60					
34										
35				PROBABILITIES AND SCORES (LEVELS)						
36		First Year Improve		Carryover Improve		Quality		New Features		
37		Prob.	Score	Prob.	Score	Prob.	Score	Prob.	Score	
38	0.05-FRACTILES:	0.185	0.39	0.185	1.46	0.185	2.10	0.185	0.20	
39	0.50-FRACTILES:	0.630	0.59	0.630	1.57	0.630	3.00	0.630	0.75	
40	0.95-FRACTILES:	0.185	0.74	0.185	1.89	0.185	3.50	0.185	1.55	
41										
42				SINGLE DIMENSIONAL VALUES						
43	0.05-FRACTILES:		0.42		0.55		0.62		0.01	
44	0.50-FRACTILES:		0.70		0.64		0.80		0.06	
45	0.95-FRACTILES:		0.91		0.91		0.90		0.15	
46										
47				SINGLE DIMENSIONAL CERTAINTY EQUIVALENT VALUES						CE Value
48			0.07		0.10		0.43		0.01	0.61
49										
50				SOLUTION TABLE						
51				Quality						
52		0.61	0.20	0.22	0.24	0.26	0.30	0.32		
53		0.20	0.322	0.377	0.414	0.448	0.511	0.543		
54		0.22	0.325	0.381	0.419	0.454	0.518	0.550		
55		0.24	0.329	0.386	0.425	0.460	0.525	0.558		
56		0.26	0.334	0.391	0.430	0.466	0.532	0.565		
57		0.28	0.339	0.397	0.437	0.472	0.539	0.573		
58	Cost	0.30	0.345	0.403	0.443	0.479	0.547	0.581		
59		0.32	0.351	0.409	0.450	0.486	0.555	0.589		
60		0.34	0.357	0.416	0.457	0.494	0.563	0.598		
61		0.36	0.363	0.423	0.464	0.502	0.571	0.606		
62		0.38	0.370	0.430	0.472	0.509	0.580	0.615		
63		0.40	0.377	0.437	0.479	0.517	0.588			

Figure 7.12   *Spreadsheet for product engineering decision*

	A	B	C	D
1				
2				
3				
4				
5				
6			First Year Improve	
7		x	Value	x
8		Low:	0.1	Low:
9		High:	0.8	High:
10		Mono:	Increasing	Mono:
11		Rho:	Infinity	Rho:
12		Weights:	0.1	
13				
14				
15			First Year Improve	
16	0.05-FRACTILES:	b	x	b
17	Low:	0.2	0.12	0.2
18	Middle:	0.3	0.24	0.3
19	High:	0.4	0.44	0.4
20	0.50-FRACTILES:			
21	Low:	0.2	0.25	0.2
22	Middle:	0.3	0.4	0.3
23	High:	0.4	0.65	0.4
24	0.95-FRACTILES:			
25	Low:	0.2	0.46	0.2
26	Middle:	0.3	0.63	0.3
27	High:	0.4	0.76	0.4
28				
29				
30			COST	QUALITY
31			0.379999997868223	0.32
32		Lower Bound:	0.2	0.2
33		Upper Bound:	0.4	0.32
34				
35				
36			First Year Improve	
37		Prob.	Score	Prob.
38	0.05-FRACTILES:	0.185	=Quad($C$31,B17:B19,C17:C19)	0.185
39	0.50-FRACTILES:	0.63	=Quad($C$31,B21:B23,C21:C23)	0.63
40	0.95-FRACTILES:	0.185	=Quad($C$31,B25:B27,C25:C27)	0.185
41				
42				
43	0.05-FRACTILES:		=ValueE(C38,C$8,C$9,C$10,C$11)	
44	0.50-FRACTILES:		=ValueE(C39,C$8,C$9,C$10,C$11)	
45	0.95-FRACTILES:		=ValueE(C40,C$8,C$9,C$10,C$11)	
46				
47				
48			=CEValue($F$3,C$12,B$38:B$40,C$43:C$45)	
49				
50				
51				
52		=SUM($C$48:$I$48)	0.2	0.22
53		0.2	=TABLE(D31,C31)	=TABLE(D31,C31)
54		0.22	=TABLE(D31,C31)	=TABLE(D31,C31)
55		0.24	=TABLE(D31,C31)	=TABLE(D31,C31)
56		0.26	=TABLE(D31,C31)	=TABLE(D31,C31)
57		0.28	=TABLE(D31,C31)	=TABLE(D31,C31)
58	Cost	0.3	=TABLE(D31,C31)	=TABLE(D31,C31)
59		0.32	=TABLE(D31,C31)	=TABLE(D31,C31)
60		0.34	=TABLE(D31,C31)	=TABLE(D31,C31)
61		0.36	=TABLE(D31,C31)	=TABLE(D31,C31)
62		0.38	=TABLE(D31,C31)	=TABLE(D31,C31)
63		0.4	=TABLE(D31,C31)	=TABLE(D31,C31)

Figure 7.13a *Spreadsheet equations for product engineering decision (part 1)*

	E	F	G
1			
2			
3	multiattribute risk tolerance:	Infinity	
4			
5			
6			Quality
7	Value	x	Value
8	0.8	0	0
9	2	1	0.4
10	Increasing	4	1
11	Infinity		
12	0.15		0.55
13			
14			
15			Quality
16	x	b	x
17	0.87	0.2	0.2
18	1.28	0.26	1.2
19	1.49	0.32	2.1
20			
21	1.1	0.2	0.75
22	1.4	0.26	1.8
23	1.6	0.32	3
24			
25	1.25	0.2	1.55
26	1.62	0.26	2.8
27	1.96	0.32	3.5
28			
29			
30	FEATURES	TOTAL	
31	=1-SUM(C31:D31)	=SUM(C31:E31)	
32	0.3	1	
33	0.6		
34			
35			
36			Quality
37	Score	Prob.	Score
38	=Quad($C$31,D17:D19,E17:E19)	0.185	=Quad($D$31,F17:F19,G17:G19)
39	=Quad($C$31,D21:D23,E21:E23)	0.63	=Quad($D$31,F21:F23,G21:G23)
40	=Quad($C$31,D25:D27,E25:E27)	0.185	=Quad($D$31,F25:F27,G25:G27)
41			
42			
43	=ValueE(E38,E$8,E$9,E$10,E$11)		=ValuePL(G38,F$8:F$10,G$8:G$10)
44	=ValueE(E39,E$8,E$9,E$10,E$11)		=ValuePL(G39,F$8:F$10,G$8:G$10)
45	=ValueE(E40,E$8,E$9,E$10,E$11)		=ValuePL(G40,F$8:F$10,G$8:G$10)
46			
47			
48	=CEValue($F$3,E$12,D$38:D$40,E$43:E$45)		=CEValue($F$3,G$12,F$38:F$40,G$43:G$45)
49			
50			
51			
52	0.24	0.26	0.3
53	=TABLE(D31,C31)	=TABLE(D31,C31)	=TABLE(D31,C31)
54	=TABLE(D31,C31)	=TABLE(D31,C31)	=TABLE(D31,C31)
55	=TABLE(D31,C31)	=TABLE(D31,C31)	=TABLE(D31,C31)
56	=TABLE(D31,C31)	=TABLE(D31,C31)	=TABLE(D31,C31)
57	=TABLE(D31,C31)	=TABLE(D31,C31)	=TABLE(D31,C31)
58	=TABLE(D31,C31)	=TABLE(D31,C31)	=TABLE(D31,C31)
59	=TABLE(D31,C31)	=TABLE(D31,C31)	=TABLE(D31,C31)
60	=TABLE(D31,C31)	=TABLE(D31,C31)	=TABLE(D31,C31)
61	=TABLE(D31,C31)	=TABLE(D31,C31)	=TABLE(D31,C31)
62	=TABLE(D31,C31)	=TABLE(D31,C31)	=TABLE(D31,C31)
63	=TABLE(D31,C31)	=TABLE(D31,C31)	=TABLE(D31,C31)

Figure 7.13b *Spreadsheet equations for product engineering decision (part 2)*

	H	I	J
1			
2			
3			
4			
5			
6		New Features	
7	x	Value	
8	Low:	0	
9	High:	4	
10	Mono:	Increasing	
11	Rho:	-1.64	SUM
12		0.2	=SUM(C12:I12)
13			
14			
15		New Features	
16	b	x	
17	0.3	0.2	
18	0.45	1.2	
19	0.6	2.1	
20			
21	0.3	0.75	
22	0.45	1.8	
23	0.6	3	
24			
25	0.3	1.55	
26	0.45	2.8	
27	0.6	3.5	
28			
29			
30			
31			
32			
33			
34			
35			
36		New Features	
37	Prob.	Score	
38	0.185	=Quad($E$31,H17:H19,I17:I19)	
39	0.63	=Quad($E$31,H21:H23,I21:I23)	
40	0.185	=Quad($E$31,H25:H27,I25:I27)	
41			
42			
43		=ValueE(I38,I$8,I$9,I$10,I$11)	
44		=ValueE(I39,I$8,I$9,I$10,I$11)	
45		=ValueE(I40,I$8,I$9,I$10,I$11)	
46			
47			CE Value
48		=CEValue($F$3,I$12,H$38:H$40,I$43:I$45)	=SUM($C$48:$I$48)
49			
50			
51			
52	0.32		
53	=TABLE(D31,C31)		
54	=TABLE(D31,C31)		
55	=TABLE(D31,C31)		
56	=TABLE(D31,C31)		
57	=TABLE(D31,C31)		
58	=TABLE(D31,C31)		
59	=TABLE(D31,C31)		
60	=TABLE(D31,C31)		
61	=TABLE(D31,C31)		
62	=TABLE(D31,C31)		
63	=TABLE(D31,C31)		

Figure 7.13c *Spreadsheet equations for product engineering decision (part 3)*

A24:I27 gives the 0.95 fractiles. As an example, these ranges show that for a budget allocation of 0.20 to cost improvement, the 0.05 fractile of the probability distribution over First Year Cost Improvement is 0.12, as shown in cell C17. Similarly, the 0.50 fractile of this distribution is 0.25, as shown in cell C21, and the 0.95 fractile is 0.46, as shown in cell C25. (The cumulative probability distribution from which these fractiles were determined is the leftmost curve in the top graph of Figure 7.10.)

Using quadratic interpolation on the 0.05, 0.50, and 0.95 fractiles, it is straightforward to make the necessary modifications to the Figure 4.9 spreadsheet to incorporate uncertainty into the evaluation of budget allocations for the product engineering decision. Range A29:F33 of Figure 7.12 shows the budget allocation with bounds on allowed allocations. This is identical to range A19:F23 of the Figure 4.9 spreadsheet.

Range A35:I40 of Figure 7.12 shows the required probabilities with corresponding evaluation measure scores necessary to use the extended Pearson-Tukey approximation. The equations for calculating evaluation measure scores are shown in Figure 7.13, and the calculation uses quadratic interpolation to determine the required evaluation measure scores. Range C42:I45 shows the single dimensional values corresponding to each score in range C38:I40, and these values are calculated using the ValuePL or ValueE function, as appropriate.

Finally, the single dimensional certainty equivalent values are calculated in range C48:I48 using the CEValue function, and the overall certainty equivalent value is calculated in cell J48 by summing the appropriate single dimensional certainty equivalent values.

The same procedure that was used in the Figure 4.9 spreadsheet to determine the preferred budget allocation is also used in the Figure 7.12 spreadsheet. A two-input Data Table is shown in range A50:H63 of Figure 7.12. This table determines the certainty equivalent values for different combinations of budget allocations to the cost improvement and quality areas in exactly the same way this was done in the Figure 4.9 spreadsheet. This table shows that the preferred allocation is still 0.32 to quality and 0.38 to cost improvement, just as it was in Figure 4.9.

For quadratic interpolation of the fractiles of probability distributions to give reasonable results, it is necessary that the assessed probability distribution not vary in an extreme manner. See Keefer (1978) and Keefer and Kirkwood (1978) for related discussion.

7.8 References

ADA Decision Systems, *DPL Standard Version User Guide,* Duxbury Press, Belmont, CA, 1995.

J. L. Corner and C. W. Kirkwood, "Decision Analysis Applications in the Operations Research Literature, 1970–1989," *Operations Research*, Vol. 39, pp. 206–219 (1991).

R. A. Howard, "Decision Analysis: Practice and Promise," *Management Science*, Vol. 34, pp. 679–695 (1988).

D. L. Keefer, "Allocation Planning for R & D with Uncertainty and Multiple Objectives," *IEEE Transactions on Engineering Management*, Vol. EM-25, pp. 8–14 (1978).

D. L. Keefer and C. W. Kirkwood, "A Multiobjective Decision Analysis: Budget Planning for Product Engineering," *Journal of the Operational Research Society*, Vol. 29, pp. 435–442 (1978).

R. L. Keeney and H. Raiffa, *Decisions with Multiple Objectives: Preferences and Value Tradeoffs*, Wiley, New York, 1976.

7.9 Review Questions

R7-1   Explain the implication of the Weak Law of Large Numbers for making decisions when there are multiple evaluation measures and also uncertainty.

R7-2   Describe the role of the multiattribute risk tolerance in making decisions when there are multiple evaluation measures and also uncertainty.

R7-3   What range of values for the multiattribute risk tolerance is seen in practical applications?

R7-4   Describe a procedure for determining whether risk aversion changes the decision for a decision problem with multiple evaluation measures and uncertainty.

R7-5   Describe a procedure for approximately determining the multiattribute risk tolerance.

7.10 Exercises

7.1   Consider once again the information presented in Exercise 4.3 of Chapter 4. Assume that all the information presented in that exercise is still valid. In addition, it is known that a decision maker is indifferent between receiving (1) an alternative that is certain to yield a Cost of $0, an Efficiency of -1, and an Effectiveness of -2, and (2) an alternative that has equal chances of yielding either of the following two outcomes: a Cost of $0, an Efficiency of 3, and an Effectiveness of 2, *or* a Cost of $1,000, an Efficiency of -1, and an Effectiveness of -2.

   Determine the multiattribute risk tolerance for a power-additive utility function over the ranges specified for the three evaluation measures in Exercise 4.3.

7.2 This is an extension of the power plant siting decision in Exercise 4.4 in Chapter 4. Assume that all the information presented in that exercise still holds except that there is uncertainty about the cost of constructing a power plant. Specifically, the cost figures given in the earlier exercise assume that it will be possible to use a new dry cooling technology at whichever of the two sites is selected. However, this technology is still under development, and it may not prove feasible to use it. This will not be known until after the site is selected. Whichever site is selected, there is a 0.3 probability that the dry cooling technology will not be feasible. If the technology is not feasible, then the cost of constructing a plant at the selected site will be ten percent higher than the figure quoted in the earlier exercise.

Determine the expected values for each of the two sites. Which site is more preferred if expected value is used as the criterion to make the decision?

7.3 A manager is considering a possible change in a batch manufacturing process. The two evaluation measures are cost per batch and number of defects per batch. The range of possible values for cost is 20 to 50 in thousands of dollars, and the range of possible values for number of defects is 30 to 60. The swing weight for cost is three times as great as the swing weight for number of defects over the specified ranges for these evaluation measures, and the single dimensional value functions over both evaluation measures are linear.

The current manufacturing process has a cost of 41 and a number of defects of 47. The proposed new alternative has uncertainty about both evaluation measures. Cumulative probability distributions have been determined for both evaluation measures for the proposed new alternative. The 0.05, 0.50, and 0.95 fractiles for cost are 25, 40, and 45, respectively. The 0.05, 0.50, and 0.95 fractiles for number of defects are 40, 51, and 58, respectively. To answer the questions below, use the extended Pearson-Tukey approximation.

(i) Confirm that the swing weights for cost and number of defects are 0.75 and 0.25, respectively.

(ii) Set up a spreadsheet model for this decision. Conduct an expected value analysis and show that the current process has an expected value of 0.33333, and the proposed process has an expected value of 0.37742. Thus, on an expected value basis, the proposed process is preferred. Include a copy of the equations for your spreadsheet with your solution.

(iii) Conduct a sensitivity analysis on the multiattribute risk tolerance ρ_m over the range $0.2 \leq \rho_m \leq$ infinity, and show that the preferred alternative changes, depending on the value of ρ_m over this range.

(iv) Consider an alternative with equal chances of yielding either a cost of 20 combined with a number of defects of 30 *or* a cost of 50 combined with a number of defects of 60. Suppose this uncertain alternative is preferred to a certain alternative that yields a cost of 50 combined with a number of defects equal to 30. Show that this implies that $\rho_m > 0.410$. Conduct a sensitivity analysis over this range of values for ρ_m, and show that the proposed process is preferred for all ρ_m in this range. Include the output from your spreadsheet analysis with your answer.

7.4 Consider once again the Zenren ElectroProducts problem in Exercise 4.6 of Chapter 4. Assume the multiobjective value function information given in that problem still holds. Also assume that the evaluation measure levels specified in that problem for the Status Quo alternative are still correct, and also that the Operational Ease evaluation measure levels given in that problem are correct for the other two alternatives.

However, now assume that there is uncertainty about some of the levels for the Cost and Number of Defects evaluation measures for the Offshore and Arizona alternatives. In particular, if the Offshore alternative is selected, then there is a 0.25 probability that the Cost will be 525, a 0.5 probability that the Cost will be 600, and a 0.25 probability that the Cost will be 800. If the Arizona alternative is selected, the Cost is certain to be 675.

If the Offshore alternative is selected, then there are equal chances that the Number of Defects will be 35 or 44. If the Arizona alternative is selected, there is a 0.6 probability that the Number of Defects will be 28 and a 0.4 probability that the Number of Defects will be 33.

Finally, it is true that a hypothetical alternative with a Cost of 500, Operational Ease of 1, and a Number of Defects of 100 is equally preferred to an alternative with equal chances of having either

a. A Cost of 500, an Operational Ease of 4, and a Number of Defects of 0, or

b. A Cost of 1000, an Operational Ease of 1, and a Number of Defects of 100.

Answer the following questions:

(i) Determine the multiattribute risk tolerance for this problem.

(ii) Specify a spreadsheet model for this problem. Include a printout of the equations for this model with your assignment.

(iii) Determine the certainty equivalent values for the three alternatives. Include a printout of the spreadsheet output with these certainty equivalent values with your assignment.

7.5 This is an extension of the MiTech Foundry problem in Exercise 4.7 of Chapter 4. Assume that the multiobjective value function information given in that problem still holds. Also assume that the evaluation measure levels given for the J941 and the Sentry 50 testers still hold. However, now assume that the OR9000 is a new product that is not yet on the market, and therefore there is uncertainty about some of its evaluation measure levels.

Probability distributions have been assessed for the cost, accuracy, and uptime evaluation measures for the OR9000. From these it has been determined that the 0.05 fractiles for the three evaluation measures are 1.2, 850, and 95, respectively. The 0.50 fractiles are 1.4, 900, and 98, respectively.

Finally, the 0.95 fractiles are 2.5, 1000, and 99, respectively. Delivery time is known to be three months for certain.

Consider a hypothetical uncertain alternative that has equal probabilities of yielding outcomes with either

   a. A cost of one hundred thousand dollars, an accuracy of 500 picoseconds, a delivery time of three months, and an uptime of 99 percent, or

   b. A cost of three hundred thousand dollars, an accuracy of 1000 picoseconds, a delivery time of six months, and an uptime of 95 percent.

This hypothetical uncertain alternative is equally preferred to a certain alternative with a cost of three hundred thousand dollars, an accuracy of 1000 picoseconds, a delivery time of three months, and an uptime of 95 percent.

   (i) Determine the parameters for the utility function for this decision. [*Hint:* In addition to the value function parameters determined in Chapter 4, you must also determine the multiattribute risk tolerance.]

  (ii) Create a spreadsheet model for this decision problem, and include a printout of the equations for this model with your homework solution. [*Hint:* Use the extended Pearson-Tukey approximation.]

 (iii) Use your spreadsheet model to calculate the certainty equivalent values for the three testers, and determine which alternative is preferred.

 (iv) Determine to two decimal points the range of the weight on cost for which the most preferred alternative found in part (iii) remains most preferred. Assume while you are varying this weight that the ratios of the other three weights remain the same. Include the spreadsheet output that you used in this sensitivity analysis with your homework solution.

7.6 **Project (Part E).** This is a continuation of the project from Chapters 1, 2, 3, and 4. In this part of the project, you extend the evaluation of your alternatives to include analysis of uncertainties using probabilities.

   (i) Extend the value function assessment from Part D of this project to assess a utility function. Present the general procedure used in the assessment but not a blow-by-blow description. Include assessed "raw data" used to determine the utility function, perhaps in a table. Show the math used to obtain the final utility function from the assessed raw data (perhaps in a figure), as well as the parameters for the final utility function.

  (ii) Present the procedure used to determine the evaluation measure scores (levels) for each alternative, along with the final evaluation measure scores. (The final scores can be presented in a table.) Reference data sources, including interviews with experts, in standard bibliography style. For evaluation measures with uncertainty, present probability distributions in figures (graphs), and give a brief description of the procedure used to determine the probability distributions. Present the raw data from which the probability distributions were determined (probably in a table or figure).

(iii) Present the expected utility calculations for the alternatives and a sensitivity analysis. Briefly describe how computations were done, but you do not have to present the actual computations if you use a spreadsheet program to do these calculations. Include a display of the equations for any spreadsheet model that you use. Conduct and present a systematic sensitivity analysis to investigate how variations in key assumptions impact the analysis results. Such a sensitivity analysis typically addresses at least (1) the impact of variations in evaluation measure weights, and (2) the impact of variations in evaluation measure scores or probabilities.

(iv) Present your conclusions based on the analysis in the preceding parts, including a qualitative discussion of the reasons that the preferred alternative is best. The goal of this discussion is that someone who does not understand the details of decision analysis methods will find your discussion to be a convincing argument for the preferred alternative. That is, the analysis should not be a mysterious procedure, but rather a way of developing insight about the key factors in the decision and how these lead to selection of the preferred alternative.

7.7 Precocious Toys, a maker of intellectually challenging toys for children, has decided to put its catalog online on the Internet. To do this, the company must purchase computer hardware and software. After screening the possible systems, two configurations from the same vendor have been selected for further consideration. These two configurations are identical except for cost and capacity. Precocious Toys is measuring cost in thousands of dollars and capacity in terms of months of useful life before the demand will become too great for the configuration, and hence a new system will be needed.

Configuration 1 costs $10,000, and Configuration 2 costs $15,000. The useful life for each of the two configurations is uncertain because the demand for an online catalog is uncertain, and it is also uncertain exactly how much load of the type generated by an online catalog can be handled by each configuration. Precocious Toys believes there is a 0.3 probability that the online catalog will be a Wild Hit and a 0.7 probability that it will be a Modest Success. These probabilities do not depend on which configuration is selected.

The probability distribution for the useful life of a configuration depends on whether the online catalog is a Wild Hit or a Modest Success. If it is a Wild Hit, then the probability distribution for the useful life of Configuration 1 has a 0.05 fractile of 12 months, a 0.50 fractile of 14 months, and a 0.95 fractile of 18 months. If the catalog is a Modest Success, then these three fractiles are 26, 32, and 36 months, respectively, for Configuration 1.

For Configuration 2, if the online catalog is a Wild Hit, then the 0.05 fractile for the useful life of the configuration is 37 months, the 0.50 fractile is 40 months, and the 0.95 fractile is 46 months. On the other hand, if the catalog is a Modest Success, Configuration 2 will have a useful life of at least 48 months. After 48 months, any configuration that is purchased will be replaced because of advances in technology.

The single dimensional value function over useful life is exponential, greater levels are more preferred, and the midvalue of the range from 12 to 48 months is 21 months. Of course, greater levels of cost are less preferred. The value increment obtained from increasing useful life from 12 to 48 months is one and a half times as great as the value increment obtained from decreasing cost from $15,000 to $10,000.

(i) Determine the parameters for an additive value function that can be used to rank the two configurations. [*Hint:* Note that in order to rank the two configurations it is not necessary to determine the values for any costs other than $10,000 and $15,000.]

(ii) Draw a decision tree to represent the information presented above for this decision, and show the values for each endpoint on this tree. Use the extended Pearson-Tukey approximation while constructing this decision tree.

(iii) Determine the expected values for the two configurations. Using expected value as the criterion to make a decision, which configuration is more preferred?

(iv) Assume that a power-additive utility function is appropriate, and show that the same configuration identified as most preferred in part (iii) is most preferred for all multiattribute risk tolerances greater than 0.2.

7.8   This is an extension of the budget allocation planning decision in Exercise 4.9 of Chapter 4. Assume that all the information given in that exercise regarding the value function for the Manufacturing Process Improvement Group is still correct. This exercise considers uncertainties about the results for various budget allocations.

Specifically, suppose that probability distributions have been assessed for (1) the throughput in parts per hour as a function of the budget allocated to throughput improvement, and (2) the number of defects per thousand parts produced as a function of the budget allocated to defect reduction. This has been done for three levels of budget allocation to each activity: (a) No budget allocated to that activity, (b) half of the budget allocated to that activity, and (c) all of the budget allocated to that activity.

If no budget is allocated to throughput improvement work, then the throughput is certain to remain at 1,000 parts per hour, and if all the budget is allocated to throughput improvement work, then the throughput is certain to improve to 2,000 parts per hour. If half of the budget is allocated to throughput improvement work, then there is uncertainty about what throughput will result. Specifically, in this case the 0.05 fractile for the probability distribution over throughput is 1,200 parts per hour, the 0.50 fractile is 1,500, and the 0.95 fractile is 1,600.

If no budget is allocated to defect reduction work, then the defects are certain to remain at 10 per thousand parts produced, and if all the budget is allocated to defect reduction work, then the defects are certain to improve to 5 defects per thousand parts produced. If half of the budget is allocated to defect reduction work, then the 0.05 fractile for the probability distribution over defects is 6 defects per thousand parts produced, the 0.50 fractile is 7.5, and the 0.95 fractile is 9.5.

For the questions below, use the extended Pearson-Tukey approximation to develop discrete approximations for the probability distributions over throughput and defects. Also, use quadratic interpolation on the fractiles of these distributions to estimate probability distributions for levels of budget other than those discussed in the exercise statement above.

(i) Develop a spreadsheet or other computer model that can be used to determine the expected value for different allocations of budget to throughput improvement and defect reduction. Include the equations for this model with your solution. [*Hint:* See part (iv) below. By developing a more general model than asked for in this part, you can save time later in solving part (iv).]

(ii) Use the model you developed in part (i) to prepare a graph of the expected value of a budget allocation as a function of the budget allocated to throughput improvement.

(iii) Use the model you developed in part (i) to determine the most preferred budget allocation. Determine this accurate to two decimal places for the fraction of available budget allocated to each activity.

(iv) Repeat part (iii), but now assume that the multiattribute risk tolerance is equal to 0.2.

CHAPTER 8

Resource Allocation

The type of decisions that are of interest in this chapter can be illustrated by considering once again the computer networking strategy decision examined in earlier chapters. To be specific, review the version of this decision that was analyzed in Chapter 4. The evaluation measure scores for the four alternatives in that decision are shown in Table 4.1. This table shows that there are three evaluation measures (Productivity Enhancement, Cost Increase, and Security), and that the four alternatives have different scores for these evaluation measures.

If you are a manager who has to meet a budget, a question regarding this decision may occur to you: What happens to the extra money for the low cost alternatives? The Cost Increase levels for the alternatives range from $0 for the Status Quo alternative to $125,000 for the High Quality/High Cost alternative. Suppose the Status Quo alternative is selected. Then there is presumably $125,000 available for other uses—money that would have been used to fund the High Quality/High Cost alternative if that alternative had been selected. It may seem that this "extra" money should be taken into account in some way in evaluating the Status Quo alternative.

When considering an isolated *project* like the computer networking strategy decision, it may be appropriate to view the monetary evaluation measure as one consideration in the decision to be traded off against others. (This was the approach taken while analyzing the computer networking decision in earlier chapters.) However, when there is a specified budget to be allocated among competing activities, then the question of how the resources not allocated to a particular activity are used becomes more important in evaluating alternatives since these resources are available to fund other activities. This type of resource allocation decision is considered in this chapter.

The next section considers a benefit/cost analysis approach to this type of decision, which can be implemented using any spreadsheet program. The section following that considers a more advanced approach using *optimization*, also called *mathematical programming*. Optimization capabilities are available in advanced spreadsheet packages, and this chapter reviews the use of the Excel Solver optimization feature.

8.1 Benefit/Cost Analysis

A widely used method for analyzing resource allocation decisions is the benefit/cost ratio approach. This method will be illustrated by considering a funding decision for routine capital projects. Many organizations allocate a specified amount of money each planning period for relatively small capital projects, for example, renovation of equipment or facilities, equipment upgrades, etc. Often, a process is followed where various organizational units submit requests for a portion of this capital budget, and the requests are prioritized and then funded in priority order until the budget is exhausted. The situations of primary interest here are those where there are a substantial number of proposed projects, perhaps several dozen or more, and only a fraction of these can be funded. In such a situation, considerable management time can be spent establishing the priority of the various requests.

A value analysis approach is useful for this type of decision because it provides a systematic method that allows everyone involved in the capital budgeting process to provide information in a clear and mutually understood way. An application from the health care industry illustrates the procedure. In this particular case, there was a routine capital budget of somewhat over ten million dollars, and the proposed projects had costs ranging from approximately $15,000 to $2,000,000, with 81 of the 133 proposed projects requiring under $100,000, and 20 requiring over $200,000.

Fourteen evaluation measures were developed covering three major evaluation areas as follows:

1. Cost benefits
   1.1 Project cost
   1.2 Employee count
   1.3 Operating expense
   1.4 Net revenue
2. Strategy
   2.1 Function (type of equipment/project)
   2.2 Quality improvement
   2.3 Physician support
   2.4 Environmental concerns
   2.5 Standardization
   2.6 System impact
3. Equipment service and support
   3.1 Maintenance cost
   3.2 Downtime
   3.3 Manufacturer support
   3.4 Life expectancy

A value function was developed to combine these evaluation measures, and each alternative was ranked using this value function. Following the completion of the multiobjective value analysis, a benefit/cost ratio approach was used to assist in establishing funding priorities for the proposed projects. The procedure is illustrated in Figure 8.1. This shows a sample of twenty-five of the proposed

	A	B	C	D	E
1		Project	Cost	Benefit	Benefit/Cost
2	1	Monitors	249000	0.75206	0.3020
3	2	Furniture&Carpet	105790	0.43634	0.4125
4	3	Mgt System	150000	0.43270	0.2885
5	4	Linear Const	2000000	0.59733	0.0299
6	5	Hyperthermia	200000	0.35885	0.1794
7	6	Brachytherapy	350000	0.42599	0.1217
8	7	CT Imaging	170000	0.33217	0.1954
9	8	Treatment Plan	50000	0.44344	0.8869
10	9	Compactor - Main	28134	0.52413	1.8630
11	10	Compactor - Small	23416	0.00000	0.0000
12	11	Comm Radio	26000	0.56600	2.1769
13	12	Carpet	15000	0.33637	2.2425
14	13	Remodel	75000	0.31954	0.4261
15	14	Twin Fetal	39000	0.45272	1.1608
16	15	Motility	35000	0.49080	1.4023
17	16	Video	140000	0.40000	0.2857
18	17	Hemoglobin	70000	0.33558	0.4794
19	18	Remodel	100000	0.41267	0.4127
20	19	Mamo	160000	0.66720	0.4170
21	20	Recorders	100000	0.34600	0.3460
22	21	Rml DVI	100000	0.40080	0.4008
23	22	ER1	245000	0.68650	0.2802
24	23	RmB	470000	0.72084	0.1534
25	24	IP Holding	60000	0.52933	0.8822
26	25	Relocation	30000	0.41267	1.3756

Figure 8.1   *Benefits and costs for capital projects*

capital projects. Each proposed project is in a separate row of this spreadsheet. An index number is given in column A, and the name of the project is in column B. The cost of the project is in column C, and the benefit of the project, as calculated using the value function with fourteen evaluation measures, is given in column D. The ratio of the benefit to cost for each project is in column E. The spreadsheet equations to calculate these benefit-to-cost ratios are shown in Figure 8.2. (In this equation, the actual benefit-to-cost ratio is multiplied by 100,000 since the ratio would otherwise have several zeros following the decimal point.) The equation to calculate the benefit-to-cost ratio is entered into cell E2, and then filled down into the range E3:E26.

Figure 8.1 shows that if there are several hundred thousand dollars available for capital projects, then it is possible to fund a number of projects. Thus, a procedure is needed to evaluate the total value of different *portfolios* made up of more than one project. The appropriate procedure for doing this depends on

	A	B	C	D	E
1		Project	Cost	Benefit	Benefit/Cost
2	1	Monitors	249000	0.75206	=(D2/C2)*100000
3	2	Furniture&Carpet	105790	0.43634	=(D3/C3)*100000
4	3	Mgt System	150000	0.4327	=(D4/C4)*100000
5	4	Linear Const	2000000	0.59733	=(D5/C5)*100000
6	5	Hyperthermia	200000	0.35885	=(D6/C6)*100000
7	6	Brachytherapy	350000	0.42599	=(D7/C7)*100000
8	7	CT Imaging	170000	0.33217	=(D8/C8)*100000
9	8	Treatment Plan	50000	0.44344	=(D9/C9)*100000
10	9	Compactor - Main	28134	0.52413	=(D10/C10)*100000
11	10	Compactor - Small	23416	0	=(D11/C11)*100000
12	11	Comm Radio	26000	0.566	=(D12/C12)*100000
13	12	Carpet	15000	0.33637	=(D13/C13)*100000
14	13	Remodel	75000	0.31954	=(D14/C14)*100000
15	14	Twin Fetal	39000	0.45272	=(D15/C15)*100000
16	15	Motility	35000	0.4908	=(D16/C16)*100000
17	16	Video	140000	0.4	=(D17/C17)*100000
18	17	Hemoglobin	70000	0.33558	=(D18/C18)*100000
19	18	Remodel	100000	0.41267	=(D19/C19)*100000
20	19	Mamo	160000	0.6672	=(D20/C20)*100000
21	20	Recorders	100000	0.346	=(D21/C21)*100000
22	21	Rml DVI	100000	0.4008	=(D22/C22)*100000
23	22	ER1	245000	0.6865	=(D23/C23)*100000
24	23	RmB	470000	0.72084	=(D24/C24)*100000
25	24	IP Holding	60000	0.52933	=(D25/C25)*100000
26	25	Relocation	30000	0.41267	=(D26/C26)*100000

Figure 8.2 *Equations for Figure 8.1 spreadsheet*

how the benefits are calculated for a combination of projects. If the value of a project does not change when it is combined with other projects, then it is appropriate to sum the benefits for the projects in a portfolio to determine the total benefit for the portfolio. Similarly, if there are no interactions between the costs of different projects, then it is appropriate to sum the costs for the projects in a portfolio to determine the total cost for the portfolio. For the remainder of this section, it will be assumed that both benefits and costs for projects add in this way.

Some thought shows that if you could select only one project, then you would obtain the highest benefit per dollar expended by selecting the project with the highest benefit-to-cost ratio. Similarly, if you want to expand from a one-project portfolio to a two-project portfolio, then you will receive the highest benefit per dollar expended if you add to the project already selected (that is, the one with the highest benefit-to-cost ratio) the project with the next highest benefit-to-cost

	A	B	C	D	E	F	G
1		Project	Cost	Benefit	Benefit/Cost	Cum Cost	Cum Ben
2	12	Carpet	15000	0.33637	2.2425	15000	0.3364
3	11	Comm Radio	26000	0.56600	2.1769	41000	0.9024
4	9	Compactor - Main	28134	0.52413	1.8630	69134	1.4265
5	15	Motility	35000	0.49080	1.4023	104134	1.9173
6	25	Relocation	30000	0.41267	1.3756	134134	2.3300
7	14	Twin Fetal	39000	0.45272	1.1608	173134	2.7827
8	8	Treatment Plan	50000	0.44344	0.8869	223134	3.2261
9	24	IP Holding	60000	0.52933	0.8822	283134	3.7555
10	17	Hemoglobin	70000	0.33558	0.4794	353134	4.0910
11	13	Remodel	75000	0.31954	0.4261	428134	4.4106
12	19	Mamo	160000	0.66720	0.4170	588134	5.0778
13	18	Remodel	100000	0.41267	0.4127	688134	5.4905
14	2	Furniture&Carpet	105790	0.43634	0.4125	793924	5.9268
15	21	Rml DVI	100000	0.40080	0.4008	893924	6.3276
16	20	Recorders	100000	0.34600	0.3460	993924	6.6736
17	1	Monitors	249000	0.75206	0.3020	1242924	7.4257
18	3	Mgt System	150000	0.43270	0.2885	1392924	7.8584
19	16	Video	140000	0.40000	0.2857	1532924	8.2584
20	22	ER1	245000	0.68650	0.2802	1777924	8.9449
21	7	CT Imaging	170000	0.33217	0.1954	1947924	9.2770
22	5	Hyperthermia	200000	0.35885	0.1794	2147924	9.6359
23	23	RmB	470000	0.72084	0.1534	2617924	10.3567
24	6	Brachytherapy	350000	0.42599	0.1217	2967924	10.7827
25	4	Linear Const	2000000	0.59733	0.0299	4967924	11.3800
26	10	Compactor - Small	23416	0.00000	0.0000	4991340	11.3800

Figure 8.3   *Cumulative benefits and costs for capital projects*

ratio. This process can be used to continue adding additional projects until the available budget is exhausted.

This procedure is easily implemented in a spreadsheet. Figure 8.3 shows the approach. In this spreadsheet, the table in range A2:E26 of Figure 8.1 is sorted in decreasing order of benefit-to-cost ratio. That is, it is sorted using column E as the sort column. Equations are entered into columns F and G to calculate the cumulative benefits and costs for the projects down to and including the one on the row of interest. The formulas to do these calculations are shown in Figure 8.4. (Note that the formulas in cells F2 and G2 are different from those in the cells below. Thus, the appropriate formula is entered into cell F2 and copied to cell G2. Then the appropriate formula is entered into cell F3 and filled into the range F3:G26.)

A graph is shown in Figure 8.5 of the cumulative benefits versus the cumulative costs as given in columns F and G of the Figure 8.3 spreadsheet. This graph

	A	B	C	D	E	F	G
1		Project	Cost	Benefit	Benefit/Cost	Cum Cost	Cum Ben
2	12	Carpet	15000	0.33637	=(D2/C2)*100000	=C2	=D2
3	11	Comm Radio	26000	0.566	=(D3/C3)*100000	=F2+C3	=G2+D3
4	9	Compactor - Main	28134	0.52413	=(D4/C4)*100000	=F3+C4	=G3+D4
5	15	Motility	35000	0.4908	=(D5/C5)*100000	=F4+C5	=G4+D5
6	25	Relocation	30000	0.41267	=(D6/C6)*100000	=F5+C6	=G5+D6
7	14	Twin Fetal	39000	0.45272	=(D7/C7)*100000	=F6+C7	=G6+D7
8	8	Treatment Plan	50000	0.44344	=(D8/C8)*100000	=F7+C8	=G7+D8
9	24	IP Holding	60000	0.52933	=(D9/C9)*100000	=F8+C9	=G8+D9
10	17	Hemoglobin	70000	0.33558	=(D10/C10)*100000	=F9+C10	=G9+D10
11	13	Remodel	75000	0.31954	=(D11/C11)*100000	=F10+C11	=G10+D11
12	19	Mamo	160000	0.6672	=(D12/C12)*100000	=F11+C12	=G11+D12
13	18	Remodel	100000	0.41267	=(D13/C13)*100000	=F12+C13	=G12+D13
14	2	Furniture&Carpet	105790	0.43634	=(D14/C14)*100000	=F13+C14	=G13+D14
15	21	Rml DVI	100000	0.4008	=(D15/C15)*100000	=F14+C15	=G14+D15
16	20	Recorders	100000	0.346	=(D16/C16)*100000	=F15+C16	=G15+D16
17	1	Monitors	249000	0.75206	=(D17/C17)*100000	=F16+C17	=G16+D17
18	3	Mgt System	150000	0.4327	=(D18/C18)*100000	=F17+C18	=G17+D18
19	16	Video	140000	0.4	=(D19/C19)*100000	=F18+C19	=G18+D19
20	22	ER1	245000	0.6865	=(D20/C20)*100000	=F19+C20	=G19+D20
21	7	CT Imaging	170000	0.33217	=(D21/C21)*100000	=F20+C21	=G20+D21
22	5	Hyperthermia	200000	0.35885	=(D22/C22)*100000	=F21+C22	=G21+D22
23	23	RmB	470000	0.72084	=(D23/C23)*100000	=F22+C23	=G22+D23
24	6	Brachytherapy	350000	0.42599	=(D24/C24)*100000	=F23+C24	=G23+D24
25	4	Linear Const	2000000	0.59733	=(D25/C25)*100000	=F24+C25	=G24+D25
26	10	Compactor - Small	23416	0	=(D26/C26)*100000	=F25+C26	=G25+D26

Figure 8.4 *Equations for Figure 8.3 spreadsheet*

shows that the first few hundred thousand dollars buy substantial benefits, but that the return on investment declines after that, and there is not much benefit for the last few projects. (In fact, the last project (in row 26 of the Figure 8.3 spreadsheet) has zero benefit. Because its cost is small, the dot for it cannot be distinguished from the adjoining dot in Figure 8.5.)

The benefit-to-cost ratio method is effective when there is a single budget constraint, and the cost of each proposed project is small relative to the entire available budget. However, there may be other budget allocations that provide a higher total value while remaining within the available budget. As an example, consider the benefit/cost analysis results shown in Figure 8.3, and suppose that the total available budget is \$750,000. Then, using the benefit/cost approach, the first twelve projects shown at the top of the spreadsheet will be funded.

However, with this allocation, there is \$750,000 − \$688,134 = \$61,866 left over from the available budget. This naturally leads to the question of whether

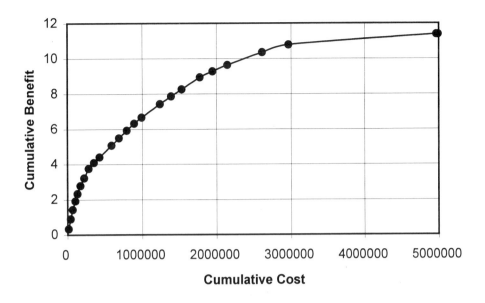

Figure 8.5 *Plot of cumulative benefits versus cumulative costs*

the available budget might be more completely utilized with a different set of projects. For example, with the $750,000 budget, the Remodel project (number 18) with a benefit of 0.41267 and a cost of $100,000 is included in the funded portfolio. However, the Furniture & Carpet project (number 2) has a benefit of 0.43634, which is higher than the benefit for the Remodel project. Since the cost for this project is $105,790, it would be possible to replace the Remodel project with this and still remain within the $750,000 budget.

This example shows that it is sometimes possible to find a portfolio of projects that has a higher value than the one selected by the benefit/cost method. However, some thought shows that any portfolio other than the one selected by the benefit/cost approach will have a lower overall benefit-to-cost ratio. The example in the preceding paragraph illustrates this. For the portfolio consisting of the first twelve projects in Figure 8.3, the overall benefit-to-cost ratio (multiplied by 100,000 as in that figure) is $100,000 \times (5.49045/688,134) = 0.7979$. If the Remodel project is replaced with the Furniture & Carpet project, then the overall benefit-to-cost ratio for the portfolio is $100,000 \times (5.5141/693,924) = 0.7946$, which is slightly lower. However, the overall value of the revised portfolio is 5.5141, which is higher than the 5.49045 obtained with the portfolio that maximizes the benefit-to-cost ratio.

This example illustrates that replacing the Remodel project with the Furniture & Carpet project increases the total value received, but it reduces the "bang for the buck" received (that is, the value received per dollar spent). When I pointed this out to a class that included many working managers, one of them promptly said, "You always spend your entire budget, or you won't get as much next year." A lively discussion continued on this point from some time.

In addition to the fact that it does not always effectively utilize the entire available budget, the cost-to-benefit ratio approach has the shortcoming that it cannot handle more than one constraint. There are often additional constraints in practical budget allocation decisions, and methods for handling these situations are considered in the next section.

Before turning to this topic, here is one additional comment that relates to the value function used to calculate project benefits. There is a question as to whether or not the cost of a project should be included in the value function used to calculate benefits. In the medical capital budgeting decision discussed in this section, the cost was included in the value function. The person in charge of that study believed that the cost should be included, although an argument can be made for leaving it out. Since the total capital budget is going to be spent on something, it seems to be "double counting" costs to include them in both the benefit and cost sides of the ratio. However, different managers who consider this type of decision problem have differing views on this.

8.2 Project Selection with Constraints

The benefit-to-cost ratio approach presented in the preceding section for resource allocation does not easily address situations with more than a single budget constraint or where there are interdependencies among the potential projects. Examples of such situations include cases where the potential projects extend over several budgeting periods or where a number of different variations are being considered for some of the proposed projects.

For example, in Figure 8.1 there are several proposed projects that address different types of remodeling or upgrading of physical facilities. Suppose that these all address the same physical space so that only one of them can be implemented. If there are only one or two of these variations, then this can be addressed manually by doing a separate benefit-to-cost ratio analysis for each variation. However, as the number of project variations grows, this manual analysis becomes less feasible. Similarly, if there are budget constraints for each of several different time periods, it can be difficult to manually analyze all of the various possibilities.

This is illustrated by the product development example in Table 8.1. There are six different possible product development projects, designated P1, P2, P3, P4, P5, and P6. Two of these, P1 and P2, each have three possible variations (designated P1.1, P1.2, P1.3 and P2.1, P2.2, P2.3, respectively), and only one of the variations for each of these projects can be implemented. The value (benefit) of each possible project is shown in the column labeled "Value." The projects would each extend over more than one year, and the required budget by year is shown for each project. The available budget for each year is shown at the bottom of the column for that year. Finally, there are two potential products, designated "Product 1" and "Product 2," which must be addressed by at least

Project	Value	Required Budget by Year					Required Work	
		1	2	3	4	5	Product 1	Product 2
P1.1	0.15	3	2	1.0	0	0	Yes	No
P1.2	0.12	2	1.5	1.5	1.0	0	Yes	No
P1.3	0.10	2	1.5	1.5	1.0	1.0	Yes	No
P2.1	0.30	4.5	5.5	6.0	6.0	5.0	No	Yes
P2.2	0.26	3.0	4.0	5.0	5.0	4.0	No	Yes
P2.3	0.18	1.5	2.5	2.5	2.0	0	No	Yes
P3	0.16	2.0	2.0	2.0	0	0	No	Yes
P4	0.19	0	1.0	1.5	1.5	1.5	Yes	No
P5	0.25	2.0	2.5	5.0	2.5	1.0	Yes	No
P6	0.21	0	1.5	3.5	1.5	0.5	No	No
Available Budget		10	10	10	10	10		

Table 8.1 *Product development portfolio decision*

one of the product development projects, and the last two columns of the table show whether each of the proposed projects addresses these products.

The preferred solution to this resource allocation decision is not immediately obvious from examining Table 8.1. Taking into account the three possible variations for P1 and P2, there are a total of ten projects to consider. Each of these could potentially either be included or not included in the selected portfolio, and so there are a total of $2^{10} = 1,024$ different portfolios to be considered. Of course, many of these do not meet the various constraints on the decision that were presented above, but it is not straightforward to be sure that you have considered all feasible combinations of the projects, let alone to find the one with the highest value.

A mathematical method called *0-1 linear integer programming* addresses this type of decision. This method can be applied using the Excel Solver. Figure 8.6a shows a spreadsheet with the information for the product development decision entered in a form that can be used by the Solver.

To use the Solver, you first assign a *decision variable* to each of the possible projects. In Figure 8.6a, the decision variables are shown in range B3:B12. For example, the decision variable for project P1.1 is in cell B3. The Excel Solver will set the values of each of these decision variables to indicate whether or not the corresponding project is included in the portfolio of projects with the highest value. Specifically, if the variable is set to zero, then the project is not included in the portfolio, and if the variable is set to one, the project is included in the portfolio.

In Figure 8.6a, the decision variables have all been set to zero to provide a starting point. Figure 8.6b shows the portfolio with the highest total value (shown in cell C13) as determined by the Excel Solver. This shows that project P1.1 is included in the portfolio since the corresponding decision variable in cell

B3 is set to one. Similarly, projects P2.3, P4, and P5 are also included in the portfolio, while the remaining projects are not included in the portfolio.

When you develop a spreadsheet for the Solver, you set the spreadsheet up as if you were going to use a trial-and-error approach to solve the problem. The Figure 8.6 spreadsheet is mostly constants with only a few equations. Specifically, the range C3:C12 shows the values for each of the potential projects, and the range D3:H12 holds the required yearly budgets for each potential project. Each of the entries in range I3:I12 is 1 if the project in that row addresses the required work on Product 1, and is zero if it does not. Similarly, the entries in range J3:J12 are each 1 if a particular potential project addresses the required work on Product 2, and zero if it doesn't. The entries of 1 in range K3:K12 show which potential projects are variations of project P1, and the entries of 1 in range L3:L12 show which potential projects are variations of project P2.

The entries in range D14:H14 are the available budgets for each of the five years. The entries of 1 in range I14:J14 show that at least one project must address Product 1, and at least one project must address Product 2. Finally, the entries of 1 in range K14:L14 show that at most one of the variations of project P1 and project P2 can be selected.

The cells in range C13:L13 are the only cells in the spreadsheet that contain equations. In each of these cells, the equation sums the product of the entry in each cell above with the corresponding decision variable in column B. These formulas are entered by first entering the formula

$$\text{SUMPRODUCT}(\$B3:\$B12,C3:C12)$$

in cell C13, and then filling this formula into range D13:L13. Note that the formula must have absolute column references for the decision variable cells in order for the reference to remain correct as the formula is filled into D13:L13.

Once this spreadsheet is set up, it is possible to attempt to find the best portfolio by trial and error. To do this, you enter some combination of zeros and ones for the decision variables in range B3:B12. After a particular set of zeros and ones is entered for the decision variables, each calculated entry in row 13 can be compared with the entry immediately below in row 14 to see whether the required condition is met. Specifically, the row 13 entries must be less than or equal to the row 14 entries for columns D through H and K through L, and the row 13 entries must be greater than or equal to the row 14 entries for columns I and J. Subject to these constraints, you want to select the set of decision variable values that maximizes the total value in cell C13.

The Excel Solver uses advanced analysis methods to quickly check all possible combinations of the decision variables and select the combination that gives the highest value for cell C13 (that is, the highest total value) while meeting all the constraints. As noted earlier, this combination is shown in Figure 8.6b, and it has a total value of 0.77, as shown in cell C13. A comparison of the entries in rows 13 and 14 shows that all the constraints are met. In fact, there is excess budget in all years except YR3.

The actual constraint conditions are not entered on the spreadsheet itself, but rather are entered from within the Solver. The Excel documentation provides

	A	B	C	D	E	F	G	H	I	J	K	L
1				Required Budget by Year					Required Work		Same Project	
2	Project	Decision	Value	YR1	YR2	YR3	YR4	YR5	Product 1	Product 2	P1	P2
3	P1.1	0	0.15	3	2	1	0	0	1	0	1	0
4	P1.2	0	0.12	2	1.5	1.5	1	0	1	0	1	0
5	P1.3	0	0.10	1.5	1.5	1.5	1	1	1	0	1	0
6	P2.1	0	0.30	4.5	5.5	6	6	5	0	1	0	1
7	P2.2	0	0.26	3	4	5	5	4	0	1	0	1
8	P2.3	0	0.18	1.5	2.5	2.5	2	0	0	1	0	1
9	P3	0	0.16	2	2	2	0	0	0	1	0	0
10	P4	0	0.19	0	1	1.5	1.5	1.5	1	0	0	0
11	P5	0	0.25	2	2.5	5	2.5	1	1	0	0	0
12	P6	0	0.21	0	1.5	3.5	1.5	0.5	0	0	0	0
13		Total:	0	0	0	0	0	0	0	0	0	0
14	CONSTRAINT:			10	10	10	10	10	1	1	1	1

a. Original spreadsheet

	A	B	C	D	E	F	G	H	I	J	K	L
1				Required Budget by Year					Required Work		Same Project	
2	Project	Decision	Value	YR1	YR2	YR3	YR4	YR5	Product 1	Product 2	P1	P2
3	P1.1	1	0.15	3	2	1	0	0	1	0	1	0
4	P1.2	0	0.12	2	1.5	1.5	1	0	1	0	1	0
5	P1.3	0	0.10	1.5	1.5	1.5	1	1	1	0	1	0
6	P2.1	0	0.30	4.5	5.5	6	6	5	0	1	0	1
7	P2.2	0	0.26	3	4	5	5	4	0	1	0	1
8	P2.3	1	0.18	1.5	2.5	2.5	2	0	0	1	0	1
9	P3	0	0.16	2	2	2	0	0	0	1	0	0
10	P4	1	0.19	0	1	1.5	1.5	1.5	1	0	0	0
11	P5	1	0.25	2	2.5	5	2.5	1	1	0	0	0
12	P6	0	0.21	0	1.5	3.5	1.5	0.5	0	0	0	0
13		Total:	0.77	6.5	8	10	6	2.5	3	1	1	1
14	CONSTRAINT:			10	10	10	10	10	1	1	1	1

b. Spreadsheet with solution

Figure 8.6   *Spreadsheet for constrained project portfolio decision*

Target Cell (Max)

Cell	Name	Original Value	Final Value
$C$13	Total: Value	0	0.77

Adjustable Cells

Cell	Name	Original Value	Final Value
$B$3	P1.1 Decision	0	1
$B$4	P1.2 Decision	0	0
$B$5	P1.3 Decision	0	0
$B$6	P2.1 Decision	0	0
$B$7	P2.2 Decision	0	0
$B$8	P2.3 Decision	0	1
$B$9	P3 Decision	0	0
$B$10	P4 Decision	0	1
$B$11	P5 Decision	0	1
$B$12	P6 Decision	0	0

Figure 8.7a   *Excel Solver answer report (part 1)*

instructions on how this is done. A report prepared by Solver, which presents the preferred portfolio and the various constraints, is shown in Figure 8.7.

This report shows that the quantity to be maximized is in cell C13 (labeled as the "target cell"), and that the decision variables (labeled as "adjustable cells") are in range B3:B12. It also shows that the decision variables are constrained to be from zero to 1, and to be integer. Thus, these variables can take only values of zero or 1. The other constraints are also shown.

This particular decision model has a special mathematical structure called *linear*. This means that the variables in the model are not raised to powers, multiplied together, or processed by functions such as the exponential. When a model is linear, some particularly efficient solution procedures are available, and the Solver allows you to specify when a model is linear so that it can use these methods.

The Solver also allows you to specify a "Tolerance," which is the allowable percentage of error in the solution to a decision problem with integer decision variables. The default setting for this is five percent, and with this setting, the value for the solution determined by the Solver could be as much as five percent worse than the best possible value. Setting the Tolerance to zero percent ensures that the Solver will find the solution with the best possible value; however, this sometimes requires an extensive amount of calculation.

This brief introduction to resource allocation has presented basic concepts and shown you how to use spreadsheet optimization to assist in resource allocation. Similar methods can be applied to a variety of resource allocation decisions. The next section reviews a specific application of this approach.

Constraints

Cell	Name	Cell Value	Formula	Status	Slack
$D$13	Total: YR1	6.5	$D$13<=$D$14	Not Binding	3.5
$I$13	Total: Product 1	3	$I$13>=$I$14	Not Binding	2
$J$13	Total: Product 2	1	$J$13>=$J$14	Binding	0
$E$13	Total: YR2	8	$E$13<=$E$14	Not Binding	2
$F$13	Total: YR3	10	$F$13<=$F$14	Binding	0
$G$13	Total: YR4	6	$G$13<=$G$14	Not Binding	4
$H$13	Total: YR5	2.5	$H$13<=$H$14	Not Binding	7.5
$K$13	Total: P1	1	$K$13<=$K$14	Binding	0
$L$13	Total: P2	1	$L$13<=$L$14	Binding	0
$B$3	P1.1 Decision	1	$B$3=integer	Binding	0
$B$4	P1.2 Decision	0	$B$4=integer	Binding	0
$B$5	P1.3 Decision	0	$B$5=integer	Binding	0
$B$6	P2.1 Decision	0	$B$6=integer	Binding	0
$B$7	P2.2 Decision	0	$B$7=integer	Binding	0
$B$8	P2.3 Decision	1	$B$8=integer	Binding	0
$B$9	P3 Decision	0	$B$9=integer	Binding	0
$B$10	P4 Decision	1	$B$10=integer	Binding	0
$B$11	P5 Decision	1	$B$11=integer	Binding	0
$B$12	P6 Decision	0	$B$12=integer	Binding	0
$B$3	P1.1 Decision	1	$B$3<=1	Binding	0
$B$4	P1.2 Decision	0	$B$4<=1	Not Binding	1
$B$5	P1.3 Decision	0	$B$5<=1	Not Binding	1
$B$6	P2.1 Decision	0	$B$6<=1	Not Binding	1
$B$7	P2.2 Decision	0	$B$7<=1	Not Binding	1
$B$8	P2.3 Decision	1	$B$8<=1	Binding	0
$B$9	P3 Decision	0	$B$9<=1	Not Binding	1
$B$10	P4 Decision	1	$B$10<=1	Binding	0
$B$11	P5 Decision	1	$B$11<=1	Binding	0
$B$12	P6 Decision	0	$B$12<=1	Not Binding	1
$B$3	P1.1 Decision	1	$B$3>=0	Not Binding	1
$B$4	P1.2 Decision	0	$B$4>=0	Binding	0
$B$5	P1.3 Decision	0	$B$5>=0	Binding	0
$B$6	P2.1 Decision	0	$B$6>=0	Binding	0
$B$7	P2.2 Decision	0	$B$7>=0	Binding	0
$B$8	P2.3 Decision	1	$B$8>=0	Not Binding	1
$B$9	P3 Decision	0	$B$9>=0	Binding	0
$B$10	P4 Decision	1	$B$10>=0	Not Binding	1
$B$11	P5 Decision	1	$B$11>=0	Not Binding	1
$B$12	P6 Decision	0	$B$12>=0	Binding	0

Figure 8.7b *Excel Solver answer report (part 2)*

8.3 Application to Proposal Evaluation

This section reviews an application of the methods in the preceding section to the selection of a portfolio of solar energy applications experiments (Golabi, Kirkwood, and Sicherman 1981). The work was conducted for the U.S. Department of Energy, which had released a Program Research and Development Announcement requesting proposals for such applications experiments. The purpose of this program was to provide an opportunity for interested parties to propose applications for concentrating solar photovoltaic experiments and, in a phased program, to design and implement the photovoltaic systems into those applications. It was expected that the awards resulting from the Program Research and Development Announcement would lead to design, implementation, and operation of between six and ten applications experiments. These were to provide technical, operations, and performance data for concentrating systems employed in on-site power generation applications and to obtain information on nontechnical issues as well.

Solar photovoltaic systems directly convert the sun's energy to electricity using photovoltaic ("solar") cells. At the time of the work reported here, it had been projected that the use of optical concentrators in conjunction with solar cells could significantly reduce photovoltaic system costs in the near term. However, operating experience with such systems was very limited.

The work was to be divided into three phases, with the number of projects supported by the Department of Energy to be reduced as work proceeded from one phase to the next. For the first phase, approximately twelve to twenty studies were to be supported at a total budget of five million dollars. After completion of this phase, it was expected that six to ten of these designs would be chosen for further funding.

The evaluation approach reviewed in this section was developed to assist in selection of projects for funding during the first phase of the work from among those proposed in response to the Program Research and Development Announcement. The selection of these projects was to proceed in two steps. The first step was an assessment of the technical worth of each proposal by a Technical Evaluation Committee of qualified professional and scientific personnel from within the government and the national laboratories. The second step was the forming of a recommendation by a Source Evaluation Panel and the actual selection by a Department of Energy Source Selection Official of proposals for negotiation. The purpose of this step was to select the portfolio of projects that would provide the maximum opportunity for advancing the overall solar photovoltaic concentrator program objectives.

It was expected that fifty to one hundred proposals, each of up to one hundred pages, would be submitted in response to the Program Research and Development Announcement. The Technical Evaluation Committee was to consist of fifteen to twenty photovoltaic specialists. Since these people would be "borrowed" from their regular positions for only a short period of time, the evaluation procedure had to be straightforward and practical to carry out. Also desired was output that would pinpoint the strengths and weaknesses of each proposal so the proposers who were not selected would clearly understand why they were not funded, should such questions arise. This was important since poor evaluation procedures might give grounds for legal or regulatory challenges that could delay the procurement.

Evaluation Procedure

The technical worth of individual proposals was evaluated using a value function approach similar to that presented in Chapter 4. A value function with twenty-two evaluation measures was used, as shown in Figure 8.8.

In addition, the programmatic issues shown in Figure 8.9 were identified as important. Constraints were developed for each of these using the approach illustrated in the preceding section. For example, the "size of system" issue was handled by defining three categories:

1 Percent of funded experiments smaller than 50 KW,
2 Percent of funded experiments between 50 and 150 KW, and
3 Percent of funded experiments larger than 150 KW.

By imposing constraints on the percentage of selected proposals allowed in each category, sufficient diversity could be guaranteed with respect to this programmatic consideration.

Implementing the Evaluation Procedure

The chair of the Technical Evaluation Committee established the twenty-two technical worth evaluation measure constructed scales, as well as the single dimensional value functions and weights.

During the actual evaluation of proposals, the seventeen members of the Technical Evaluation Committee did not need to be concerned with the computation of values. A computer program was used to handle the mathematics necessary to determine the overall value of each proposal, and the evaluators had only to fill out an evaluation form with the scores for the twenty-two evaluation measures for each proposal.

Each proposal was read by five evaluators. For each evaluation measure, the mean of the ratings for the five evaluators was used in the computations. However, in cases where the ratings of the five evaluators differed significantly, this was indicated by the computer program so the differences could be reviewed.

The computer program printed out the values for each proposal, a rank-ordered list of the proposals, and also supplementary information, such as the individual evaluation measure levels for each proposal. Thus, the reasons for a particular proposal rating could be easily traced from this output.

The results of the technical worth evaluation were then used as input to an optimization program that took the programmatic considerations into account. This program was available to run during the deliberations of the Source Evaluation Panel. Thus, the Panel could see the impact on the technical worth of the selected portfolio resulting from changes to constraint levels.

I. Conceptual design and system analysis

   A. Application identification
     — Load characterization
     — Matchup of array output and load

   B. System conceptual design
     — System conceptual design description

   C. Analysis, optimization, and tradeoff studies
     — Consideration of system design options
     — Parameter identification and optimization approach

II. Technical performance and integration

   A. Component specification plan
     — Specification of array
     — Specification of other major components

   B. System control and interfacing
     — Understanding of system control and operation
     — Interfaces with other energy sources

   C. Evaluation of potential performance
     — Evaluation of potential performance

   D. Development of major components
     — Development of major components

III. Implementation plan

     — Definition of work tasks for Phase I
     — Identification of team members for Phase I
     — General Phase II and III plans
     — Program management for Phase I

IV. Proposer's capabilities

     — Experience of firm(s)
     — Experience of personnel assigned to project
     — Disciplines of personnel

V. Other characteristics

     — Accessibility to technical community and visibility to public
     — Potential for low cost
     — Percent of load met by photovoltaic system
     — Institutional considerations

Figure 8.8 *Technical worth evaluation considerations*

1. Cost concerns
   a. Phase I cost
   b. Phase II cost
   c. Small and minority business participation
   d. Amount of cost sharing
2. Type of application
3. Size of system
4. Location
5. Load type
6. Load condition
7. Degree of utility cooperation
8. Whether proposal was a total energy system or not
9. Type of optics
10. Type of tracking
11. Concentration ratio
12. Type of solar cell used
13. Potential for low cost
14. Market potential

Figure 8.9 *Programmatic considerations*

In use, the optimization program might initially be run with only budgetary constraints. After inspecting the resulting optimal portfolio, the Source Evaluation Panel might, for example, decide that the distribution of selected proposals by location was unacceptable and impose constraints forcing certain minimum fractions of the portfolio to be in different parts of the country. The optimization program would then be run again and a new optimal portfolio determined. To meet the newly imposed constraints, proposals of lower technical worth would have to be selected, and the total technical worth of the selected portfolio would drop. The Panel could compare the technical worth of the selected portfolio with that of the portfolio selected previously (using fewer constraints) and decide if the increased locational diversity justified the decrease in technical worth. This type of interactive variation of constraints could be continued until a portfolio was found that had satisfactory technical worth and was also satisfactory with regard to programmatic issues.

Results of Applying the Approach

Using specially prepared data entry forms, the seventeen-member Technical Evaluation Committee completed its assessment of the seventy-seven submitted proposals in about two weeks. The results of this evaluation were then summarized for the Source Evaluation Panel. Members of this panel used the portfolio optimization procedure in the interactive manner discussed above to prepare a series of displays showing the consequences of varying certain cost and programmatic diversity constraints.

Using this procedure, about three days were required for the Source Evaluation Panel to recommend, and the Source Selection Official to select, a portfolio of seventeen proposals. This contrasted with a recent similar procurement where the activities of the Source Evaluation Panel and the Source Selection Official had required about a month.

The evaluation method presented here was important to this successful result; however, proper management and adequate computer support were also needed. In a procurement effort of the size discussed here, it is vital that easily used computer software be available to carry out the required calculations, and that adequate support staff be available to ensure that interaction with the computer goes smoothly.

8.4 Decisions with Continuous Decision Variables

Optimization approaches can also be used in decisions where alternatives are specified by continuous decision variables. The procedure is illustrated in this section using the product engineering budget allocation decision, which has previously been considered in Sections 4.8 and 7.7. Figure 8.10 shows a spreadsheet to analyze this decision using Solver, and the equations for this spreadsheet are shown in Figure 8.11. This spreadsheet is almost the same as the portion of the spreadsheet in Figure 7.12 above row 49. Specifically, the only difference is that in the Figure 8.11 spreadsheet there is no equation in cell E31 to calculate the budget allocation to New Features and Models work in terms of the allocations to Cost Improvement and to Quality.

Instead, Solver is used to determine the preferred budget allocation, subject to the constraints on the budget allocations for each of the three areas, as well as the constraint that the sum of the three budget allocations must be 1. The three "adjustable cells" used by Solver are in the range C31:E31, the "target cell" is cell J48, and the constraints on the budget allocations are in range C32:E33.

The budget allocation that the product engineering group had been using prior to this study is shown in range C31:E31 of the Figure 8.10 spreadsheet. The preferred allocation, as determined by Solver, is shown in the same cells of the Figure 8.12 spreadsheet. Note that this allocation has a certainty equivalent value of 0.61 (as shown in cell J48), while the allocation that had previously been used had a certainty equivalent value of only 0.51, as shown in Figure 8.10.

The Solver answer report for this decision is shown in Figure 8.13. This shows the lower and upper bounds on the three decision variables, as well as the constraint on the total budget allocation.

It was noted in Section 8.2 that when a decision problem has a linear structure, then Solver can use a particularly efficient solution procedure, and that there is an option within Solver that allows you to specify that a problem is linear. Note that the mathematical structure for the product engineering problem is *not* linear. Therefore, the linear option should not be selected when solving this decision problem.

The solution procedure for the product engineering decision using a Data Table, which was presented in Sections 4.8 and 7.7, works well with two independent decision variables, but it does not generalize to more than two variables because there is no "three-input" Data Table. The procedure presented in this section

	A	B	C	D	E	F	G	H	I	J
1				PRODUCT ENGINEERING BUDGET ALLOCATION						
2										
3				multiattribute risk tolerance:	Infinity					
4										
5				VALUE FUNCTIONS						
6		First Year Improve		Carryover Improve		Quality		New Features		
7		x	Value	x	Value	x	Value	x	Value	
8		Low:	0.10	Low:	0.80	0.00	0.00	Low:	0.00	
9		High:	0.80	High:	2.00	1.00	0.40	High:	4.00	
10		Mono:	Increasing	Mono:	Increasing	4.00	1.00	Mono:	Increasing	
11		Rho:	Infinity	Rho:	Infinity			Rho:	-1.64	SUM
12		Weights:	0.10		0.15		0.55		0.20	1.00
13										
14				RESPONSE FUNCTION PARAMETERS						
15		First Year Improve		Carryover Improve		Quality		New Features		
16	0.05-FRACTILES:	b	x	b	x	b	x	b	x	
17	Low:	0.20	0.12	0.20	0.87	0.20	0.20	0.30	0.20	
18	Middle:	0.30	0.24	0.30	1.28	0.26	1.20	0.45	1.20	
19	High:	0.40	0.44	0.40	1.49	0.32	2.10	0.60	2.10	
20	0.50-FRACTILES:									
21	Low:	0.20	0.25	0.20	1.10	0.20	0.75	0.30	0.75	
22	Middle:	0.30	0.40	0.30	1.40	0.26	1.80	0.45	1.80	
23	High:	0.40	0.65	0.40	1.60	0.32	3.00	0.60	3.00	
24	0.95-FRACTILES:									
25	Low:	0.20	0.46	0.20	1.25	0.20	1.55	0.30	1.55	
26	Middle:	0.30	0.63	0.30	1.62	0.26	2.80	0.45	2.80	
27	High:	0.40	0.76	0.40	1.96	0.32	3.50	0.60	3.50	
28										
29				BUDGET ALLOCATION						
30				COST	QUALITY	FEATURE	TOTAL			
31				0.30	0.28	0.42	1.00			
32		Lower Bound:		0.20	0.20	0.30	1.00			
33		Upper Bound:		0.40	0.32	0.60				
34										
35				PROBABILITIES AND SCORES (LEVELS)						
36		First Year Improve		Carryover Improve		Quality		New Features		
37		Prob.	Score	Prob.	Score	Prob.	Score	Prob.	Score	
38	0.05-FRACTILES:	0.19	0.24	0.19	1.28	0.19	1.51	0.19	1.01	
39	0.50-FRACTILES:	0.63	0.40	0.63	1.40	0.63	2.18	0.63	1.58	
40	0.95-FRACTILES:	0.19	0.63	0.19	1.62	0.19	3.09	0.19	2.59	
41										
42				SINGLE DIMENSIONAL VALUES						
43	0.05-FRACTILES:		0.20		0.40		0.50		0.08	
44	0.50-FRACTILES:		0.43		0.50		0.64		0.15	
45	0.95-FRACTILES:		0.76		0.68		0.82		0.37	
46										
47				SINGLE DIMENSIONAL CERTAINTY EQUIVALENT VALUES						CE Value
48				0.04		0.08		0.36		0.04	0.51

Figure 8.10 *Spreadsheet for product engineering decision*

	A	B	C	D
1				
2				
3				
4				
5				
6				
7		x	Value	x
8		Low:	0.1	Low:
9		High:	0.8	High:
10		Mono:	Increasing	Mono:
11		Rho:	Infinity	Rho:
12		Weights:	0.1	
13				
14				
15				
16	0.05-FRACTILES:	b	x	b
17	Low:	0.2	0.12	0.2
18	Middle:	0.3	0.24	0.3
19	High:	0.4	0.44	0.4
20	0.50-FRACTILES:			
21	Low:	0.2	0.25	0.2
22	Middle:	0.3	0.4	0.3
23	High:	0.4	0.65	0.4
24	0.95-FRACTILES:			
25	Low:	0.2	0.46	0.2
26	Middle:	0.3	0.63	0.3
27	High:	0.4	0.76	0.4
28				
29				
30			COST	QUALITY
31			0.3	0.28
32		Lower Bound:	0.2	0.2
33		Upper Bound:	0.4	0.32
34				
35				
36				
37		Prob.	Score	Prob.
38	0.05-FRACTILES:	0.185	=Quad($C$31,B17:B19,C17:C19)	0.185
39	0.50-FRACTILES:	0.63	=Quad($C$31,B21:B23,C21:C23)	0.63
40	0.95-FRACTILES:	0.185	=Quad($C$31,B25:B27,C25:C27)	0.185
41				
42				
43	0.05-FRACTILES:		=ValueE(C38,C$8,C$9,C$10,C$11)	
44	0.50-FRACTILES:		=ValueE(C39,C$8,C$9,C$10,C$11)	
45	0.95-FRACTILES:		=ValueE(C40,C$8,C$9,C$10,C$11)	
46				
47				
48			=CEValue($F$3,C$12,B$38:B$40,C$43:C$45)	

Figure 8.11a *Equations for Figure 8.10 spreadsheet (part 1)*

	E	F	G
1			
2			
3	multiattribute risk tolerance:	Infinity	
4			
5			
6			Quality
7	Value	x	Value
8	0.8	0	0
9	2	1	0.4
10	Increasing	4	1
11	Infinity		
12	0.15		0.55
13			
14			
15			Quality
16	x	b	x
17	0.87	0.2	0.2
18	1.28	0.26	1.2
19	1.49	0.32	2.1
20			
21	1.1	0.2	0.75
22	1.4	0.26	1.8
23	1.6	0.32	3
24			
25	1.25	0.2	1.55
26	1.62	0.26	2.8
27	1.96	0.32	3.5
28			
29			
30	FEATURES	TOTAL	
31	0.42	=SUM(C31:E31)	
32	0.3	1	
33	0.6		
34			
35			
36			Quality
37	Score	Prob.	Score
38	=Quad($C$31,D17:D19,E17:E19)	0.185	=Quad($D$31,F17:F19,G17:G19)
39	=Quad($C$31,D21:D23,E21:E23)	0.63	=Quad($D$31,F21:F23,G21:G23)
40	=Quad($C$31,D25:D27,E25:E27)	0.185	=Quad($D$31,F25:F27,G25:G27)
41			
42			
43	=ValueE(E38,E$8,E$9,E$10,E$11)		=ValuePL(G38,F$8:F$10,G$8:G$10)
44	=ValueE(E39,E$8,E$9,E$10,E$11)		=ValuePL(G39,F$8:F$10,G$8:G$10)
45	=ValueE(E40,E$8,E$9,E$10,E$11)		=ValuePL(G40,F$8:F$10,G$8:G$10)
46			
47			
48	=CEValue($F$3,E$12,D$38:D$40,E$43:E$45)		=CEValue($F$3,G$12,F$38:F$40,G$43:G$45)

Figure 8.11b   *Equations for Figure 8.10 spreadsheet (part 2)*

	H	I	J
1			
2			
3			
4			
5			
6			
7	x	Value	
8	Low:	0	
9	High:	4	
10	Mono:	Increasing	
11	Rho:	-1.64	SUM
12		0.2	=SUM(C12:I12)
13			
14			
15			
16	b	x	
17	0.3	0.2	
18	0.45	1.2	
19	0.6	2.1	
20			
21	0.3	0.75	
22	0.45	1.8	
23	0.6	3	
24			
25	0.3	1.55	
26	0.45	2.8	
27	0.6	3.5	
28			
29			
30			
31			
32			
33			
34			
35			
36			
37	Prob.	Score	
38	0.185	=Quad($E$31,H17:H19,I17:I19)	
39	0.63	=Quad($E$31,H21:H23,I21:I23)	
40	0.185	=Quad($E$31,H25:H27,I25:I27)	
41			
42			
43		=ValueE(I38,I$8,I$9,I$10,I$11)	
44		=ValueE(I39,I$8,I$9,I$10,I$11)	
45		=ValueE(I40,I$8,I$9,I$10,I$11)	
46			
47			CE Value
48		=CEValue($F$3,I$12,H$38:H$40,I$43:I$45)	=SUM(C48:I48)

Figure 8.11c *Equations for Figure 8.10 spreadsheet (part 3)*

	A	B	C	D	E	F	G	H	I	J
1				PRODUCT ENGINEERING BUDGET ALLOCATION						
2										
3				multiattribute risk tolerance:	Infinity					
4										
5					VALUE FUNCTIONS					
6			First Year Improve		Carryover Improve		Quality		New Features	
7			x	Value	x	Value	x	Value	x	Value	
8			Low:	0.10	Low:	0.80	0.00	0.00	Low:	0.00	
9			High:	0.80	High:	2.00	1.00	0.40	High:	4.00	
10			Mono:	Increasing	Mono:	Increasing	4.00	1.00	Mono:	Increasing	
11			Rho:	Infinity	Rho:	Infinity			Rho:	-1.64	SUM
12			Weights:	0.10		0.15		0.55		0.20	1.00
13										
14					RESPONSE FUNCTION PARAMETERS					
15			First Year Improve		Carryover Improve		Quality		New Features	
16	0.05-FRACTILES:		b	x	b	x	b	x	b	x	
17	Low:	0.20	0.12	0.20	0.87	0.20	0.20	0.30	0.20	
18	Middle:	0.30	0.24	0.30	1.28	0.26	1.20	0.45	1.20	
19	High:	0.40	0.44	0.40	1.49	0.32	2.10	0.60	2.10	
20	0.50-FRACTILES:									
21	Low:	0.20	0.25	0.20	1.10	0.20	0.75	0.30	0.75	
22	Middle:	0.30	0.40	0.30	1.40	0.26	1.80	0.45	1.80	
23	High:	0.40	0.65	0.40	1.60	0.32	3.00	0.60	3.00	
24	0.95-FRACTILES:									
25	Low:	0.20	0.46	0.20	1.25	0.20	1.55	0.30	1.55	
26	Middle:	0.30	0.63	0.30	1.62	0.26	2.80	0.45	2.80	
27	High:	0.40	0.76	0.40	1.96	0.32	3.50	0.60	3.50	
28										
29				BUDGET ALLOCATION						
30				COST	QUALITY	FEATURE	TOTAL			
31				0.38	0.32	0.30	1.00			
32		Lower Bound:		0.20	0.20	0.30	1.00			
33		Upper Bound:		0.40	0.32	0.60				
34										
35					PROBABILITIES AND SCORES (LEVELS)					
36			First Year Improve		Carryover Improve		Quality		New Features	
37			Prob.	Score	Prob.	Score	Prob.	Score	Prob.	Score	
38	0.05-FRACTILES:		0.19	0.39	0.19	1.46	0.19	2.10	0.19	0.20	
39	0.50-FRACTILES:		0.63	0.59	0.63	1.57	0.63	3.00	0.63	0.75	
40	0.95-FRACTILES:		0.19	0.74	0.19	1.89	0.19	3.50	0.19	1.55	
41										
42					SINGLE DIMENSIONAL VALUES					
43	0.05-FRACTILES:			0.42		0.55		0.62		0.01	
44	0.50-FRACTILES:			0.70		0.64		0.80		0.06	
45	0.95-FRACTILES:			0.91		0.91		0.90		0.15	
46										
47					SINGLE DIMENSIONAL CERTAINTY EQUIVALENT VALUES					CE Value
48				0.07		0.10		0.43		0.01	0.61

Figure 8.12   *Spreadsheet with preferred budget allocation*

Target Cell (Max)

Cell	Name	Original Value	Final Value
$J$48	CE Value	0.513193254	0.614888355

Adjustable Cells

Cell	Name	Original Value	Final Value
$C$31	COST	0.3	0.379999998
$D$31	QUALITY	0.28	0.32
$E$31	FEATURES	0.42	0.3

Constraints

Cell	Name	Cell Value	Formula	Status	Slack
$F$31	TOTAL	0.999999998	$F$31=$F$32	Binding	0
$C$31	COST	0.379999998	$C$31>=$C$32	Not Binding	0.179999998
$D$31	QUALITY	0.32	$D$31>=$D$32	Not Binding	0.12
$E$31	FEATURES	0.3	$E$31>=$E$32	Binding	0
$C$31	COST	0.379999998	$C$31<=$C$33	Not Binding	0.020000002
$D$31	QUALITY	0.32	$D$31<=$D$33	Binding	0
$E$31	FEATURES	0.3	$E$31<=$E$33	Not Binding	0.3

Figure 8.13   *Excel Solver answer report for product engineering decision*

works with a larger number of variables, and, in fact, the spreadsheet shown in Figure 8.10 actually uses three decision variables (in cells C31, D31, and E31).

Finally, a technical note is in order if you plan to use an approach like this on your own decision problems: An optimization approach will not always find the most preferred solution when the mathematical structure for a decision problem is not linear. In particular, the specific solution that is found may depend on what "starting point" you give it—that is, what initial budget allocation you specify. In practical situations, you should try a variety of different starting points. If Solver finds the same solution for these different starting points, then you can be fairly sure that it has found the best possible solution.

8.5 Reference

K. Golabi, C. W. Kirkwood, and A. Sicherman, "Selecting a Portfolio of Solar Energy Projects Using Multiattribute Preference Theory," *Management Science*, Vol. 27, pp. 174–189 (1981).

8.6 Review Question

R8-1   Discuss the pros and cons of including cost as an evaluation measure in the value function used to evaluate alternatives for possible inclusion in a portfolio when there is a budget constraint.

8.7 Exercises

8.1   Suppose there are three possible activities to include in a portfolio, and that these have the following values and costs:
   a. Activity 1: A value of 0.35 and a cost of 700,
   b. Activity 2: A value of 0.22 and a cost of 500, and
   c. Activity 3: A value of 0.18 and a cost of 450.
There is a total budget of 1,000, and the value of a portfolio is equal to the sum of the values of the activities in the portfolio.
   (i) Determine the preferred portfolio subject to the budget constraint if benefit-to-cost ratio is used to select activities.
   (ii) Determine which portfolio yields the greatest value subject to the budget constraint.
   (iii) Discuss the reasons for the differences between the answers in parts (i) and (ii).

8.2   BrandZip Industries has a product improvement group that works on improvements to existing consumer products sold by the company. For the coming budget year, fifteen potential improvement projects have been identified, and BrandZip needs to select the portfolio of projects that will be implemented. A value function with four evaluation measures (first year cost improvement, carryover cost improvement, new features, and quality improvement) has been developed to evaluate the value of each project. Assume that the total benefit for a portfolio of projects is equal to the sum of the values for each project in the portfolio.

   The value function has been used to determine the benefit (value) of each project; in addition, the cost of each project has been determined. The results are shown in the following table (costs are in thousands of dollars):

	Project	Cost	Benefit
1.	Digital display A	200	0.13
2.	Digital display B	490	0.21
3.	Fluid line reroute	425	0.17
4.	Fluid coupler upgrade	325	0.11
5.	Sense switch field upgrade	125	0.07
6.	Thrust bearing lubricator	275	0.14
7.	Bearing tolerance improvement	525	0.39
8.	Control line multiplexer	1,425	0.77
9.	Air filter mesh reduction	75	0.10
10.	Coolant seal field upgrade	400	0.22
11.	Direct drive	1,325	0.67
12.	Manufacturing simplification A	1,500	0.83
13.	Manufacturing simplification B	1,700	0.91
14.	Packaging cost reduction A	875	0.57
15.	Packaging cost reduction B	700	0.49

(i) Plot a curve of cumulative benefit versus cumulative cost for these projects, with the projects sorted in order of decreasing benefit-to-cost ratio. Using benefit-to-cost ratio as the criterion for selecting projects, determine the preferred portfolio if a total budget of $5 million is available. Determine how much of the available budget is not expended with this portfolio.

(ii) Determine the portfolio of projects that provides the largest benefit subject to the $5 million budget constraint. Determine the differences between this portfolio and the one found in part (i). Discuss how this portfolio is better than the one found in part (i), and how it is not as good.

(iii) The two packaging cost reduction projects (numbers 14 and 15 in the list above) both require specialized expertise that is in short supply within the product improvement group. Therefore, the group can undertake at most one of these two projects. Determine the portfolio of projects with the largest benefit subject to this constraint in addition to the $5 million budget constraint.

8.3 The programming support group for the Northeast Division of Cirrus Industries has 9100 hours of programming time available to undertake new projects during the next budget year. A value analysis has been conducted on the fourteen proposed new projects, and the following table shows the time requirements (in hours) and benefit (value) for each of the projects. Assume that the total benefit for a portfolio of projects is equal to the sum of the values for each project in the portfolio.

	Project	Time	Benefit
1.	Mail list upgrade	200	0.08
2.	Contact reporting	750	0.20
3.	Com fix	400	0.19
4.	Router bug	1000	0.21
5.	Accounting interface	1500	0.27
6.	Scheduler A	600	0.15
7.	Scheduler B	2500	0.37
8.	On-line manuals	3000	0.44
9.	WAN upgrade	2700	0.40
10.	Print queue fix	350	0.32
11.	EIS prototype	4000	0.50
12.	GDSS prototype	4200	0.51
13.	Contact management	2700	0.45
14.	Real-time sales	4500	0.76
	CONSTRAINT:	9100	

(i) Plot a curve of cumulative benefit versus cumulative time required for these projects, with the projects sorted in order of decreasing benefit-to-time ratio. Using benefit-to-time ratio as the criterion for selecting projects, determine the preferred portfolio, assuming that a total of 9100 hours of programming time is available. Determine how much of the available programming time is not expended with this portfolio.

(ii) Determine the portfolio of projects that provides the largest benefit subject to the 9100-hour constraint. Determine the differences between this portfolio and the one found in part (i). Discuss how this portfolio is better than the one found in part (i), and how it is not as good.

8.4 This is a variation on the preceding exercise. The head of the programming support group realizes that not all of the programmers in the group can work on each of the fourteen proposed projects. Specifically, three types of programming are required for the projects: Excel programming, Access programming, and C programming. Not all of the programmers know each of these types of programming.

The following table shows the number of hours of each type of programming required for each project. There is a total of 9000 potential hours of Excel programming available, a total of 7200 potential hours of Access programming available, and a total of 3600 potential hours of C programming available. In addition, 9100 hours of total programming time are available to distribute across the three types of programming. Assume that the total benefit for a portfolio of projects is equal to the sum of the values for each project in the portfolio.

	Project	Benefit	Excel	Access	C
				Time Requirements	
1.	Mail list upgrade	0.08		200	
2.	Contact reporting	0.20		500	250
3.	Com fix	0.19			400
4.	Router bug	0.21			1000
5.	Accounting interface	0.27	1000		500
6.	Scheduler A	0.15	600		
7.	Scheduler B	0.37	1500		1000
8.	On-line manuals	0.44		3000	
9.	WAN upgrade	0.40			2700
10.	Print queue fix	0.32			350
11.	EIS prototype	0.50	1000	3000	
12.	GDSS prototype	0.51	2000	1000	1200
13.	Contact management	0.45		2700	
14.	Real-time sales	0.76		3000	1500
	CONSTRAINTS:		9000	7200	3600

Determine the portfolio of projects that provides the largest benefit subject to the constraints on the amount of programming time of each type that is available in addition to the constraint on the total number of hours of programming time that is available.

8.5 Solve parts (ii) and (iv) of Exercise 4.9 again, but this time use the Excel Solver.

8.6 Solve parts (i), (iii), and (iv) of Exercise 7.8 again, but this time use the Excel Solver.

Multiattribute Preference Theory

This chapter summarizes the multiattribute preference theory that is most relevant to the methods presented in earlier chapters. Specifically, a number of relevant definitions, theorems, and assessment procedures are given. While some theorem proofs are presented, these should be regarded more as "plausibility arguments" than formal proofs. In particular, the conditions of continuity, differentiability, etc., that are required for the theorems to hold are often not specified. The required conditions do not cause significant problems for realistic applications.

The presentation in this chapter assumes that the evaluation measures (attributes) and alternatives have been specified for the decision problem. These tasks are considered in detail in Chapters 2 and 3. Other relevant references are Buede (1986), Keeney (1981), Keeney (1988), and Keller and Ho (1988). Corner and Kirkwood (1991) survey applications of multiattribute preference theory.

The results presented below are the most relevant for actual multiobjective decision problems. There is an enormous literature on preference theory, some of which is not applications oriented. The interested reader should consult the references to pursue this literature. The mathematical level of this chapter is higher than that of other chapters. You do not need to be familiar with this chapter to use the methods presented in the rest of the book.

9.1 Some Notation

The attributes (evaluation measures) for a decision problem are signified by $X = (X_1, X_2, \ldots, X_n)$, where X indicates the entire set of attributes. A specific *level* or *score* for attribute X_i is specified by x_i. For example, X_1 might represent the attribute cost, measured in dollars, and x_1 might represent the level \$1,000 for this attribute. Alternatives are specified by $a_1, a_2, \ldots, a_m$. We will restrict attention to situations where preferences are to be analyzed over the rectangular region $x_i^L \leq x_i \leq x_i^H$, $i = 1, 2, \ldots, n$, where x_i^L is the lowest level of X_i that is of interest, and x_i^H is the highest level of X_i that is of interest. In some cases, it will be useful to specify the *least preferred* level of X_i, and this will be signified by x_i^o. The *most preferred* level of X_i will be signified by x_i^*. With this notation, $x^o = (x_1^o, x_2^o, \ldots, x_n^o)$ designates the least preferred point in the rectangular region of interest, and $x^* = (x_1^*, x_2^*, \ldots, x_n^*)$ designates the most preferred point in that region.

Finally, $(x_i; \bar{x}_i)$ represents levels for all the attributes with particular attention called to the level of X_i. For example, $(x_i; \bar{x}_i^o) = (x_1^o, x_2^o, \ldots, x_{i-1}^o, x_i, x_{i+1}^o, \ldots, x_n^o)$. That is, $(x_i; \bar{x}_i^o)$ has all attributes except x_i at their least preferred levels. Analogously, $(x_i, x_j; \bar{x}_{ij})$ represents levels for all the attributes with particular attention called to the levels of X_i and X_j.

Definition 9.1. Monotonicity: *Preferences over X_i are* monotonic *if either higher levels of an attribute are always more preferred, or higher levels are always less preferred, regardless of the levels of the other attributes.*

While many of the results below hold in more general situations, it will suffice for our purposes to consider only cases where preferences are monotonic over each X_i for any fixed levels of the other attributes. (With this restriction, either $x_i^o = x_i^L$ and $x_i^* = x_i^H$, or $x_i^o = x_i^H$ and $x_i^* = x_i^L$.)

Definition 9.2. Indifference Curve: *An* indifference curve *is a set of points in $x_1, x_2, \ldots, x_n$ that are equally preferred. An indifference curve is also called an* isopreference curve.

9.2 Preferences under Certainty

This section considers situations where there is no uncertainty about the outcomes of the decision alternatives, and thus it presents theory relevant for the application methods in Chapter 4. If there is only one attribute, and if preferences over this attribute are monotonic, then there is not much difficulty making a decision: Simply pick the alternative with the highest level of the attribute if preferences are monotonically increasing over the attribute, or pick the alternative with the lowest level of the attribute if preferences are monotonically decreasing over the attribute.

However, if there are two or more attributes, the decision may be difficult even when there is no uncertainty about the outcomes of the decision alternatives. This is because you may have to consider *tradeoffs* among the attributes. For example, suppose you are considering possible changes to a semiconductor component manufacturing process, and the evaluation attributes are cost and percent yield of acceptable components from the process. It is likely that higher cost alternatives will also have a higher percent yield. The alternative that you prefer may vary, depending on how much you are willing to pay to reduce the defective component rate. That is, different tradeoffs between cost and percent yield may lead to different preferred alternatives.

Some Preliminary Notation and Results

Definition 9.3. Dominance: *An alternative a_1 dominates a second alternative a_2 if a_1 is at least as preferred as a_2 with respect to all the attributes and more preferred with respect to at least one attribute.*

If a_1 dominates a_2, then clearly a_1 is preferred to a_2. The fact that a_1 is preferred to a_2 will be designated by $a_1 \succ a_2$. Similarly, if a_1 and a_2 are equally preferred, this will be designated by $a_1 \sim a_2$.

Definition 9.4. Efficient Set: *The* efficient set *for a decision problem is the subset of the decision alternatives consisting of all the alternatives that are not dominated by another alternative. The efficient set is also called the* efficient frontier *or the* Pareto optimal set.

Clearly, the most preferred alternative for a decision problem will be in the efficient set.

Definition 9.5. Value Function: *A function $v(x)$ is a* value function *if it is true that $v(x') > v(x'')$ if and only if $x' \succ x''$, where x' and x'' are specified but arbitrary levels of x.*

Definition 9.6. Strategic Equivalence: *Two value functions are* strategically equivalent *if they give the same rank ordering for any set of alternatives. (A* rank ordering *of a set of alternatives is a list of the alteratives in decreasing order of preference. That is, the first alternative in the list is more preferred than all the rest, the second is more preferred than all the others except the first, and so forth.)*

It follows directly from the definitions that two strategically equivalent value functions must have the same indifference curves.

Theorem 9.7. Monotonicity of Strategically Equivalent Value Functions: *Value function $v''(x)$ is strategically equivalent to value function $v'(x)$ if and only if $v''(x)$ is a positive monotonic transformation of $v'(x)$.*

Proof. A positive monotonic transformation is a function $f(z)$ over the scalar z such that for any two levels z_1 and z_2, $z_1 > z_2$ implies that $f(z_1) > f(z_2)$. When $v''(x)$ is a positive monotonic transformation of $v'(x)$, it is clearly true that the two functions yield the same ranking of alternatives. The converse can be established by contradiction. Suppose that the transformation $v''(x) = f[v'(x)]$ between two value functions $v'(x)$ and $v''(x)$ is not monotonic. Then there must be values of x (designated x' and x'') such that $v'(x') > v'(x'')$ but $f[v'(x')] \leq f[v'(x'')]$. However, since $v''(x) = f[v'(x)]$, this means that $v''(x') \leq v''(x'')$, and hence v'' does not give the same ranking as v'. This is a contradiction of the theorem conditions, and hence f must be monotonic. ∎

Definition 9.8. Additive Value Function: *Value function $v(x)$ is called an additive value function if it is strategically equivalent to a value function of the form*

$$v(x) = \sum_{i=1}^{n} \lambda_i v_i(x_i) \qquad (9.8.1)$$

for some functions $v_i(x_i)$ and constants λ_i.

Note from Theorem 9.7 that for any additive value function of the form (9.8.1), it is possible to construct a strategically equivalent additive value function with $v(x^o) = 0$, $v(x^*) = 1$, $v_i(x_i^o) = 0$, and $v_i(x_i^*) = 1$, for all i. When the function is in this form, it must be true that $\lambda_i = v(x_i^*; \bar{x}_i^o)$, and $\sum_{i=1}^{n} \lambda_i = 1$. This additive value function is what is most commonly used in practice, and, in particular, this is the form used in Chapter 4. [The λ_i are called *scaling constants*, *swing weights*, or *weights*. The $v_i(x_i)$ are called *single attribute value functions* or *single dimensional value functions*.]

Definition 9.9. Normalized Additive Value Function: *An additive value function is called* normalized *if it is in the form specified in the preceding paragraph.*

The theoretical basis for this form of value function is developed during the remainder of this section.

The Two-attribute Case

Once a value function has been specified, the evaluation of alternatives is straightforward: For each alternative, determine levels for each attribute, and then calculate the value for each alternative using the value function. The alternative with the highest value is most preferred. A difficulty in implementing this procedure is in determining the value function. As shown below, this function must be determined by questioning the decision maker. Even with only two evaluation attributes, the process of determining a general function $v(x_1, x_2)$ can be difficult. With more than two attributes, it is almost impossible without making some assumptions that simplify the form of the value function.

The value function form that has been generally used in practice is the additive form $v(x_1, x_2) = \lambda_1 v_1(x_1) + \lambda_2 v_2(x_2)$. Another form that might seem intuitively appealing is a multiplicative form $v(x_1, x_2) = v_1'(x_1) v_2'(x_2)$. However, there is always an additive form that is strategically equivalent to any multiplicative form. Simply apply the positive monotonic transformation $f(z) = \log(z)$ to the multiplicative value function to obtain a strategically equivalent additive value function.

This section presents the condition necessary for an additive value function to be valid when there are two evaluation attributes. Then a procedure is presented for assessing an additive value function. The condition necessary for an additive value function when there are two attributes is somewhat obscure. The proof that this condition leads to an additive value function is even more obscure. The condition requires examination of a variety of different levels for X_1 and X_2. In order to keep the discussion of these different levels straight without resorting

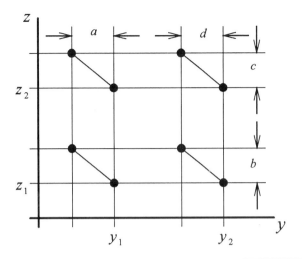

Figure 9.1   *Corresponding tradeoffs condition*

to subscripts on subscripts, the notation $Y = X_1$ and $Z = X_2$ will be used for the remainder of this section. Note in particular that y_1, y_2, etc., will designate different levels of attribute Y rather than different attributes. Similarly, z_1, z_2, etc., will designate different levels of attribute Z.

Definition 9.10. Corresponding Tradeoffs Condition: *Two attributes Y and Z obey the* corresponding tradeoffs condition *if the following is true: For any y_1, y_2, z_1, and z_2, if $(y_1, z_1) \sim (y_1 - a, z_1 + b)$ and $(y_2, z_1) \sim (y_2 - d, z_1 + b)$, then for the c such that $(y_1, z_2) \sim (y_1 - a, z_2 + c)$ it is true that $(y_2, z_2) \sim (y_2 - d, z_2 + c)$.*

Figure 9.1 illustrates the corresponding tradeoffs condition, and we will shortly show that this condition must hold to have an additive value function. However, showing that this condition implies an additive value function is not straightforward. We will restrict ourselves to showing that this is true for a grid of points over $\{Y, Z\}$, which are defined by the combinations of $y = y_i, i = 0, 1, 2, \ldots, n$, and $z = z_j, j = 0, 1, 2, \ldots, n$. As we shall see, y_0, z_0, and either y_1 or z_1 can be specified arbitrarily. The remainder of the points in the grid are then determined by the procedure shown below. By making y_1 as close to y_0 as desired (or z_1 as close to z_0 as desired), it is possible to make this grid of points as fine as desired. Thus, we can show that the corresponding tradeoffs condition implies that an additive value function holds for an arbitrarily dense set of points over $\{Y, Z\}$. Showing that this implies an additive value function for all (y, z) requires some continuity conditions on $v(y, z)$. See Keeney and Raiffa (1976), Section 3.4, for additional discussion and references.

Lemma 9.11. Corresponding Tradeoffs: *For y_0, z_0, y_1, and z_1 such that $(y_1, z_0) \succ (y_0, z_0)$ and $(y_0, z_1) \sim (y_1, z_0)$, if the corresponding tradeoffs condition holds, then there exist $y_2, y_3, \ldots, y_n$, and $z_2, z_3, \ldots, z_n$, for any n, such that*

$$(y_i, z_{k-i}) \sim (y_{i+1}, z_{k-i-1}); \quad i = 0, 1, \ldots, k-1; \; k = 2, 3, \ldots, n \qquad (9.11.1)$$

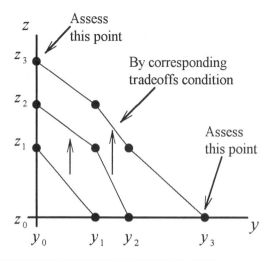

Figure 9.2 *Corresponding tradeoffs lemma*

Proof. The proof proceeds by induction, and is illustrated in Figure 9.2. Arbitrarily select y_1 and z_1 as specified in the lemma statement. Then determine the y_2 and z_2 such that $(y_2, z_0) \sim (y_1, z_1) \sim (y_0, z_2)$. With this definition of y_2 and z_2, condition 9.11.1 holds for $n = 2$. We will now show that if condition 9.11.1 holds for $k = 2, 3, \ldots, n$, it also holds for $k = n + 1$. This suffices to prove the desired result.

Replace i with $i - 1$ and k with $n - 1$ in condition 9.11.1. Then it follows by substitution into condition 9.11.1 that

$$(y_{i-1}, z_{(n-1)-(i-1)}) \sim (y_i, z_{(n-1)-(i-1)-1}); \quad i = 1, 2, \ldots, n - 1 \qquad (9.11.2)$$

(Note that $i = 0$ does not yield a valid relationship in condition 9.11.1.) Replace k with n in condition 9.11.1. Then it follows by substitution into condition 9.11.1 that

$$(y_i, z_{n-i}) \sim (y_{i+1}, z_{n-i-1}); \quad i = 0, 1, \ldots, n - 1 \qquad (9.11.3)$$

Finally, replace i with $i - 1$ and k with n in condition 9.11.1. Then it follows by substitution into condition 9.11.1 that

$$(y_{i-1}, z_{n-(i-1)}) \sim (y_i, z_{n-(i-1)-1}); \quad i = 1, 2, \ldots, n - 1 \qquad (9.11.4)$$

However, condition 9.11.4 is just the same as condition 9.11.2 except that (1) z has one higher index, and (2) the range of indices for condition 9.11.2 goes only to $i = n - 1$. Thus, it follows from the corresponding tradeoffs condition that if condition 9.11.3 has z shifted to one higher index, it must still hold for $i = 1, 2, \ldots, n - 1$. Hence,

$$(y_i, z_{(n+1)-i}) \sim (y_{i+1}, z_{(n+1)-i-1}); \quad i = 1, 2, \ldots, n - 1 \qquad (9.11.5)$$

However, this is just condition 9.11.1 with $k = n+1$, except that the upper limit on i in condition 9.11.5 needs to be n rather than $n - 1$, and the lower limit needs to be 0 rather than 1. To fill in the cases where $i = n$ and $i = 0$, find the y_{n+1} such that $(y_{n+1}, z_o) \sim (y_n, z_1)$ and the z_{n+1} such that $(y_o, z_{n+1}) \sim (y_1, z_n)$. Such values will exist, and hence the limits on i in condition 9.11.6 can be extended to $i = 0, 1, \ldots, n$. This establishes the induction. ∎

Theorem 9.12.   Additive Value Function: *The corresponding tradeoffs condition holds if and only if there exists an additive value function* $v(y, z) = v_Y(y) + v_Z(z)$.

Proof. It is easy to show that the corresponding tradeoffs condition holds for an additive value function, The converse is less straightforward. However, this has essentially been established by Lemma 9.11. We will use the result of that lemma to construct an additive value function. Let $v_Y(y_i) = i$ and $v_Z(z_j) = j$. Then the value function $v(y, z) = v_Y(y_i) + v_Z(z_j)$ yields the indifference relations shown in condition 9.11.1. (Note that while this is an additive value function, there may be nonadditive value functions that are strategically equivalent to it.) This proof applies only at the grid of points considered in condition 9.11.1. However, we can make this grid arbitrarily dense over $\{Y, Z\}$. ∎

Assessing a Two-attribute Additive Value Function

In concept, the procedure used to prove Lemma 9.11 could be used to determine an additive value function. This assessment procedure is called the *Lock-Step Procedure*, and Keeney and Raiffa (1976), Section 3.4.6, discuss it in further detail. However, in practice, an alternative approach called the *Midvalue Splitting Technique* is more commonly used. A preliminary definition and result are useful to present this procedure.

This section shifts back to our standard notation and uses X_1 and X_2 to designate the two attributes in the decision problem.

Definition 9.13.   Midvalue: *When an additive value function is valid, the midvalue of an interval* $[x'_i, x''_i]$ *is the level* x^m_i *such that, starting from a specified level of another attribute, the decision maker would give up the same amount of that other attribute to improve* x_i *from* x'_i *to* x^m_i *as to improve* x_i *from* x^m_i *to* x''_i. *(Note that this definition implicitly assumes that preferences are monotonically increasing over* x_i. *If preferences are monotonically decreasing, then the same amount would be given up to improve* x_i *from* x''_i *to* x^m_i *as from* x^m_i *to* x'_i.*)*

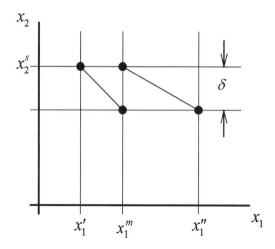

Figure 9.3   *Midvalue*

Theorem 9.14. Value Averaging: *If x_i^m is the midvalue of $[x_i', x_i'']$, then it is true that $v_i(x_i^m) = [v_i(x_i') + v_i(x_i'')]/2$.*

Proof. For simplicity, assume that we are interested in the midvalue for attribute X_1. The proof can easily be extended to consider X_2. Let x_2^s designate the "specified starting level" of X_2 that is used in the definition of the midvalue. From the definition of the midvalue, it follows that there exists a δ such that $(x_1', x_2^s) \sim (x_1^m, x_2^s - \delta)$ and $(x_1^m, x_2^s) \sim (x_1'', x_2^s - \delta)$. (See Figure 9.3.) Thus,

$$\lambda_1 v_1(x_1') + \lambda_2 v_2(x_2^s) = \lambda_1 v_1(x_1^m) + \lambda_2 v_2(x_2^s - \delta)$$
$$\lambda_1 v_1(x_1^m) + \lambda_2 v_2(x_2^s) = \lambda_1 v_1(x_1'') + \lambda_2 v_2(x_2^s - \delta)$$

Subtracting one of these equations from the other leads to the desired result. Note that the specified starting point x_2^s does not affect the result, and hence can be selected arbitrarily. ∎

A normalized additive value function can be assessed as follows. [See Keeney and Raiffa (1976), Section 3.4.8, for a more extensive discussion of this assessment procedure.]

Step 1   Determine the attributes X_1 and X_2 together with x_1^o, x_1^*, x_2^o, and x_2^*. In what follows, we assume that (x_1^*, x_2^o) is at least as preferred as (x_1^o, x_2^*). The attributes can always be defined so this is true, and it is assumed for notational simplicity in presenting the results.

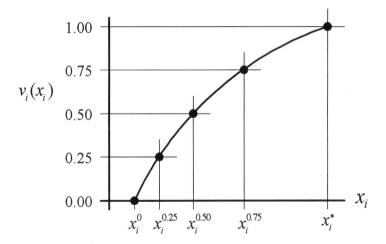

Figure 9.4   *Single attribute value function*

Step 2   Confirm or assume that the corresponding tradeoffs condition holds, so that an additive value function is valid.

Step 3   Determine the midvalue $x_i^{0.50}$ of $[x_i^o, x_i^*]$, the midvalue $x_i^{0.25}$ of $[x_i^o, x_i^{0.50}]$, and the midvalue $x_i^{0.75}$ of $[x_i^{0.50}, x_i^*]$, for $i = 1, 2$. From Definition 9.9 for the normalized additive value function, it follows that $v_i(x_i^o) = 0$ and $v_i(x_i^*) = 1$. Then, from Theorem 9.14, it follows that $v_i(x_i^{0.25}) = 0.25$, $v_i(x_i^{0.50}) = 0.50$, and $v_i(x_i^{0.75}) = 0.75$. This gives five points for the function $v_i(x_i)$. The entire function can generally be adequately approximated by drawing a curve through these points. (See Figure 9.4.) [An alternative to Step 3, which yields an analytic representation for $v_i(x_i)$, is given below following Theorem 9.16.]

Step 4   Determine the level x_1' such that $(x_1', x_2^o) \sim (x_1^o, x_2^*)$. (See Figure 9.5.) Then, from the definition of the normalized additive value function, $\lambda_1 v_1(x_1') = \lambda_2$. This equation, together with the condition that $\lambda_1 + \lambda_2 = 1$, determines the λ_i.

The steps above yield the value function. However, the single attribute value functions $v_i(x_i)$ are available only in graphical form. For computational ease, it is sometimes useful to have analytic forms for these functions. Kirkwood and Sarin (1980) have investigated conditions that lead to analytic forms for $v_i(x_i)$, and one of these conditions is discussed now.

Definition 9.15. Constant Tradeoff Attitude: *Assume that an additive value function is valid. Constant tradeoff attitude holds for attribute X_i if it is true that whenever x_i^m is the midvalue of $[x_i', x_i'']$, then $x_i^m + \delta$ is the midvalue of $[x_i' + \delta, x_i'' + \delta]$, for any δ. [Kirkwood and Sarin (1980) call this condition the delta property.]*

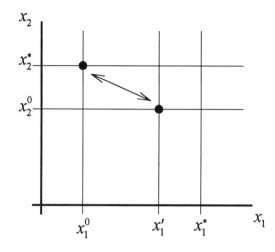

Figure 9.5   *Assessing the* λ_i

Theorem 9.16. Exponential Single Attribute Value Functions: *Assume that an additive value function is valid. Constant tradeoff attitude holds for X_i if and only if, for monotonically increasing preferences over X_i,*

$$v_i(x_i) = \begin{cases} \dfrac{1 - \exp\left[-(x_i - x_i^L)/\rho_i\right]}{1 - \exp\left[-(x_i^H - x_i^L)/\rho_i\right]}, & \rho_i \neq \text{Infinity} \\[2ex] \dfrac{x_i - x_i^L}{x_i^H - x_i^L}, & \text{otherwise} \end{cases} \tag{9.16.1}$$

or for monotonically decreasing preferences over X_i,

$$v_i(x_i) = \begin{cases} \dfrac{1 - \exp\left[-(x_i^H - x_i)/\rho_i\right]}{1 - \exp\left[-(x_i^H - x_i^L)/\rho_i\right]}, & \rho_i \neq \text{Infinity} \\[2ex] \dfrac{x_i^H - x_i}{x_i^H - x_i^L}, & \text{otherwise} \end{cases} \tag{9.16.2}$$

where x_i^L is the lowest level of X_i that is of interest, x_i^H is the highest level of interest, and ρ_i is called the exponential constant *for the single attribute value function. The value function is scaled so that it varies between 0 and 1 over the range from $x_i = x_i^L$ to $x_i = x_i^H$. That is, for monotonically increasing preferences, $v_i(x_i^L) = 0$ and $v_i(x_i^H) = 1$, while for monotonically decreasing preferences $v_i(x_i^L) = 1$ and $v_i(x_i^H) = 0$.*

Proof. It is easy to show that the specified forms lead to constant tradeoff attitude. The following argument shows that constant tradeoff attitude leads to the specified exponential forms.

If constant tradeoff attitude holds, then from Theorem 9.14,

$$v_i(x_i^m + \delta) = [v_i(x_i' + \delta) + v_i(x_i'' + \delta)]/2 \tag{9.16.3}$$

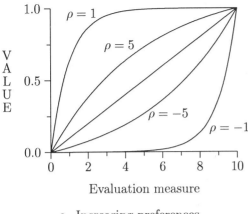

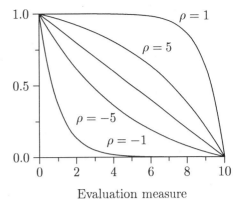

a. Increasing preferences

b. Decreasing preferences

Figure 9.6   *Example exponential value functions*

For notational convenience, let $x = (x'_i + x''_i)/2$, $h = x''_i - x = x - x'_i$, and $\pi = x - x^m_i$. Then equation 9.16.3 can be rewritten as

$$v_i(x - \pi + \delta) = [v_i(x - h + \delta) + v_i(x + h + \delta)]/2 \qquad (9.16.4)$$

Now consider the case where x'_i and x''_i approach x, thereby forcing h and π to approach zero in the limit. Performing a Taylor series expansion of equation 9.16.4 around $x + \delta$, and keeping only the first terms that do not cancel out, results in $\pi = -(1/2)[v''_i(x + \delta)/v'_i(x + \delta)]h^2$, where v'_i and v''_i represent the first and second derivatives, respectively. Substituting $x_i = x + \delta$ leads to $\pi = -(1/2)[v''_i(x_i)/v'_i(x_i)]h^2$. Since π must be a constant when constant tradeoff attitude holds, this differential equation yields the equations 9.16.1 and 9.16.2 when the scaling conventions for $v_i(x_i)$ are imposed. ∎

Examples of exponential value functions are shown in Figure 9.6. This theorem can be used to replace Step 3 in the above assessment procedure with the following:

Step 3′   Determine or assume that constant tradeoff attitude holds for x_i. Then determine the midvalue $x^{0.50}_i$ of $[x^o_i, x^*_i]$. The value of ρ_i can be determined by solving $v_i(x^{0.50}_i) = [v_i(x'_i) + v_i(x''_i)]/2$ using either equation 9.16.1 or equation 9.16.2, as appropriate, for $v_i(x_i)$.

It follows directly from equations 9.16.1 and 9.16.2 that when preferences are monotonically increasing over x_i, then

1 $x^m_i < (x'_i + x''_i)/2$ implies that $\rho_i > 0$.
2 $x^m_i = (x'_i + x''_i)/2$ implies that $\rho_i =$ Infinity.
3 $x^m_i > (x'_i + x''_i)/2$ implies that $\rho_i < 0$.

When preferences are monotonically decreasing over x_i, the direction of the first inequalities in (1) and (3) are reversed.

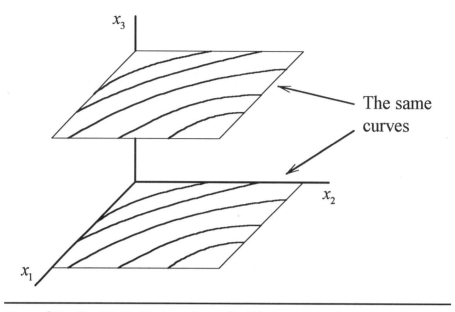

Figure 9.7   *Preferential independence (indifference curves)*

The N-attribute Case

The corresponding tradeoffs condition required for the additive value function to hold in the two-attribute case is not particularly intuitive. In situations with more than two attributes, the conditions required for an additive value function are more intuitively appealing. However, the proofs that lead to the additive value function form are even less obvious than in the two-attribute case.

Definition 9.17. Preferential Independence: *Suppose that Y and Z are a partition of $\{X_1, X_2, \ldots, X_n\}$. (That is, each X_i is included in exactly one of Y or Z.) Then Y is* preferentially independent *of Z if the rank ordering of alternatives that have common levels for all attributes in Z does not depend on these common levels. (The common levels do not have to be the same for different attributes, but the level of each X_i in Z is the same for all alternatives.)*

Note that when preferential independence holds, then the indifference curves over Y for fixed levels of Z will be the same regardless of the level of Z. Figure 9.7 illustrates this for a three-attribute case where $\{X_1, X_2\}$ is preferentially independent of $\{X_3\}$.

Definition 9.18. Mutual Preferential Independence: *A set of attributes $\{X_1, X_2, \ldots, X_n\}$ displays* mutual preferential independence *if Y is preferentially independent of Z for every partition $\{Y, Z\}$ of $\{X_1, X_2, \ldots, X_n\}$.*

Theorem 9.19.   Pairwise Preferential Independence: *Let $Y_k = \{X_1, X_k\}, k = 2, 3, \ldots, n$, and let Z_k be all the attributes that are not in Y_k. Then $\{X_1, X_2, \ldots, X_n\}$ will have mutual preferential independence if Y_k is preferentially independent of Z_k for $k = 2, 3, \ldots, n$.*

The proof of this result is not straightforward. It follows from Keeney and Raiffa (1976), Theorem 3.7, and the interested reader should consult that result and the references given there. Also, see Section 9.8 for a proof in the three-attribute case.

Theorem 9.19 is useful in practice because it shows that determining whether mutual preferential independence holds requires only testing the preferential independence of $n - 1$ pairs of attributes. Thus, the required number of tests grows only linearly with the number of attributes. Theorem 9.19 says that if preferential independence holds for the specified pairs of attributes, then mutual independence must hold. That is, there is no possible value function that obeys the specified pairwise preferential independence conditions that does not also obey mutual preferential independence.

The reason that preferential independence is important is shown by the following theorem:

Theorem 9.20. Additive Value Function: *If $\{X_1, X_2, \ldots, X_n\}, n \geq 3$, have mutual preferential independence, then $v(x_1, x_2, \ldots, x_n)$ is additive.*

The proof of this is not intuitive. Keeney and Raiffa (1976) give a plausibility discussion for the three-attribute case in Section 3.5.3 and state the result in Theorem 3.6. Note also that Aczél (1966) discusses several different results from functional equation theory that imply Theorem 9.20 for the three-attribute case. See Section 9.7 for additional discussion of this approach to proving the theorem.

Theorem 9.20 shows that you do not have to consider the corresponding tradeoffs condition to establish that a value function is additive when there are three or more attributes. However, it is straightforward to show by substitution into the additive value function that condition 9.11.1 is true with an additive value function, and hence the corresponding tradeoffs condition holds between any two attributes. That is, the preferential independence conditions that are required to establish an additive value function also establish that the corresponding tradeoffs condition holds.

Assessing an N-attribute Value Function

The normalized n-attribute additive value function can be assessed by the following procedure. [See Keeney and Raiffa (1976), Section 3.7, for a more detailed discussion.]

Step 1   Determine the attributes $X_i, i = 1, 2, \ldots, n$, together with x_i^o and x_i^*. In what follows, assume that $(x_i^*; \bar{x}_i^o)$ is at least as preferred as $(x_j^*; \bar{x}_j^o)$ for all $j > i$. The attributes can always be defined so this is true, and it is assumed for notational simplicity in presenting the results.

Step 2   Confirm or assume that $\{X_1, X_2, \ldots, X_n\}$ obey mutual preferential independence, so that an additive value function is valid. From Theorem 9.19, it suffices to consider only pairs of attributes $\{X_1, X_j\}, j = 2, 3, \ldots, n$, to establish this result. The usual test of this condition for any specified $\{X_1, X_j\}$ proceeds as follows: Specify convenient levels $\bar{x}_{1j}$ for all the $X_i, i \neq 1, j$. Then determine the

x_1' such that $(x_1', x_j^o; \bar{x}_{1j}) \sim (x_1^o, x_j^*; \bar{x}_{1j})$ Repeat this process for several different levels of $\bar{x}_{1j}$. If the same x_1' is found in each case, then this is taken as establishing that $\{X_1, X_j\}$ is preferentially independent of the remaining attributes.

Step 3   Determine the midvalue $x_i^{0.50}$ of $[x_i^o, x_i^*]$, the midvalue $x_i^{0.25}$ of $[x_i^o, x_i^{0.50}]$ and the midvalue $x_i^{0.75}$ of $[x_i^{0.50}, x_i^*]$. From the definition of the normalized additive value function (Definition 9.9), it follows that $v_i(x_i^o) = 0$ and $v_i(x_i^*) = 1$. Then, from Theorem 9.14, it follows that $v_i(x_i^{0.25}) = 0.25$, $v_i(x_i^{0.50}) = 0.50$, and $v_i(x_i^{0.75}) = 0.75$. This gives five points for the function $v_i(x_i)$. The entire function can generally be adequately approximated by drawing a curve through these points. [As an alternative to this step, constant tradeoff attitude, as discussed earlier, can be assumed or tested to establish that $v_i(x_i)$ has an exponential form. Although Theorems 9.14 and 9.16 were established only for the two-attribute case, it is straightforward to show that these also hold for the n-attribute case.]

Step 4   Determine the levels x_i' such that $(x_i', x_{i+1}^o; \bar{x}_{ij}) \sim (x_i^o, x_{i+1}^*; \bar{x}_{ij}), i = 1, 2, \ldots, n-1$. Then, from the definition of the normalized additive value function, $\lambda_i v_i(x_i') = \lambda_{i+1}, i = 1, 2, \ldots, n-1$. These equations, with the condition that $\sum_{i=1}^n \lambda_i = 1$, determine the λ_i.

The additive value function has been widely used in practice. However, the formal assessment procedure specified above is not often used. [An exception is the Somerstown case discussed in Keeney and Raiffa (1976), Section 7.2.] More informal procedures usually include two steps: (1) The decision maker is asked to directly draw the functions $v_i(x_i)$, or these are assumed to be linear, and (2) the decision maker is asked to numerically specify the "relative importance" of swinging each attribute from its least preferred to its most preferred level. See von Winterfeldt and Edwards (1986), Chapter 11, for a discussion of the reasons that these informal assessment procedures often give adequately accurate value functions. The theory of *measurable value functions*, which is reviewed below, provides a formal basis for these informal procedures.

One point about these approximate procedures is particularly important: The scaling constants (weights) λ_i depend on the *ranges of variation* $[x_i^L, x_i^H]$ for each attribute. Thus, if these ranges are changed, *even if there is no change in the decision maker's preferences,* the scaling constants will change. An intuitive demonstration of this is given by the following example: Consider a production decision that will impact cost and quality. In the abstract, cost may be more important than quality. However, suppose that all the alternatives have exactly the same cost. Then clearly cost has no impact on the decision. If the range of variation in cost is small, it may still have a limited impact on the decision, while if the range of variation in cost is large, it may have a large impact.

Here is a simple example that illustrates how weights change with changes in the range of attributes. Suppose that there are two attributes X_1 and X_2, each having a range from 0 to 10 and monotonically increasing preferences. Suppose further that the value functions are linear over this range, so that $v_i(x_i) = x_i/10$. Finally, suppose that $\lambda_1 = \lambda_2 = 0.5$, so that $v(x_1, x_2) = 0.50 \times (x_1/10) + 0.50 \times (x_2/10)$.

Suppose that it is discovered during the process of collecting data for the actual alternatives in the decision problem that, for these actual alternatives, X_2 never exceeds 5. Therefore, it is decided to change the range for X_2 to go from 0 to 5 with no changes in the preferences of the decision maker. Then, from the scaling conventions for a normalized additive value function, the new single attribute value function over X_2 is $v_2'(x_2) = x_2/5$. What happens to the λ_i? From the original value function, we know that $(5,0) \sim (0,5)$. Therefore, from Step 4 of the value function assessment procedure given above, if the rescaled value function is used, $\lambda_1' v_1(5) = \lambda_2'$ or $\lambda_1' \times (5/10) = \lambda_2'$. Hence, $0.5\lambda_1' = \lambda_2'$, so that $\lambda_1' = 2/3$ and $\lambda_2' = 1/3$. Therefore, the rescaled value function is $v'(x_1, x_2) = (2/3) \times (x_1/10) + (1/3) \times (x_2/5)$.

This shows that changing the range of variation for X_2 has changed the scaling constants, even though there has been no change in the decision maker's preferences. [To emphasize that the scaling constants depend on the range, these constants are sometimes referred to as *swing weights*. This terminology comes from the fact that λ_i is equal to the increase in $v(x_1, x_2, \ldots, x_n)$ that results from moving ("swinging") x_i from x_i^o to x_i^*.]

Measurable Value Functions

It seems intuitively reasonable to make a statement like "the increase in value when going from an asset position of \$0 to an asset position of \$1 million is greater than the increase when going from \$1 million to \$2 million." (That is, the first million dollars has a greater value than the second million dollars. You may not agree with this sentiment, but the statement seems to make sense.)

The value functions presented so far do not allow us to analyze such *differences* in value between two outcomes. A value function $v(x)$ will show that \$2 million is preferred to \$1 million, which is in turn preferred to \$0, but $v(x)$ does not say anything about the *differences* in values between these different amounts of money. This is because, as Theorem 9.7 shows, a value function is determined only to within a positive monotonic transformation. Thus, by applying the appropriate positive monotonic transformation, we can convert the difference in value $v(x') - v(x'')$ between any two outcomes x' and x'' into any number that is desired, as long as the transformed value has the same sign as $v(x') - v(x'')$. Hence, differences in value as measured by $v(x)$ are not meaningful.

Dyer and Sarin (1979) have studied conditions necessary for differences in value to have meaning, and this section presents some results that are most relevant for decision analysis practice. The presentation is informal, and the interested reader should consult Dyer and Sarin's article for discussion of some technical matters related to proving the results given below. For notational convenience, $s = (s_1, s_2, \ldots, s_n)$, $t = (t_1, t_2, \ldots, t_n)$, $y = (y_1, y_2, \ldots, y_n)$, and $z = (z_1, z_2, \ldots, z_n)$ will represent distinct outcomes. Further, $st \succ^* yz$ will be used to mean that the value difference between t and s is greater than the value difference between z and y, where $s \succ t$ and $y \succ z$. Equivalently, we can say that the *strength of preference* for s over t is greater than the strength of preference for y over z. Similarly, $st \sim^* yz$ will be used to mean that the strength of preference for s over t is equal to the strength of preference for y over z.

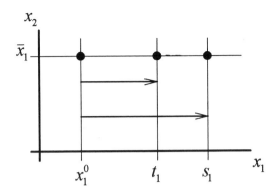

Figure 9.8   *Difference consistency*

Dyer and Sarin note that it is possible to define a set of preference axioms such that if $s \succ t$ and $y \succ z$, then $st \succ^* yz$ if and only if $\overset{m}{v}(s) - \overset{m}{v}(t) > \overset{m}{v}(y) - \overset{m}{v}(z)$ for some function $\overset{m}{v}(x)$, where $\overset{m}{v}(x)$ is called a *measurable value function*. Furthermore, $\overset{m}{v}(x)$ is determined to within a positive linear transformation, rather than only to a positive monotonic transformation as is the case for the value functions we have studied in earlier sections. That is, if $\overset{m}{v}(x)$ is a measurable value function, then another measurable value function $\overset{m}{v}'(x)$ will give the same rank ordering of value differences as $\overset{m}{v}(x)$ only if $\overset{m}{v}'(x) = a\overset{m}{v}(x) + b$ for some constants $a > 0$ and b.

We will now examine the conditions under which $\overset{m}{v}(x)$ is additive, and the relationship between $\overset{m}{v}(x)$ and the value function $v(x)$ that we that already studied. Finally, we will examine how measurable value functions can be used in practice.

Definition 9.21. Difference Consistency: *The mutually preferentially independent attributes* $X_1, X_2, \ldots, X_n$ *are* difference consistent *if, for* $i = 1, 2, \ldots, n$, $(s_i; \bar{x}_i) \succ (t_i; \bar{x}_i)$ *if and only if* $(s_i; \bar{x}_i)(x_i^o; \bar{x}_i) \succ^* (t_i; \bar{x}_i)(x_i^o; \bar{x}_i)$ *for all* $s_i, t_i, \bar{x}_i$.

Informally, this condition says that the preferability of different levels of an attribute X_i can be ranked by comparing the value differences between each level and the least desirable level x_i^o. (See Figure 9.8.) Dyer and Sarin comment that this condition seems so intuitively appealing that it can be assumed to hold in most practical applications without testing.

Definition 9.22. Difference Independence: *An attribute* X_i *is* difference independent *of the remaining attributes if it is true that when* $(s_i; \bar{s}_i) \succ (t_i; \bar{s}_i)$ *for some* $\bar{s}_i$, *then* $(s_i; \bar{s}_i)(t_i; \bar{s}_i) \sim^* (s_i; \bar{x}_i)(t_i; \bar{x}_i)$ *for any* $\bar{x}_i$.

That is, this condition says that the value difference between two levels t_i and s_i of attribute X_i does not depend on the levels of the other attributes. When this condition holds, the form of the measurable value function is particularly simple, as the following theorem shows. (See Figure 9.9.)

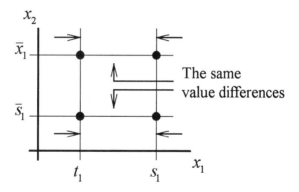

Figure 9.9   *Difference independence*

Theorem 9.23. Additive Measurable Value Function: *Assume X_1, X_2, $\ldots$, X_n, $n \geq 3$, are mutually preferentially independent and difference consistent. Assume further that X_1 is difference independent of the remaining attributes. Then $\overset{m}{v}(x)$ can be written in the normalized additive form*

$$\overset{m}{v}(x) = \sum_{i=1}^{n} \lambda_i \overset{m}{v}_i(x_i) \tag{9.23.1}$$

where $\overset{m}{v}(x^o) = 0$, $\overset{m}{v}(x^*) = 1$, $\overset{m}{v}_i(x_i^o) = 0$, and $\overset{m}{v}_i(x_i^*) = 1$, for all i, and furthermore, $\overset{m}{v}(x)$ can be used to rank-order outcomes.

Proof. This is implied by Theorem 1 in Dyer and Sarin (1979). ∎

When the conditions for Theorem 9.23 hold, then it is possible to assess an additive value function using value difference questions. These questions seem to be more intuitively understandable for many people than the questions required for the assessment procedure given earlier. Using the value difference approach, Steps 2 through 4 of that earlier procedure are replaced by the following.

Step 2′   Confirm or assume that $X_1, X_2, \ldots, X_n$ obey mutual preferential independence and difference consistency, and also confirm or assume that X_1 is difference independent of the other attributes. We have already discussed tests for preferential independence. Difference consistency can be assumed without testing in most cases. A test for difference independence is to consider the value difference between the two extreme outcomes x_1^o and x_1^* for several different levels of $\bar{x}_1$. If the value difference is the same for the various levels of $\bar{x}_1$, then this can be taken as establishing difference independence. (Note that while it is necessary to test only for the value independence of X_1, it follows from Theorem 9.23 that all the X_i must be value independent of the remaining attributes when the conditions of the theorem hold.)

Step 3′   Determine the $\overset{m}{v}_i(x_i)$. As an alternative to the midvalue splitting technique discussed earlier, the relative value increments between different attribute scale levels can be directly assessed and used to determine $\overset{m}{v}_i(x_i)$. For example, suppose that X_i has the possible levels 0, 1, and 2, and that higher levels are more preferred. If the value increment going from 0 to 1 is twice as great as the value increment going from 1 to 2, then it must be true that $\overset{m}{v}_i(1) = 2/3$ since $\overset{m}{v}_i(0) = 0$ and $\overset{m}{v}_i(2) = 1$.

Step 4′   Determine the λ_i. As an alternative to the procedure given earlier, direct assessments can be made of the relative value increments going from the least preferred level to the most preferred level of each attribute. These can then be used to determine the λ_i. As an example, for a three-attribute problem, assume that the value increment between x_1^o and x_1^* is equal to the value increment between x_2^o and x_2^*, and that this value increment is twice as great as the value increment between x_3^o and x_3^*. Then it follows from the normalized additive value function form that $\lambda_1 = \lambda_2$ and $\lambda_2 = 2 \times \lambda_3$. Since $\lambda_1 + \lambda_2 + \lambda_3 = 1$, then $\lambda_1 = \lambda_2 = 0.4$ and $\lambda_3 = 0.2$.

Based on my own experience with this type of questioning, it is clear that people can answer the questions in Steps 2′, 3′, and 4. However, someone who gives careful thought to these questions may be left a little uneasy about what the answers mean. For example, what does it mean to say that the value increment between x_2^o and x_2^* is twice as great as the value increment between x_3^o and x_3^*? Neither of these increments is anything that someone can actually obtain in a decision problem—in a decision problem, you obtain *outcomes*, rather than increments between outcomes.

There is not so much of a conceptual difficulty with the questions for the previous procedure. All of those questions relate to outcomes that are at least conceptually obtainable, and a decision maker faced with one of the situations addressed in the questions could actually implement the answers he or she had given to the questions. On the other hand, suppose that someone says the value increment between $0 and $1,000 is greater than the value increment between $1 million and $1 million plus $10. While this may be true, it is also probably true that the decision maker would prefer to have either the $1 million or the $1 million plus $10, rather than either the $0 or the $1,000. Thus, what is the operational meaning of the answer to the value increment question in an actual decision situation?

Nonetheless, the questions seem meaningful to practical decision makers who are not familiar with the issues raised in the last couple of paragraphs. The answers they give to the questions seem to provide reasonable information about their preferences. Even if these answers are viewed as only approximations to the answers that would be provided to the assessment questions in the earlier procedure, they are useful approximations because they reduce the need to address the relatively complex questions in that earlier procedure.

Dyer and Sarin (1979) present a number of other value difference conditions that lead to forms of measurable value functions other than the additive. These do not appear to have been applied to date.

9.3 Preferences under Uncertainty

This section reviews theory for evaluating alternatives in decisions where there is uncertainty about the specific consequence that will result from selecting a particular alternative. This theory underlies the practical analysis methods presented in Chapters 6 and 7. This theory assumes that expected utility is used as the criterion to rank-order alternatives under uncertainty. Section 9.8 presents the theoretical basis for using expected utility as a decision criterion.

Using expected utility as a decision criterion requires that a *utility function* be assessed over the evaluation attributes. Even with only one attribute, the direct assessment of a utility function can take some effort. With the addition of more attributes, it becomes increasingly difficult to directly assess a utility function. To reduce this difficulty, conditions are examined that restrict the form of the utility function in a way that allows straightforward assessment of the function.

Some Preliminary Results

Definition 9.24. Strategic Equivalence: *Two utility functions are strategically equivalent if they give the same rank ordering for any set of alternatives.*

Theorem 9.25. Linear Transformation Property for Strategic Equivalence: *Two utility functions* $u'(x)$ *and* $u''(x)$ *over the consequences* x *for a decision problem are strategically equivalent if and only if*

$$u''(x) = au'(x) + b \qquad (9.25.1)$$

for some constants $a > 0$ *and* b. *In the decision analysis literature, this is referred to as the* linear transformation property *of utility functions. (Note, however, that the formal mathematical name for the transformation in equation 9.25.1 is a* positive affine transformation. *The special case of this where* $b = 0$ *is a* linear transformation.)

The proof of this follows directly from Theorem 9.44 in Section 9.9. See also Keeney and Raiffa (1976), Section 4.3.2.

Definition 9.26. Certainty Equivalent: *For a decision problem with a single attribute* Z, *the* certainty equivalent *for an uncertain alternative is the certain amount of* Z *that is equally preferred to the uncertain alternative. The certainty equivalent is also sometimes called the* certain equivalent.

Definition 9.27. Constant Risk Aversion: *For a decision problem with a single evaluation attribute, preferences are* constantly risk averse *if it is true that whenever all the possible consequences of an uncertain alternative are changed by the same amount, the certainty equivalent for the alternative also changes by that same amount. (Constant risk aversion is also sometimes called the* delta property.)

Theorem 9.28. Exponential Utility Function: *For a decision problem with a single evaluation attribute Z, if preferences are constantly risk averse, then when preferences are monotonically increasing over Z,*

$$u(z) = \begin{cases} \dfrac{1 - \exp\left[-(z - z^L)/\rho\right]}{1 - \exp\left[-(z^H - z^L)/\rho\right]}, & \rho \neq \text{Infinity} \\[2ex] \dfrac{z - z^L}{z^H - z^L}, & \text{otherwise} \end{cases} \tag{9.28.1}$$

while if preferences are monotonically decreasing over Z,

$$u(z) = \begin{cases} \dfrac{1 - \exp\left[-(z^H - z)/\rho\right]}{1 - \exp\left[-(z^H - z^L)/\rho\right]}, & \rho \neq \text{Infinity} \\[2ex] \dfrac{z^H - z}{z^H - z^L}, & \text{otherwise} \end{cases} \tag{9.28.2}$$

where "z^L" is the lowest level of z that is of interest, "z^H" is the highest level of interest, and ρ is called the risk tolerance *for the utility function. The utility functions in equations 9.28.1 and 9.28.2 are scaled so that they vary between 0 and 1, inclusive, over the range from $z = z^L$ to $z = z^H$. That is, for monotonically increasing preferences, $u(z^L) = 0$ and $u(z^H) = 1$, while for monotonically decreasing preferences, $u(z^L) = 1$ and $u(z^H) = 0$. (Some presentations of the exponential utility function use the* risk aversion coefficient *c rather than the risk tolerance ρ. The two are related by the equation $c = 1/\rho$.)*

Proof. Showing that a linear or exponential utility function form implies that constant risk aversion holds is straightforward. We will show the converse, which is that constant risk aversion implies that one of these forms holds. Only the case where preferences are monotonically increasing with respect to the evaluation measure will be considered. (The proof for the monotonically decreasing case is analogous.)

Additional notation is useful for this proof. The *risk premium* for an alternative is the difference between the expected value $\bar{z}$ for the alternative and its certainty equivalent CE. Specifically, for an evaluation measure with monotonically increasing preferences, the risk premium π is given by $\pi = \bar{z} - \text{CE}$, while with monotonically decreasing preferences, it is $\pi = \text{CE} - \bar{z}$. Thus, the risk premium is positive for a risk averse decision maker, zero for one who is risk neutral, and negative for one who is risk seeking.

Consider two alternatives related in the manner given in the definition of constant risk aversion. That is, the second alternative differs from the first only by having the same amount either added to or subtracted from each outcome. It follows directly from the definition of the risk premium π that the risk premiums must be the same for the two alternatives if constant risk aversion holds. Furthermore, if w is defined by $w = z - \bar{z}$, then w has the same probability distribution for both alternatives.

From the definition of the certainty equivalent, it follows that if the certainty equivalent of an uncertain alternative is CE, then it must be true that $u(\text{CE}) = \text{E}[u(z)]$. This can be rewritten in terms of π and w as $u(\bar{z} - \pi) = \text{E}[u(\bar{z} + w)]$.

Taylor expand both sides of this equation around $\bar{z}$. The left side becomes

$$u(\bar{z} - \pi) = u(\bar{z}) - \frac{du(\bar{z})}{dz}\pi + (1/2)\frac{d^2u(\bar{z})}{dz^2}\pi^2 + \cdots$$

and the right side becomes

$$\text{E}[u(\bar{z} + w)] = \text{E}[u(\bar{z}) + \frac{du(\bar{z})}{dz}w + (1/2)\frac{d^2u(\bar{z})}{dz^2}w^2 + \cdots]$$

$$= u(\bar{z}) + \frac{du(\bar{z})}{dz}\text{E}(w) + (1/2)\frac{d^2u(\bar{z})}{dz^2}\text{E}(w^2) + \cdots$$

Since $\bar{z}$ is the expected value for the alternative and $w = z - \bar{z}$, then $\text{E}(w) = 0$ and $\text{E}(w^2) = \sigma^2$, where σ^2 is the variance for the alternative.

Using these results, equating the right-hand and left-hand Taylor expansions, and dropping common terms leads to

$$-\frac{du(\bar{z})}{dz}\pi + (1/2)\frac{d^2u(\bar{z})}{dz^2}\pi^2 + \cdots = (1/2)\frac{d^2u(\bar{z})}{dz^2}\sigma^2 + \cdots$$

Now assume a situation with "small" risk aversion so that $\pi \ll \sigma$ and where the uncertainty is small enough that only terms through second order need be considered in the Taylor expansion. Then the equation above reduces to

$$-\frac{du(\bar{z})}{dz}\pi = (1/2)\frac{d^2u(\bar{z})}{dz^2}\sigma^2$$

or

$$\pi = -(1/2)\frac{d^2u(\bar{z})/dz^2}{du(\bar{z})/dz}\sigma^2 \tag{9.28.3}$$

With constant risk aversion, π and σ will not change when a constant amount is added to each possible outcome of an alternative. However, $\bar{z}$ will change by the constant amount. Thus, for equation 9.28.1 to hold, it must be true that

$$\frac{d^2u(z)/dz^2}{du(z)/dz} = -\frac{1}{\rho}$$

for some constant ρ. (Otherwise, the right-hand side of equation 9.28.1 would vary as $\bar{z}$ changes, and hence constant risk aversion would not hold.) This is a second-order linear constant-coefficient differential equation, and the solution is

$$u(z) = \begin{cases} a + b\exp(-z/\rho), & \rho \neq \text{Infinity} \\ a + bz, & \rho = \text{Infinity} \end{cases}$$

where a and b are undetermined constants.

From Theorem 9.25, it follows that the values of a and b do not matter except that b must have the correct sign so that preferences either increase or decrease as is appropriate for the evaluation measure of interest. Set these constants so the most preferred end of the range of interest has a utility of 1, and the least preferred end of the range has a utility of zero. The result is equations 9.28.1 and 9.28.2. ∎

To determine ρ, consider an alternative with equal chances of yielding z' or z''. If z_{ce} is the certainty equivalent for this alternative, then it follows directly from equation 9.28.1 that when preferences are monotonically increasing over z,

1 $z_{ce} < (z' + z'')/2$ implies that $\rho > 0$.
2 $z_{ce} = (z' + z'')/2$ implies that $\rho = $ Infinity.
3 $z_{ce} > (z' + z'')/2$ implies that $\rho < 0$.

When preferences are monotonically decreasing over x_i, the direction of the first inequalities in (1) and (3) are reversed. It must be true that $0.5u(z') + 0.5u(z'') = u(z_{ce})$, and this equation can be solved for ρ. In general, the solution cannot be determined in closed form, and Section 6.6 presents methods for finding the solution. Keeney and Raiffa (1976), Chapter 4, discuss additional utility function forms. Bell (1995) presents a generalization of the exponential utility function.

Definition 9.29. Additive Utility Function: *Utility function $u(x)$ is additive if it is strategically equivalent to a utility function of the form*

$$u(x) = \sum_{i=1}^{n} k_i u_i(x_i) \tag{9.29.1}$$

for some functions $u_i(x_i)$ and constants k_i.

From Theorem 9.25 it follows that for any additive utility function in the form of equation 9.29.1, it is possible to construct a strategically equivalent additive utility function with $u(x^o) = 0$, $u(x^*) = 1$, $u_i(x_i^o) = 0$, and $u_i(x_i^*) = 1$, for all i. When the function is in this form, it must be true that $k_i = u(x_i^*; \bar{x}_i^o)$, where $\sum_{i=1}^{n} k_i = 1$. This is the most commonly used additive utility function in practice, and it is the form used in Chapter 7. [The $u_i(x_i)$ are called *single attribute utility functions* or *single dimensional utility functions*.]

Definition 9.30. Normalized Additive Utility Function: *An additive utility function is called normalized if it is in the form specified in the preceding paragraph.*

Additive Independence

The condition on preferences that leads to the additive utility function form is restrictive, and is often not met in practice. However, it is useful to study this condition to see how restrictive it is necessary to be for an additive utility function form to be valid.

Definition 9.31. Additive Independence: *Preferences over* $X_1, X_2, \ldots, X_n$ *are* additive independent *if the rank ordering for any set of alternatives depends only on the marginal probability distributions over the attributes for each alternative.*

Theorem 9.32. Additive Utility Function: *Preferences are additive independent if and only if the utility function is additive.*

> **Proof.** Showing that an additive utility function implies additive independence is straightforward by direct substitution into the utility function. We prove the converse by induction. (See Figure 9.10 for an illustration of the approach to the proof for two attributes.) Consider two uncertain alternatives, where one has equal chances of yielding $(x_1, x_2, \ldots, x_n)$ or $(x_1^o, x_2^o, \ldots, x_n^o)$, while the other has equal chances of yielding $(x_1, x_2, \ldots, x_i, x_{i+1}^o, \ldots, x_n^o)$ or $(x_1^o, x_2^o, \ldots, x_i^o, x_{i+1}, \ldots, x_n)$. Because these two alternatives have the same marginal probability distributions for all attributes, they must be equally preferred when additive independence holds, and hence they must have equal expected utilities. Therefore,
>
> $$0.5 \times [u(x_1, x_2, \ldots, x_n) + u(x_1^o, x_2^o, \ldots, x_n^o)]$$
> $$= 0.5 \times [u(x_1, x_2, \ldots, x_i, x_{i+1}^o, \ldots, x_n^o) + u(x_1^o, x_2^o, \ldots, x_i^o, x_{i+1}, \ldots, x_n)]$$
> $$(9.32.1)$$
>
> By Theorem 9.25, it can be assumed without loss of generality that $u(x^o) = 0$, and therefore equation 9.32.1 reduces to
>
> $$u(x_1, x_2, \ldots, x_n) = u(x_1, x_2, \ldots, x_i, x_{i+1}^o, \ldots, x_n^o) + u(x_1^o, x_2^o, \ldots, x_i^o, x_{i+1}, \ldots, x_n)$$
> $$(9.32.2)$$
>
> For the induction, we assume that
>
> $$u(x_1, x_2, \ldots, x_n) = \sum_{k=1}^{i-1} u(x_k; \bar{x}_k^o) + u(x_1^o, x_2^o, \ldots, x_{i-1}^o, x_i, \ldots, x_n) \quad (9.32.3)$$
>
> and then show that this must also hold when i is replaced by $i + 1$. To establish this, set $x_k = x_k^o, k = i+1, i+2, \ldots, n$ in both equation 9.32.2 and equation 9.32.3 and equate the right-hand sides of the two equations. This establishes that equation 9.32.3 holds with i replaced by $i + 1$. But when $i = n$ in equation 9.32.3, we have established the additive utility function form
>
> $$u(x_1, x_2, \ldots, x_n) = \sum_{k=1}^{n} u(x_k; \bar{x}_k^o) \quad (9.32.4)$$

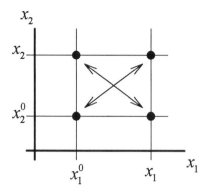

Figure 9.10   *Additive independence*

All that remains to complete the induction is to establish that equation 9.32.3 holds for $i = 1$ to start the induction. But for $i = 1$, equation 9.32.3 holds trivially. ∎

The additive utility function form is relatively simple to use because it requires consideration of functions over only one variable at a time. However, the additive independence condition is fairly restrictive. For example, consider a decision problem with two attributes, cost and quality. Suppose alternative a_1 has equal chances of yielding consequences with cost and quality at either (1) their best levels together or (2) their worst levels together. Similarly, suppose that alternative a_2 has equal chances of yielding consequences with either (3) cost at its best level and quality at its worst level or (4) cost at its worst level and quality at its best level. If additive independence holds, these two alternatives must be equally preferred since they have the same marginal probability distributions over cost and quality. Most people would prefer not to have to select either of these alternatives, but if they had to select one, they would usually have a preference between them. However, when additive independence holds, the two alternatives must be equally preferred. See Richard (1975) for further related discussion.

Utility Independence

This section investigates a condition that is not as restrictive as additive independence but which still leads to a relatively simple form of utility function. This condition has been tested in many applications and has been found to often hold (Corner and Kirkwood 1991).

Definition 9.33. Utility Independence: *Suppose Y and Z are a partition of $\{X_1, X_2, \ldots, X_n\}$. Then Y is utility independent of Z if the rank ordering of any set of alternatives with uncertainty about the outcomes for attributes in Y and common specified levels for the attributes in Z does not depend on the specified levels of the attributes in Z. (The common levels do not have to be the same for different attributes, but the level of each X_i in Z is the same for all alternatives, and there is no uncertainty about this level.)*

The Power-additive Utility Function

Theorem 9.34. Power-Additive Utility Function: *Assume that an additive value function $v(x_1, x_2, \ldots, x_n)$ holds over the attributes $\{X_1, X_2, \ldots, X_n\}$, and that X_1 is utility independent of $\{X_2, X_3, \ldots, X_n\}$. Then*

$$u(x_1, x_2, \ldots, x_n) = \begin{cases} \dfrac{1 - \exp[-v(x_1, x_2, \ldots, x_n)/\rho_m]}{1 - \exp(-1/\rho_m)}, & \rho_m \neq \text{Infinity} \\ v(x_1, x_2, \ldots, x_n), & \text{otherwise} \end{cases}$$

(9.34.1)

where ρ_m is a constant called the multiattribute *risk tolerance.*

Proof. Since an additive value function holds, we know that

$$v(x_1, x_2, \ldots, x_n) = \sum_{i=1}^{n} \lambda_i v_i(x_i)$$

(9.34.2)

Because a utility function is also a value function, we know from Theorem 9.7 that $u(x_1, x_2, \ldots, x_n)$ must be some positive monotonic transformation of $v(x_1, x_2, \ldots, x_n)$. That is,

$$u(x_1, x_2, \ldots, x_n) = f[\textstyle\sum_{i=1}^{n} \lambda_i v_i(x_i)]$$

(9.34.3)

for some function f. To complete the proof, we need to show that f is either exponential or linear.

Consider an alternative with specified levels for $X_2, X_3, \ldots, X_n$ and uncertainty about the level of X_1. For any such alternative, it follows from equation 9.34.3 that there is a certain level x_{ce} of X_1 such that

$$f[\lambda_1 v_1(x_{ce}) + \textstyle\sum_{i=2}^{n} \lambda_i v_i(x_i)] = E\{f[\textstyle\sum_{i=1}^{n} \lambda_i v_i(x_i)]\}$$

(9.34.4)

Define $z = \lambda_1 v_1(x_1)$, $\delta = \sum_{i=2}^{n} \lambda_i v_i(x_i)$, and $z_{ce} = \lambda_1 v_1(x_{ce})$. Then (9.34.4) can be rewritten as

$$f(z_{ce} + \delta) = E[f(z + \delta)]$$

(9.34.5)

Since x_1 is utility independent of the other attributes, equation 9.34.5 must hold regardless of the value of δ. But from the definition of constant risk aversion (Definition 9.27), this means that z is constantly risk averse. Thus, Theorem 9.28 implies that f must be exponential or linear, and hence equation 9.34.1 holds. ■

Examples of the relationship between the value function and the power-additive utility function are shown in Figure 9.11. It is easy to show by direct calculation that it is possible to distinguish between the two alternatives discussed in the cost-quality example presented earlier when the exponential form of equation 9.34.1 holds. In particular, if $\rho_m > 0$, then $a_2 \succ a_1$, while if $\rho_m < 0$, then $a_1 \succ a_2$. A decision maker who prefers a_2 to a_1 is called *multiattribute*

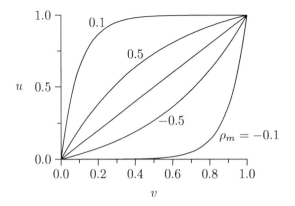

Figure 9.11   *Power-additive utility function*

risk averse; a decision maker who prefers a_1 to a_2 is called *multiattribute risk seeking*; and a decision maker who finds the two alternatives equally preferred is called *multiattribute risk neutral*.

To completely specify the power-additive utility function, it is necessary to find only one additional piece of information once the additive value function has been found—namely, the multiattribute risk tolerance ρ_m. In order to determine ρ_m, it is necessary for the decision maker to find the certainty equivalent for one uncertain alternative. From this, ρ_m can be determined numerically. A procedure to determine ρ_m is presented in Section 7.3. The required utility independence condition appears to be fairly weak; however, the following discussion shows that it implies much stronger conditions.

Definition 9.35.   Mutual Utility Independence: *The attributes $\{X_1, X_2, \ldots, X_n\}$ are mutually utility independent if every subset of the attributes is utility independent of the remaining attributes.*

Theorem 9.36. Utility Independence: *If the conditions hold for an additive value function and one attribute is utility independent of the remaining attributes, then the attributes are mutually utility independent.*

> **Proof.**   This follows from Theorem 9.34. With the specified conditions in Theorem 9.36, then equation 9.34.1 holds. It is straightforward to show that this utility function form implies that any subset of attributes is utility independent of the remaining attributes by direct substitution into the utility function. ∎

Theorem 9.37. Additive Form: *If the conditions for Theorem 9.34 hold, then the additive form of equation 9.34.1 holds if and only if there exist x'_1, x''_1, x'_2, and x''_2 such that for some fixed levels $\bar{x}_{12}$ of attributes $X_3, X_4, \ldots, X_n$, an alternative with equal chances of yielding either $(x'_1, x'_2, \bar{x}_{12})$ or $(x''_1, x''_2, \bar{x}_{12})$ is equally preferred to an alternative with equal chances of yielding either $(x'_1, x''_2, \bar{x}_{12})$ or $(x''_1, x'_2, \bar{x}_{12})$.*

> **Proof.**   This is the corollary on page 291 of Keeney and Raiffa (1976). ∎

If the stated indifference condition holds for any one $\bar{x}'_{12}$, then the additive utility function form implies that it must hold for all $\bar{x}'_{12}$. Furthermore, if it holds for any one set of x'_1, x''_1, x'_2, and x''_2, it must hold for all sets.

The Multiplicative Utility Function

An alternate version of the power-additive utility function, called the multiplicative utility function, has seen wide use in practice. After the theory is presented for this form, a comparison will be made of the power-additive and multiplicative forms.

Theorem 9.38. Multiplicative Utility Function: *A mathematically equivalent form for the power-additive utility function $u(x_1, x_2, \ldots, x_n)$ in equation 9.34.1, which requires the same preferential and utility independence conditions, is the multiplicative form*

$$u(x_1, x_2, \ldots, x_n) = \begin{cases} \{\prod_{i=1}^{n} [kk_i u_i(x_i) + 1] - 1\}/k, & -1 < k \neq 0 \\ \sum_{i=1}^{n} k_i u_i(x_i), & \text{otherwise} \end{cases} \tag{9.38.1}$$

where

$$1 + k = \prod_{i=1}^{n} (kk_i + 1), \tag{9.38.2}$$

both u and the u_i are scaled to lie between zero and one, and $0 \leq k_i \leq 1$.

Proof. This is proved by determining the multiplicative utility function that is equivalent to any power-additive utility function. If the additive form in equation 9.34.1 holds, then the corresponding multiplicative utility function is also additive with $\lambda_i = k_i$ and $v_i(x_i) = u_i(x_i)$. Otherwise, the conversion from the power-additive form to the multiplicative form is made with the following substitutions into the exponential form of equation 9.34.1:

$$\rho_m = -1/\ln(1 + k) \tag{9.38.3}$$

$$\lambda_i = \ln(kk_i + 1)/\ln(k + 1) \tag{9.38.4}$$

$$v_i(x_i) = \ln[kk_i u_i(x_i) + 1]/\ln(kk_i + 1) \tag{9.38.5}$$

∎

A multiplicative utility function can be determined by first assessing a power-additive utility function and then using equations 9.38.3, 9.38.4, and 9.38.5 to convert it to the equivalent multiplicative utility function. However, there is not much point in doing this conversion since the power-additive function can be used directly to rank alternatives. In practice, when the multiplicative utility function is used, it is determined by direct assessment. Keeney and Raiffa (1976), Section 6.6, present an assessment procedure, and here is a summary:

Step 1   Determine the attributes $X_i, i = 1, 2, \ldots, n$, together with x_i^o and x_i^*. In what follows, we assume that $(x_i^*; \bar{x}_i^o)$ is at least as preferred as $(x_j^*, \bar{x}_j^o)$ for all $j > i$. The attributes can always be defined so this is true, and it is assumed for notational simplicity in presenting the results.

Step 2   Confirm or assume that the relevant conditions hold for an additive value function. (The required condition is the corresponding tradeoffs condition for a two-attribute problem, or preferential independence for more than two attributes.) The required tests were discussed earlier.

Also, confirm or assume that X_1 is utility independent of the other attributes. The usual test is to consider an uncertain alternative with specified levels for all attributes except X_1, and equal chances of yielding the two extreme outcomes x_1^o and x_1^*. Determine the certainty equivalent $x_1 = x_1^{ce}$ for this alternative that has the same fixed levels for the attributes other than X_1 as the uncertain alternative. Try different fixed levels for the attributes other than X_1 and see if x_1^{ce} changes. If it does not, then this is taken as establishing that X_1 is utility independent of the other attributes.

Step 3   Determine the utility functions $u_i(x_i), i = 1, 2, \ldots, n$, using standard single attribute utility assessment procedures. [Since $u_i(x_i)$ is a utility function over x_i scaled so that $u_i(x_i^o) = 0$ and $u_i(x_i^*) = 1$, conventional single attribute utility function assessment procedures can be used to determine this function.]

Step 4   Determine the level x_i' such that $(x_i', x_{i+1}^o; \bar{x}_{ij}) \sim (x_i^o, x_{i+1}^*; \bar{x}_{ij})$. Then, from equation 9.38.1 it follows that $k_i u_i(x_i') = k_{i+1}, i = 1, 2, \ldots, n - 1$. This provides $n - 1$ equations to solve for the n scaling constants k.

Step 5   One final equation must be determined in order to solve for the k_i and k. (Note that while there are a total of $n+1$ constants in the multiplicative utility function, equation 9.38.2 provides one of the required equations.) The simplest question as far as carrying out the mathematics is concerned is to ask for the probability p such that a certainty equivalent for an uncertain alternative with a probability p of yielding x^* and a probability $1 - p$ of yielding x^o is $(x_1^*; \bar{x}_i^o)$. It is easy to show from equation 9.38.1 that $k_1 = p$. Then the remaining k_i can be determined using the results of Step 4. (If there is some reason to think that additive independence might hold, then Theorem 9.37 can be used to check directly for this.)

Step 6   Calculate k by solving equation 9.38.2. Note that this equation is an nth-order polynomial and hence may have multiple roots. Keeney and Raiffa (1976), Appendix 6B, show that if $\sum_{i=1}^n k_i > 1$, then $-1 < k < 0$; if $\sum_{i=1}^n k_i = 1$, then $k = 0$ (that is, the additive version of equation 9.38.1 holds); and if $\sum_{i=1}^n k_i < 1$, then $0 < k$. Keeney and Raiffa also show that if the search for a solution to equation 9.38.2 is restricted to the region specified in the preceding sentence, then there is a unique solution to the equation. Except in special cases, the solution must be found numerically.

Power-additive versus Multiplicative Utility Functions

Theorem 9.38 shows that the power-additive and multiplicative utility functions are mathematically equivalent. However, the assessment procedures for the two types of utility functions require somewhat different information, and thus one of them may be more appropriate in a particular decision problem. The power-additive form incorporates attitude toward risk taking in an intuitively appealing way. First, the decision maker assesses a value function that encodes returns-to-scale effects for each attribute through the $v_i(x_i)$ functions, as well as tradeoffs among the attributes through the λ_i. After this process is completed, attitude toward risk taking is addressed through the multiattribute risk tolerance ρ_m. This corresponds to our intuitive view that attitude toward risk taking is somehow separate from tradeoffs or returns-to-scale effects.

With the multiplicative utility function, on the other hand, single attribute utility functions must be assessed over each attribute. Thus, we must address attitude toward risk taking multiple times (that is, for each attribute). This generally requires answering substantially more lottery-type questions than with the power-additive utility function. Furthermore, in practice, it is sometimes true that different individuals assess the functions over each attribute. (For example, in a decision involving environmental and cost issues, ecologists and project engineers might be most knowledgeable about attributes in their areas of expertise. Hence, they might separately determine the single attribute utility functions for attributes in their areas.) There is no reason to think that these different individuals will have the same attitude toward risk taking, nor that their attitudes will be the same as that of the ultimate decision maker.

This is somewhat intuitively disconcerting, since it seems that a utility function should encode a single consistent attitude toward risk taking. Furthermore, the single attribute utility functions assessed for each attribute (with their possibly differing attitudes toward risk taking) affect the scaling constants k_i, as shown in Step 4 of the assessment procedure given earlier. Hence, changing the assessed risk attitudes for each attribute can change the scaling constants. Thus, if people with different attitudes toward risk taking assess the single attribute utility functions for different attributes, their differing attitudes toward risk taking can affect not only the single attribute utility functions, but also the scaling constants that encode tradeoffs among the attributes.

This discussion seems to argue for using the power-additive utility function form. However, this form has one disadvantage relative to the multiplicative form. The midvalue splitting technique, which can be used to determine the $v_i(x_i)$, requires that it be possible to find midvalues for all the required intervals, as discussed in the assessment procedure given earlier. However, if a particular attribute scale is discrete, then the midvalue for an interval may lie between two of the defined scale levels. Hence, it may not be possible to find midvalues for some intervals. While there are some conceptually correct ways to address this problem, they require answering questions that are difficult.

This may not be such a problem with the multiplicative utility function form, since a utility function can be assessed over such a scale, although it may require finding certainty equivalents for other than the standard 50:50 lotteries that are usually used for such assessments.

There has been research done on the degree to which changes in different aspects of a value or utility function affect evaluation results using the value/utility function. This work indicates that the impact of small errors in either $v_i(x_i)$ or $u_i(x_i)$ is often relatively small compared with the impact of errors in assessing the scaling constants. [See Keefer and Pollock (1980) and Barron, von Winterfeldt, and Fischer (1984) for further discussion of related matters.] Thus, it may be adequate to approximate $v_i(x_i)$ in situations where the midvalue does not fall exactly on a defined level of the scale. (For example, if the midvalue for an interval falls between levels of 2 and 3 on a particular attribute scale, it may be adequate to round the midvalue to either 2 or 3, or to take the midvalue as 2.5.) Alternatively, the $v_i(x_i)$ can be assessed using the measurable value function approach discussed earlier.

The Two-attribute Case

Of the various preference conditions presented in earlier sections, the corresponding tradeoffs condition seems to be the most difficult to test. Explaining this condition can be difficult, and obtaining a believable assessment of whether or not it is true also seems difficult. Thus, it would be useful to have an alternative test that could be done to determine whether the power-additive or multiplicative utility function holds in the two-attribute case. In fact, such a test exists.

Theorem 9.39. Two-attribute Multiplicative Utility Function: *For a decision problem with two attributes, the multiplicative utility function in equation 9.38.1, or the equivalent power-additive utility function in equation 9.34.1, holds if each of the attributes is utility independent of the other attribute.*

Proof. It follows directly from Theorem 9.25 that because of the utility independence conditions

$$u(x_1, x_2) = a_1(x_2)u(x_1, x_2^o) + b_1(x_2) \tag{9.39.1}$$
$$u(x_1, x_2) = a_2(x_1)u(x_1^o, x_2) + b_2(x_1) \tag{9.39.2}$$

where the functions $a_1(x_2) > 0$, $a_2(x_1) > 0$, $b_1(x_2)$, and $b_2(x_1)$ are arbitrary except for the designated sign restriction.

Without loss of generality, assume that $u(x_1^o, x_2^o) = 0$ and $u(x_1^*, x_2^*) = 1$. Then successively set $(x_1, x_2) = (x_1^o, x_2^o)$, $(x_1, x_2) = (x_1^*, x_2^o)$, $(x_1, x_2) = (x_1^o, x_2^*)$, and $(x_1, x_2) = (x_1^*, x_2^*)$ in equation 9.39.1 to establish four equations that can be solved for $a_1(x_2^o)$, $a_1(x_2^*)$, $b_1(x_2^o)$, and $b_1(x_2^*)$. Solving these gives

$$u(x_1, x_2^o) = k_1 u_1(x_1) \tag{9.39.3}$$
$$u(x_1, x_2^*) = k_2 + (1 - k_2)u_1(x_1) \tag{9.39.4}$$

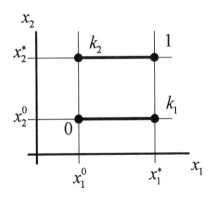 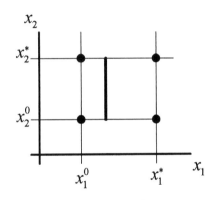

a. Equations 9.39.3 and 9.39.4   b. Equation 9.39.5

Figure 9.12   *Two-attribute multiplicative utility function*

where $k_1 = u(x_1^*, x_2^o)$, $k_2 = u(x_1^o, x_2^*)$, and $u_1(x_1) = u(x_1, x_2^o)/u(x_1^*, x_2^o)$. (See Figure 9.12a.)

In equation 9.39.2, successively set $(x_1, x_2) = (x_1, x_2^o)$ and $(x_1, x_2) = (x_1, x_2^*)$, and solve for $a_2(x_1)$ and $b_2(x_1)$. This gives

$$u(x_1, x_2) = [u(x_1, x_2^*) - u(x_1, x_2^o)]u_2(x_2) + u(x_1, x_2^o) \tag{9.39.5}$$

where $u_2(x_2) = u(x_1^o, x_2)/u(x_1^o, x_2^*)$. (See Figure 9.12b.)

Substitute into equation 9.39.5 for $u(x_1, x_2^*)$ from equation 9.39.4, and substitute into equation 9.39.5 for $u(x_1, x_2^o)$ from equation 9.39.3. This yields

$$u(x_1, x_2) = k_1 u_1(x_1) + k_2 u_2(x_2) + (1 - k_1 - k_2)u_1(x_1)u_2(x_2) \tag{9.39.6}$$

Define $k = (1 - k_1 - k_2)/(k_1 k_2)$. Then equation 9.39.6 is equivalent to

$$u(x_1, x_2) = \{[kk_1 u_1(x_1) + 1][kk_2 u_2(x_2) + 1] - 1\}/k \tag{9.39.7}$$

if $k \neq 0$; and, of course, if $k = 0$, then equation 9.39.6 reduces to an additive utility function. But equation 9.39.7 is the multiplicative utility function, and hence the theorem is proved. ■

Some Final Results about Utility Independence

Theorem 9.36 shows how a combination of preferential independence and utility independence (for situations with at least three attributes) implies that the attributes are mutually utility independent. It is also possible to specify seemingly less restrictive utility independence conditions that imply mutual utility independence without requiring consideration of preferential independence.

Theorem 9.40. Pairwise Utility Independence: *If* $\{X_1, X_i\}, i = 2, 3, \ldots, n,$ *are utility independent of the remaining attributes, then* $\{X_1, X_2, \ldots, X_n\}$ *are mutually utility independent.*

Proof. This follows from Keeney and Raiffa (1976), Theorem 6.7. ∎

Thus, if one of the attributes in combination with each of the other attributes is utility independent of the remaining attributes, then the attributes are mutually utility independent. What about the situation when each attribute alone is utility independent of the other attributes? Does this also imply that the attributes are mutually utility independent? The answer is no as the following theorem shows.

Theorem 9.41. Multilinear Utility Function: *If* $X_i, i = 1, 2, \ldots, n,$ *is utility independent of the remaining attributes, then*

$$
\begin{aligned}
u(x) = {} & \sum_{i=1}^{n} k_i u_i(x_i) + \sum_{i=1}^{n}\sum_{j>i} k_{ij} u_i(x_i) u_j(x_j) \\
& + \sum_{i=1}^{n}\sum_{j>i}\sum_{\ell>j} k_{ij\ell} u_i(x_i) u_j(x_j) u_\ell(x_\ell) \\
& + \cdots + k_{123\cdots n} u_1(x_1) u_2(x_2) \cdots u_n(x_n)
\end{aligned}
\tag{9.41.1}
$$

where the various subscripted k's are constants.

Proof. The proof is given in Keeney and Raiffa (1976), Theorem 6.3. That theorem statement also gives detailed expressions for the subscripted k's. ∎

Note that more constants are required to determine the $u(x_1, x_2, \ldots, x_n)$ in this situation than when there is mutual utility independence. (In fact, there are $2^n - 2$ independent constants in this case versus n independent constants when there is mutual utility independence.) However, the utility function in equation 9.41.1 still requires only the single attribute utility functions $u_i(x_i), i = 1, 2, \ldots, n$. Thus, while more constants are required than with mutual utility independence, it is still necessary to determine only single attribute utility functions.

Finally, some readers may be wondering whether having X_i utility independent of the remaining attributes implies that those remaining attributes are utility independent of X_i. (This would be analogous to the situation for probabilistic independence where having one random variable probabilistically independent of a second random variable implies that the second random variable is also probabilistically independent of the first.)

However, we can see that this is not true for utility independence by examining equation 9.39.5 in Theorem 9.39. That equation holds if X_2 is utility independent of X_1. But it is straightforward to construct examples that obey equation 9.39.5 but that do not obey the linear transformation equation 9.39.1, which must hold if X_1 is utility independent of X_2. (See Keeney and Raiffa 1976, Section 5.6.) Therefore, having X_2 utility independent of X_1 does not imply that X_1 is utility independent of X_2.

9.4 Other Approaches to Multiobjective Decision Making

The examples discussed in earlier chapters, as well as numerous applications (Corner and Kirkwood 1991), demonstrate the usefulness of the decision analysis methods presented in this book. However, other approaches have also been developed for multiobjective decision making. This section reviews the two most widely studied of these approaches—the Analytic Hierarchy Process and multiple criteria decision making—and provides references for the interested reader.

The Analytic Hierarchy Process

The Analytic Hierarchy Process (AHP) was developed by Thomas Saaty in the early 1970s, and it has an active community of researchers and users. Saaty (1986) presents an axiomatic basis for the methods used in the process. Later Saaty writings (1994b, 1995) are recent presentations of the approach. There has been a spirited discussion in the scholarly literature of the advantages and disadvantages of the AHP relative to the decision analysis methods presented in this book. Articles by Dyer (1990a, 1990b), Harker and Vargas (1990), and Saaty (1990) give a flavor of this discussion. [Note that the Analytic Hierarchy Process does not use probabilities to address uncertainties, but instead relies on scenario analysis (Forman 1993). Thus, applications of the AHP to decision making are comparable in scope to what is presented in Chapters 1 through 4 of this book, with possible augmentation using the scenario methods in Appendix B.]

With the AHP approach, the value function assessment process and the ranking of alternatives are not as distinctly separated as with the methods presented in this book. The decision maker does not first assess a value function and then apply it to rank alternatives. Instead, he or she answers questions of the following type (Vargas 1990): Given an evaluation consideration and two alternatives A and B, which alternative performs better with respect to that consideration and by how much? While not required by the theory, this comparison is often done using a nine-point scale (Saaty 1994b, Table 1), where the points on this scale are defined qualitatively and then translated using a standard scheme into numerical measures of the relative degree of performance of one alternative with respect to the other. A mathematical procedure is then used to estimate a value function based on the various pairwise comparisons, and this value function is used to rank the alternatives.

This procedure may require a considerable number of pairwise comparisons if there are a substantial number of alternatives and/or evaluation measures. Thus, given the limitations of human cognitive processes, it is not surprising that there can be inconsistencies among the decision maker's responses to the pairwise comparison questions. The AHP computation procedure provides a measure of this inconsistency as part of its output.

With some AHP approaches, the addition of a new alternative can change the ranking of existing alternatives, even though the new alternative does not influence the desirability of existing alternatives (that is, the new alternative is "irrelevant" for ranking the earlier alternatives). Dyer (1990a, 1990b) and other commentators view that as a significant flaw in the approach. Forman (1993) notes that this characteristic is not present in some variations of the AHP.

The Analytic Hierarchy Process is "a theory of measurement concerned with deriving dominance priorities from paired comparisons of homogeneous elements with respect to a common criterion or attribute" (Saaty 1994a). Thus, the AHP is directed at a broader range of issues than just making decisions. It may have advantages with respect to these broader issues, but I agree with Winkler's (1990) assessment that decision analysis methods are more appealing for aiding decision making. Even ignoring the theoretical objections that have been raised to the AHP, the approach seems overly complex with its need for sometimes extensive pairwise comparisons of alternatives and extensive mathematical calculations to determine rankings. These characteristics seem to obscure, rather then illuminate, the tradeoffs involved in making decisions with multiple objectives.

In contrast, the methods of this book can be implemented using a standard spreadsheet program, and the specific factors that lead to a particular ranking of alternatives can be traced in a straightforward manner. The separation of value assessment and scoring of alternatives that is characteristic of decision analysis methods makes it straightforward to determine whether disagreements among stakeholders to a decision are with regard to values or the estimated performance of the alternatives. This aids in tracing the primary factors that lead to a particular ranking of alternatives, and also provides a clear audit trail for the decision process.

In addition, the methods presented in the first four chapters of this book generalize to decisions with uncertainties in a straightforward manner using probabilities, as presented in Chapters 5 through 7. Probability analysis has been under development for several hundred years, and hence the use of probabilities in decision analysis allows us to bring to bear the extensive tools that have resulted from this long development for decisions requiring this level of analysis.

Multiple Criteria Decision Making

The phrase *multiple criteria decision making* (MCDM) refers to a set of related approaches for analyzing multiobjective decisions using mathematical programming (optimization) methods. Thus, MCDM methods are an alternative to the procedures presented in Chapter 8 for analyzing decisions with sufficiently many alternatives that it is not feasible to explicitly enumerate them all. MCDM methods do not usually include analysis of uncertainty with probabilities. Dyer et al. (1992) review the field and discuss future research directions. Yoon and Hwang (1995) present a variety of different MCDM methods.

Multiple criteria decision making methods differ from the approaches presented in Chapter 8 in that a value or utility function is not explicitly assessed with MCDM. Instead, MCDM approaches usually apply mathematical programming methods to identify the efficient set (that is, the alternatives that are not dominated), and then these approaches use an interactive computer-based procedure to help the decision maker explore the various alternatives in the efficient set in order to select the one that is preferred.

Thus, MCDM methods (like the AHP) do not separate the assessment of a value or utility function from the ranking of alternatives. Instead, the decision maker examines various alternatives and provides information that a mathematical programming system can use to propose different alternatives that may be preferred to the ones that have been previously examined. This interactive process between the decision maker and the mathematical programming system generally requires specialized computer software to do the necessary calculations and provide appropriate information to the decision maker.

Both the Analytic Hierarchy Process and multiple criteria decision making methods are implicitly based on the view that explicitly assessing a value or utility function in the manner presented in this book is either difficult or undesirable, and hence to be avoided. However, the substantial number of decision analysis applications over the last twenty-five years demonstrates that value/utility function assessment as presented in Chapters 4, 6, and 7 is not particularly difficult from a conceptual standpoint for most decision makers. That is, these decision makers understand the meaning of the questions that are asked during the assessment and also understand that answering such questions can be critical to making an informed decision.

It is, however, true that decision makers sometimes find it difficult to answer the questions because these decision makers are unclear about their values or preferences. It seems to be desirable to articulate these values/preferences rather than to avoid directly doing this, as in the AHP or MCDM procedures. An explicit value assessment helps to clarify the role of values/preferences in the decision. Given the significant impact of values on many decisions, this clarification can be useful. Thus, it is worth some effort to examine these values for decisions that are important enough to warrant the quantitative analysis of decision analysis or the other methods discussed in this section.

Finally, it is worth noting that practical issues associated with implementing the Analytic Hierarchy Process or multiple criteria decision making methods can be significant. This book has presented detailed instructions for implementing multiobjective decision analysis approaches using standard spreadsheet techniques. You do not need complex algorithms or proprietary computer software, as you typically do for most AHP or MCDM approaches.

9.5 References

J. Aczél, *Lectures on Functional Equations and Their Applications*, Academic Press, New York, 1966.

F. H. Barron, D. von Winterfeldt, and G. W. Fischer, "Empirical and Theoretical Relationships Between Value and Utility Functions," *Acta Psychologica*, Vol. 56, pp. 233–244 (1984).

D. E. Bell, "A Contextual Uncertainty Condition for Behavior under Risk," *Management Science*, Vol. 41, pp. 1145–1150 (1995).

D. M. Buede, "Structuring Value Attributes," *Interfaces*, Vol. 16, No. 2, pp. 52–62 (March-April 1986).

J. L. Corner and C. W. Kirkwood, "Decision Analysis Applications in the Operations Research Literature, 1970-1989," *Operations Research*, Vol. 39, pp. 206–219 (1991).

J. S. Dyer, "Remarks on the Analytic Hierarchy Process," *Management Science*, Vol. 36, pp. 249–258 (1990a).

J. S. Dyer, "A Clarification of 'Remarks on the Analytic Hierarchy Process,'" *Management Science*, Vol. 36, pp. 274–275 (1990b).

J. S. Dyer, P. C. Fishburn, R. E. Steuer, J. Wallenius, and S. Zionts, "Multiple Criteria Decision Making, Multiattribute Utility Theory: The Next Ten Years," *Management Science*, Vol. 38, pp. 645–654 (1992).

J. S. Dyer and R. K. Sarin, "Measurable Multiattribute Value Functions," *Operations Research*, Vol. 27, pp. 810–822 (1979).

P. Fishburn and P. Wakker, "The Invention of the Independence Condition for Preferences," *Management Science*, Vol. 41, pp. 1130–1144 (1995).

E. H. Forman, "Facts and Fictions about the Analytic Hierarchy Process," *Mathematical and Computer Modelling*, Vol. 17, No. 4/5, pp. 19–26 (1993).

P. T. Harker and L. G. Vargas, "Reply to 'Remarks on the Analytic Hierarchy Process' by J. S. Dyer," *Management Science*, Vol. 36, pp. 269–273 (1990).

D. L. Keefer and S. M. Pollock, "Approximations and Sensitivity in Multiobjective Resource Allocation," *Operations Research*, Vol. 28, pp. 114–128 (1980).

R. L. Keeney, "Measurement Scales for Quantifying Attributes," *Behavioral Science*, Vol. 26, pp. 29–36 (1981).

R. L. Keeney, "Structuring Objectives for Problems of Public Interest," *Operations Research*, Vol. 36, pp. 396–405 (1988).

R. L. Keeney and H. Raiffa, *Decisions with Multiple Objectives: Preferences and Value Tradeoffs*, Wiley, New York, 1976.

L. R. Keller and J. L. Ho, "Decision Problem Structuring: Generating Options," *IEEE Transactions on Systems, Man, and Cybernetics*, Vol. 18, pp. 715–728 (1988).

C. W. Kirkwood and R. K. Sarin, "Preference Conditions for Multiattribute Value Functions," *Operations Research*, Vol. 28, pp. 225–232 (1980).

J. W. Pratt, H. Raiffa, and R. O. Schlaifer, "The Foundations of Decision under Uncertainty: An Elementary Exposition," *Journal of the American Statistical Association*, Vol. 59, pp. 353–375 (1964).

S. F. Richard, "Multivariate Risk Aversion, Utility Independence and Separable Utility Functions," *Management Science*, Vol. 22, pp. 12–21 (1975).

T. L. Saaty, "Axiomatic Foundation of the Analytic Hierarchy Process," *Management Science*, Vol. 32, pp. 841–855 (1986).

T. L. Saaty, "An Exposition of the AHP in Reply to the Paper 'Remarks on the Analytic Hierarchy Process,'" *Management Science*, Vol. 36, pp. 259–268 (1990).

T. L. Saaty, "Highlights and Critical Points in the Theory and Application of the Analytic Hierarchy Process," *European Journal of Operational Research*, Vol. 74, pp. 426–447 (1994a).

T. L. Saaty, "How to Make a Decision: The Analytic Hierarchy Process," *Interfaces*, Vol. 24, No. 6, pp. 19–43 (November–December 1994b).

T. L. Saaty, *Decision Making for Leaders: The Analytic Hierarchy Process for Decisions in a Complex World*, Third Edition, RWS Publications, Pittsburgh, PA, 1995.

L. G. Vargas, "An Overview of the Analytic Hierarchy Process and its Applications," *European Journal of Operational Research*, Vol. 48, pp. 2–8 (1990).

D. von Winterfeldt and W. Edwards, *Decision Analysis and Behavioral Research*, Cambridge University Press, Cambridge, England, 1986.

R. L. Winkler, "Decision Modeling and Rational Choice: AHP and Utility Theory," *Management Science*, Vol. 36, pp. 247–248 (1990).

K. P. Yoon and C. Hwang, *Multiple Attribute Decision Making: An Introduction*, Sage, Thousand Oaks, CA, 1995.

9.6 Review Questions

R9-1 Define dominance, and explain its significance.

R9-2 Define the corresponding tradeoffs condition, and explain its significance.

R9-3 Define constant tradeoff attitude, and explain its significance.

R9-4 Define mutual preferential independence, and explain its significance.

R9-5   Define difference consistency and difference independence, and explain their significance.

R9-6   Review the process for assessing an additive measurable value function.

R9-7   Define constant risk aversion, and explain its significance.

R9-8   Define additive independence, and explain its significance.

R9-9   Define utility independence, and explain its significance.

R9-10   Explain the relationship between the power-additive and multiplicative utility functions.

9.7 Exercises

9.1   Consider a decision problem with three evaluation measures X, Y, and Z, where the range of possible levels for each evaluation measure is from 0 to 10, inclusive. Preferences are monotonically increasing for X and Y, while preferences are monotonically decreasing for Z. Furthermore, $\{X,Y\}$ is preferentially independent of Z, and $\{X,Z\}$ is preferentially independent of Y.

(i) Assume:
  a. The midvalue of any interval on X, Y, or Z is the average of the upper and lower bounds of the interval,
  b. $(10,0,5)$ is indifferent to $(0,10,5)$, and
  c. $(10,4,10)$ is indifferent to $(0,4,6)$.

Determine a value function $v(x,y,z)$.

(ii) Assume that in addition to the conditions in (i), X is utility independent of $\{Y,Z\}$ with $(4,6,3)$ indifferent to a 50:50 lottery between $(10,6,3)$ and $(0,6,3)$. Determine $u(x,y,z)$.

(iii) Assume that in addition to the conditions in (i), X is utility independent of $\{Y,Z\}$. Determine the values of the multiattribute risk tolerance ρ_m for which a 50:50 lottery between $(10,10,0)$ and $(0,0,10)$ is preferred to receiving $(5,5,5)$ for certain.

9.2   Consider an alternative with a 0.8 probability of winning 50 and a 0.2 probability of losing 10. Also, consider another alternative that has equal chances of winning 10 or 25.

(i) Suppose a decision maker is constantly risk averse with monotonically increasing preferences and a risk tolerance of 20. Show by direct calculation that if the two alternatives are probabilistically independent, then the certainty equivalent for the sum of these two alternatives is equal to the sum of the certainty equivalents for the alternatives.

(ii) Now assume that $u(x) = \sqrt{x + 10}$. Show that the certainty equivalent for the sum of the two probabilistically independent alternatives is *not* equal to the sum of the certainty equivalents for the alternatives.

9.3 Show that if an alternative has a normal probability distribution and preferences are constantly risk averse and monotonically increasing, then the certainty equivalent is given by $\mathrm{CE} = \bar{x} - \sigma^2/(2\rho)$.

9.4 The *selling price* of an alternative is the minimum amount for which a decision maker who owns the alternative would sell it, while the *buying price* is the maximum amount that a decision maker who does not own the alternative would pay to buy it. From the definitions, it follows that the selling price is equal to the certainty equivalent and that the buying price is the amount which, when subtracted from each outcome of the alternative, yields a certainty equivalent of zero for the alternative.

(i) Show that when constant risk aversion holds, a decision maker's buying price for any alternative is equal to his or her selling price.

(ii) Consider an alternative with equal chances of yielding 0 and 10. Show that if $u(x) = \sqrt[3]{x/10}$, then the buying price for this alternative is not equal to the selling price.

9.5 Suppose that the value function over four attributes X, Y, Z, and W is given by

$$v(x, y, z, w) = x^2 y + z^3 \log_{10}(w)$$

over the inclusive range 1 to 10 for each evaluation measure. Determine whether this value function is consistent with mutual utility independence of X, Y, Z, and W.

9.6 Suppose that a decision maker's preferences over two evaluation measures X and Y are mutually utility independent over the range $0 \le x \le 10$ and $1000 \le y \le 2000$, and that preferences are monotonically increasing over x and monotonically decreasing over y. Suppose further that the decision maker is risk neutral for lotteries over X for $y = 1000$, and that the decision maker is constantly risk averse with a risk tolerance of 3000 for lotteries over Y for $x = 10$. Finally, suppose that $u(0, 2000) = 0$ and $u(10, 1000) = 1$.

(i) Determine $u(x, y)$ over the range $0 \le x \le 10$ and $1000 \le y \le 2000$ to the extent that $u(x, y)$ can be determined given the information above.

(ii) Suppose that in addition to the information given above, it is also known that $(0, 1000)$ is equally preferred to a 50:50 lottery between $(10, 1000)$ and $(0, 2000)$, and also that $(10, 2000)$ is equally preferred to $(0, 1500)$. Determine $u(x, y)$ over the range $0 \le x \le 10$ and $1000 \le y \le 2000$.

9.7 Consider a decision problem with three evaluation measures X, Y, and Z. Suppose that all indifference curves over X and Y with Z held fixed at any level have the equation $xy = $ constant, and that all indifference curves over X and Z with Y held fixed at any level have the equation $x + z^2 = $ constant. Further, the certainty equivalent (x, y', z') for any lottery with uncertainty only in X and certain levels y' and z' for Y and Z, respectively, does not depend on the levels y' and z'.

(i) Determine whether or not X, Y, and Z are mutually utility independent.

(ii) Suppose that $u(1,3,1) = 0.1$ and $u(5,3,1) = 0.7$. Determine the expected utility for a 50:50 lottery between $(3,1,1)$ and $(2,3,2)$.

9.8 Consider a decision problem with five evaluation measures X_1, X_2, X_3, X_4, and X_5. Suppose that Y_1 and Y_2 are mutually utility independent, where $Y_1 = \{X_1, X_2\}$ and $Y_2 = \{X_3, X_4, X_5\}$. Show that the utility functions

$$u(x_1, x_2, x_3', x_4', x_5') \text{ and } u(x_1', x_2', x_3, x_4, x_5)$$

where x_1', x_2', x_3', x_4' and x_5' are arbitrarily specified levels of the evaluation measures, are sufficient to determine $u(x_1, x_2, x_3, x_4, x_5)$ if $u(x_1', x_2', x_3', x_4', x_5') = 0$.

9.9 Consider a utility function $u(w, x, y, z)$ over four evaluation measures, where the evaluation measures are mutually utility independent and preferences are monotonically increasing for each evaluation measure over the range from 0 to 100, inclusive. Suppose further that when all the other evaluation measures are set at 50, risk neutrality holds for lotteries over W, and when all the other evaluation measures are set at 50, risk neutrality holds for lotteries over X. On the other hand, constant risk aversion holds for lotteries over Y with all the other evaluation measures set at 50, and constant risk aversion also holds for lotteries over Z with all the other evaluation measures set at 50. (The risk tolerance for Y is 50, and the risk tolerance for Z is 25.) Finally, suppose
   a. $u(100, 100, 100, 100) = 1$ and $u(0, 0, 0, 0) = 0$,
   b. $u(100, 0, 0, 0) = u(0, 100, 0, 0) = 0.4$,
   c. $(80, x, 0, z) \sim (0, x, 100, z)$ for any x and z, and
   d. $(0, 0, 0, 60)$ is equally preferred to a 50:50 lottery between $(100, 100, 100, 100)$ and $(0, 0, 0, 0)$.
Determine $u(w, x, y, x)$ over the inclusive range from 0 to 100 on each evaluation measure.

9.8 Supplement: Additive Value Proof (Three Attributes)

The proofs of Theorems 9.19 and 9.20 for the three-attribute case follow directly from theorems in Aczél (1966). Specifically, see Theorem 1 on page 311 of that book, and the theorem and corollary on page 329. While Theorem 9.20 follows for the three-attribute case directly from these results in Aczél, the proof of Aczél's results is not straightforward, and he does not appear to give generalizations to situations with more than three attributes.

Here is an informal statement of Aczél's page 329 corollary, and a demonstration of how it is used to prove Theorem 9.20 for the three-attribute case. (The required continuity, monotonicity, and differentiability conditions for this theorem should be obeyed for realistic value functions.)

Theorem 9.42. Additive Decomposition: *Under appropriate continuity, monotonicity, and differentiability conditions, when*

$$\Phi(x_1, x_2, x_3) = F[G(x_1, x_2), x_3] = H[x_1, K(x_2, x_3)] \qquad (9.42.1)$$

then it is also true that

$$\Phi(x_1, x_2, x_3) = L[M(x_3, x_1), x_2] \qquad (9.42.2)$$

and

$$\Phi(x_1, x_2, x_3) = h[k(x_1) + m(x_2) + g(x_3)] \qquad (9.42.3)$$

The connection between this theorem and Theorems 9.19 and 9.20 is made by recognizing that the following conditions hold for $v(x_1, x_2, x_3)$:

1 $\{X_1, X_2\}$ preferentially independent of X_3 implies that

$$v(x_1, x_2, x_3) = F[G(x_1, x_2), x_3]$$

for some functions F and G, and

2 $\{X_1, X_3\}$ preferentially independent of X_2, implies that

$$v(x_1, x_2, x_3) = H[x_2, K(x_1, x_3)]$$

for some functions H and K.

The first of these two conditions follows from the fact that under the stated preferential independence condition, the rank ordering of alternatives that differ only in X_1 and X_2 cannot depend on the common level of X_3. Thus, if we have the value function $v(x_1, x_2, x_3^o)$ for some fixed x_3^o, it follows from Theorem 9.7 that the value function for any other level of X_3 must be a positive monotonic transformation of $v(x_1, x_2, x_3^o)$. Thus, $v(x_1, x_2, x_3)$ can depend on x_1 and x_2 only through $v(x_1, x_2, x_3^o)$. Hence, if we set $G(x_1, x_2) = v(x_1, x_2, x_3^o)$, the equation shown for the first condition above holds. The argument for the second condition is analogous.

Thus, these two conditions together correspond to equation 9.42.1 if we set $\Phi(x_1, x_2, x_3) = v(x_1, x_2, x_3)$. Hence, equation 9.42.3 implies that $\{X_2, X_3\}$ is preferentially independent of X_1, and thus that the three attributes display mutual preferential independence, as Theorem 9.19 asserts they must. Finally, equation 9.42.3 implies that $v(x_1, x_2, x_3)$ is additive in agreement with Theorem 9.20, since we know from Theorem 9.7 that a strategically equivalent version of the Φ in equation 9.42.3 can be constructed with h set equal to $h(z) = z$.

9.9 Supplement: Two Fundamental Theorems

This appendix presents the axioms of consistent choice and two fundamental theorems that result from these axioms. In this appendix, the symbol "$\succ$" means "is preferred to," and the *consequences* of a decision are designated $c_1, c_2, \ldots, c_n$. Note that these consequences may themselves be uncertain alternatives. These axioms assume that probabilities exist and that the rules of probability apply. Pratt, Raiffa, and Schlaifer (1964) present a more extensive set of axioms that develops probability from first principles. See also Fishburn and Wakker (1995). Here are the axioms of consistent choice:

1 (Transitivity) If $c_i \succ c_j$ and $c_j \succ c_k$, then $c_i \succ c_k$.
2 (Reduction) If the standard rules of probability can be used to show that two alternatives have the same probability for each c_i, then the two alternatives are equally preferred.
3 (Continuity) If $c_i \succ c_j \succ c_k$, then there is a p such that an alternative with a probability p of yielding c_i and a probability $1 - p$ of yielding c_k is equally preferred to c_j.
4 (Substitution) If two consequences are equally preferred, then one can be substituted for the other in any decision without changing the preference ordering of alternatives.
5 (Monotonicity) For two alternatives that each yield either c_i or c_j, where $c_i \succ c_j$, then the first alternative is preferred to the second if it has a higher probability of yielding c_i.

If these conditions hold, then it is possible to prove the following theorem.

Theorem 9.43. Expected Utility: *If the axioms of consistent choice hold, then there exists a function $u(c_i)$ such that alternative A is preferred to alternative B if*

$$\sum_{i=1}^{n} p(c_i|A)u(c_i) > \sum_{i=1}^{n} p(c_i|B)u(c_i) \qquad (9.43.1)$$

where $p(c_i|A)$ is the probability of c_i if A is selected, and $p(c_i|B)$ is the probability of c_i if B is selected.

Proof. The following steps demonstrate the desired result:

1 Using the Transitivity Axiom, the consequences can be rank-ordered in terms of preferability. Suppose the consequences are labeled so that $c_1 \succ c_2 \succ \cdots \succ c_n$.
2 By the Reduction Axiom, there exists for any uncertain alternative an equally preferred alternative that directly yields the outcomes $c_1, c_2, \ldots, c_n$. Suppose the equally preferred alternative for A has probabilities $p(c_1|A), p(c_2|A), \ldots, p(c_n|A)$ of yielding $c_1, c_2, \ldots, c_n$, respectively, and the equally preferred alternative for B has probabilities $p(c_1|B), p(c_2|B), \ldots, p(c_n|B)$ of yielding $c_1, c_2, \ldots, c_n$. Then by the Substitution Axiom, the original alternatives can be replaced by their equally preferred reduced equivalents. Make this replacement.

3 By the Continuity Axiom, there is a number $u(c_i)$ such that c_i is equally preferred to an alternative with a probability $u(c_i)$ of yielding c_1 and a probability $1 - u(c_i)$ of yielding c_n. Thus, by the Substitution Axiom, each c_i can be replaced by the equally preferred alternative that has a probability $u(c_i)$ of yielding c_1 and a probability $1 - u(c_i)$ of yielding c_n. Make this substitution.

4 By the Reduction Axiom, A is equally preferred to an alternative with a probability $\sum_{i=1}^{n} p(c_i|A)u(c_i)$ of yielding c_1 and a probability $1 - \sum_{i=1}^{n} p(c_i|A)u(c_i)$ of yielding c_n. Similarly, B is equally preferred to an alternative with a probability $\sum_{i=1}^{n} p(c_i|B)u(c_i)$ of yielding c_1 and a probability $1 - \sum_{i=1}^{n} p(c_i|B)u(c_i)$ of yielding c_n. Thus, by the Substitution Axiom, A and B can be replaced by these alternatives, which have outcomes that include only c_1 and c_n. Make this substitution.

5 Thus, by the Monotonicity Axiom, $A \succ B$ if

$$\sum_{i=1}^{n} p(c_i|A)u(c_i) > \sum_{i=1}^{n} p(c_i|B)u(c_i).$$

This is the relationship in equation 9.43.1. ∎

The function $u(c_i)$ is called a *utility function*, and the decision criterion in Theorem 9.43 says that expected utility must be used as a decision criterion if the axioms of consistent choice are to be obeyed. These axioms were originally postulated as a model of unaided human decision making behavior. Many experiments have been done to test whether unaided human decision making naturally obeys the axioms. The results have shown that unaided human decision making does *not* obey the axioms of consistent choice. This has led to some questioning of whether the axioms are a good basis for decision analysis procedures, and a number of alternative axiom sets have been developed to better describe unaided human decision making.

However, there is a difference between describing how unaided decision making processes work and using analysis to make better decisions. Our focus here is on making better decisions. From this perspective, each of the axioms of consistent choice is reasonable, and it is hard to give any of them up as logical principles that we would want our reasoning to obey.

On a more practical level, we are all aware of limitations in human reasoning. Very few of us would trust ourselves to accurately add a column of 100 numbers in our head. Many decisions under uncertainty are more complex than adding 100 numbers. Why should we trust our unaided reasoning processes to be more accurate at analyzing these decisions than at adding 100 numbers? Thus, the fact that unaided decision making does not obey the axioms of consistent choice is not a convincing argument that these axioms should not be used as a basis for decision making.

Another theorem related to utility functions is useful. The proof of Theorem 9.43 shows that $u(c_i)$ must be between 0 and 1, inclusive, since it is defined as a probability. However, it is not necessary that all utility functions be in that range, as the following theorem demonstrates.

Theorem 9.44. Linear Transformation: *Given a utility function $u(c_i)$, then another function $u'(c_i)$ is guaranteed to give the same ranking of alternatives if and only if*

$$u'(c_i) = au(c_i) + b \tag{9.44.1}$$

for some constants $a > 0$ and b.

Proof. Suppose that $\sum_{i=1}^{n} p(c_i|A)u(c_i) > \sum_{i=1}^{n} p(c_i|B)u(c_i)$. Then it is also true that $a \sum_{i=1}^{n} p(c_i|A)u(c_i) > a \sum_{i=1}^{n} p(c_i|B)u(c_i)$ for any constant $a > 0$, and hence $a \sum_{i=1}^{n} p(c_i|A)u(c_i) + b > a \sum_{i=1}^{n} p(c_i|B)u(c_i) + b$ for any constant b.

However, since $a \sum_{i=1}^{n} p(c_i|A)u(c_i) + b = \sum_{i=1}^{n} p(c_i|A)[au(c_i) + b]$ and $a \sum_{i=1}^{n} p(c_i|B)u(c_i) + b = \sum_{i=1}^{n} p(c_i|B)[au(c_i) + b]$, which is the same as

$$\sum_{i=1}^{n} p(c_i|A)u'(c_i) > \sum_{i=1}^{n} p(c_i|B)u'(c_i)$$

Therefore, the desired result is demonstrated. ∎

Case: Computer Networking Strategy

George Whingard frowned as he paced back and forth in front of his office window, occasionally looking out at the sunny spring day.[1] "Just my luck to be stuck here while everyone else is out playing golf," he muttered to himself. Whingard, the General Manager of the Avtex division of General Micro Components, had been wrestling with a request from his engineering manager that had turned into a fight among his engineering manager, division business operations manager, and computer operations manager. He had decided to come into the office on a Saturday to think it through without interruptions.

It had all started a couple of weeks ago when Alice Benson, the engineering manager, had come in to discuss a proposed upgrade to the personal computers (PCs) used by her engineers. She wanted to add a computer network tying together the personal computers used by the engineers. "As you know, we use the PCs for computer-aided design work, which usually involves several people working on the same project. In addition, we have an extensive database of previous designs that we consult and crib from when we do new designs. As it is now, we have to carry things around on floppy disks, and it is very time consuming to coordinate and find previous designs."

"In addition," Benson continued, "different engineers often end up with different, and often outdated, versions of something they are working on. This happens when they make a floppy disk copy of something, and then the original gets updated. Of course, we try to keep track of who has copies, but it is difficult."

Whingard asked how much the network would cost. "Well," Benson said, "it depends on how we decide to do it. As you know, we are still using the mainframe for a lot of the heavy-duty stuff. It would be helpful to tie the network into the mainframe so we could access mainframe files directly. In fact, it would probably be best to use the mainframe as a file server to hold the master

---

[1] This case uses analysis methods presented in Chapters 2 and 3, as well as Sections 4.1 through 4.7 of Chapter 4.

copies of everything. That way, there wouldn't be any question about who had the most up-to-date version."

Whingard said, "Can you give me a ballpark number?"

Benson replied, "Depending on whether or not we tie into the mainframe, it will be from \$20,000 to \$60,000. I'm including operating costs in those figures."

Whingard said, "Well, that is a pretty small part of our budget. I don't see any problem with doing it if you think it would be useful."

A.1 The Trouble Begins

Whingard frowned as he recalled the conversation. If only he had known! After further discussion with Benson, the two of them had decided that they would need to discuss the proposal for a network with Alvin Barrett, the computer operations manager. But Benson said as she left Whingard's office, "You know how those mainframe guys are—they don't want to have anything to do with PCs."

Whether or not that was true, it was certainly true that Barrett was not enthused about tying the engineering PCs to the mainframe. As Whingard, Benson, and Barrett started their meeting, he commented, "We have a lot of confidential data on that machine. I'm not happy about the prospect of letting any old engineer on a PC into that. Besides, those PC networks are real flaky. If the network goes down and takes the mainframe with it, we'll all be in trouble."

The meeting went downhill from there, and Whingard finally told Barrett and Benson that he had another meeting and showed them out of his office.

Alas, there were even more problems. Yolanda Galvez, the division's business operations manager, came in later the same day. She commented, "Well, those techies have got my people all stirred up. Several of our sales reps heard about the proposal by engineering to network their PCs, and started saying how much a PC network would help their work. Now everyone in business operations is talking about what a great improvement having a network would be."

* * *

Whingard looked out at the sunshine again. "Well," he recalled, "Barrett was even less interested in having the business people tied to his mainframe than the engineers. I guess it's time to call in the management consultants. When you've got to tick off some important people, it's nice to have outside confirmation ...."

A.2 First Meeting with the Management Consultant

George Whingard looked across his desk at Deborah Abramson. He said, "That about brings you up to date on this whole mess. What do you think?"

Abramson, a senior consultant with A. J. Erperts and Associates, said, "Let me see if I have it straight. The money is not a major issue. Even the highest estimates put the total at less than $100,000, taking into account operating costs. The problem is that your computer operations people don't like the idea."

Whingard replied, "That's not the only issue, of course, but it is an important one. Barrett has been feeding me all sorts of horror stories about system security problems caused by connecting PC networks to mainframes. When I suggested that we just go ahead with a network that isn't connected to the mainframe, he didn't like that either. He said that they would just download stuff from the mainframe and then put it on the PC network."

"I asked him," Whingard continued, "why they couldn't do that now with the current PCs. He said they could, but that at least without a network the data is limited to one PC, so that security isn't such an issue.

"But, to tell you the truth, I found another of his arguments more convincing. He showed me some studies that have been done that indicate that adding a PC network actually *decreases* worker productivity. According to these studies, the state of the art for PC networks is still not very good, and using a network consumes a lot of people's time. If this is true, then I question the value of doing a network for anybody. The problem is that this has become a 'quality of work environment' issue for both the engineers and business office staff. If I don't give them what they want, they will let me know how unhappy they are. It won't be pleasant, and we could even lose some key people over it—I understand that some of our competitors are putting in networks."

Abramson said, "It sounds like you need to establish a *process* that everyone can agree on to make this decision. If we get everyone to agree on the process, and they have input to it, then that may decrease the controversy."

"Perhaps," Whingard commented. "At least if they can see that their views were taken into account in the process."

Value Analysis

Abramson continued, "Value analysis could be a good approach for this decision. With this approach, you establish a *value tree* that includes all the evaluation concerns for the decision. Then you set up an *evaluation measure* for each concern that allows you to specify in numerical form how each alternative performs with respect to each evaluation concern. If you make sure to include everybody's concerns in the value tree, then it will be clear that you have considered the things that are important for all parties."

"Maybe," Whingard commented, "but I think the problem is that the engineers and the computer operations people attach considerably different importance to, for example, system security."

"Well, if that is true, then we may not be able to reach a decision that everybody likes," said Abramson. "However, using a quantitative approach like value analysis allows you to investigate the impact of varying the importance attached to different evaluation concerns. This is called *sensitivity analysis*. Often we find that for reasonable variations in the *weights* attached to different evaluation concerns, the preferred alternative doesn't change."

"All these people who want the network have been talking about how useful having access to more numbers would be," noted Whingard. "They ought to be willing to try some procedure like this that uses numbers. Also, if they start beefing to my boss, at least I'll have a good record of what I did to show her that I didn't do something crazy."

"That's true," Abramson replied. "An advantage of this approach is that it leaves a good audit trail."

"I'm not sure that is always an advantage, but it should be in this case," Whingard noted. "How will you proceed?"

"Let me go around and talk to people to see if I can develop a value tree and maybe even evaluation measures for each of the concerns in the tree. I'll get back to you after I do that."

A.3 Second Meeting: Value Tree and Evaluation Measures

"I've made good progress," Abramson noted as she started her second meeting with Whingard. "I met with Barrett, Benson, and Galvez separately, and I also met with several of the people that work for each of them. From this I developed an initial value tree. Then I met with the three managers and several of their key people in a single meeting for about three hours. In that meeting, we agreed on a value tree and got preliminary evaluation measures. I wrote this up and ran it by everyone. They are all in agreement with what I am going to show you."

"That's amazing!" Whingard commented. "I figured you wouldn't get any agreement from them."

"I generally find that it isn't too tough to get agreement on a value tree and evaluation measures. After all, if you just include everything that anyone thinks is important, you can get agreement. The problems are likely to come later on when we look at the *tradeoffs* among the various evaluation considerations. We are a lot less likely to have agreement on those.

"Here is the value tree," continued Abramson. (See Figure A.1.) "Also, here are more detailed descriptions of each evaluation concern. (See Figure A.2.) In addition, let me go over the evaluation measures with you." (See Section A.7.)

Whingard and Abramson discussed this information for a while. Whingard said, "This looks good. What is the next step?"

"I am going to work on getting the possible alternatives sorted out."

Whingard noted, "It seems like there will be a lot of alternatives. Every time someone comes in to discuss this with me, they start talking about RAM and LANs and stuff like that. There seems to be a lot of technical stuff that has to be considered."

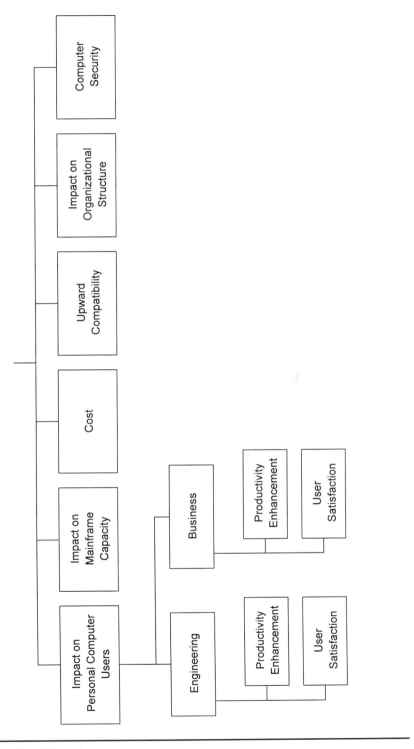

Figure A.1   *The value tree*

1. **Impact on Personal Computer Users:** The degree to which addition of a specified personal computer network changes the work environment of PC users. Since the impact may be different for engineering and business users, each group is considered separately.

   a. **Productivity Enhancement:** Change in the quantity and quality of the work output of users due to the presence of the network.

   b. **User Satisfaction:** The degree to which personal computer users perceive that their work environment has been improved or degraded due to the presence of the network.

2. **Impact on Mainframe Capacity:** Changes in utilization of the mainframe due to the presence of the network.

3. **Cost:** The capital and operating costs to support the specified network.

4. **Upward Compatibility:** The compatibility of the specified network with potential future expanded network uses.

5. **Impact on Organizational Structure:** The degree and type of changes in organizational structure required for successful operation of the network.

6. **Computer Security:** The impact of the specified network on data and software integrity and security.

Figure A.2 *Description of evaluation concerns*

"That's true if you want to decide on exactly what hardware and software to buy, but you are more interested in what *strategy* to follow with regard to a personal computer network. That requires less specification of the network details, and I don't think there are that many different strategic alternatives for this decision. I'll get back to you with the results."

A.4 Third Meeting: Alternatives

"We got good agreement on the alternatives that need to be considered," Abramson noted when she returned to see Whingard. "I approached this by developing a *strategy generation table*. This breaks the alternatives down into their parts, and then we put the parts together in ways that make sense to develop alternatives. After I made it clear that we are interested in a networking strategy, rather than the details of exactly what network to select, it was fairly straightforward to obtain agreement on alternatives.

"Here is the strategy generation table for this decision. (See Figure A.3.) As you can see from this, the elements of the decision are

1 What capabilities are provided to the engineering staff,
2 What capabilities are provided to the business operations staff, and
3 Whether or not the network is connected to the mainframe.

For either the engineering or business staff, the basic issue is whether they have analysis and/or document processing capabilities. Of course, exactly what each

Engineering Capabilities	Business Capabilities	Network Type
1. No network	1. No network	1. No mainframe connection
2. Analysis	2. Analysis	2. Networked to mainframe
3. Document processing	3. Document processing	
4. Analysis and document processing	4. Analysis and document processing	

Figure A.3 *Strategy generation table*

of these terms means is somewhat different for the two groups. I have a more complete specification here if you would like to see it."

"That's all right. I'll read it in your report," Whingard replied.

"If you examine the strategy generation table, you will see that there are four possibilities for the capabilities that could be provided to each of the two groups (no network, analysis only, document processing only, or both analysis and document processing); and then, of course, the network is either connected to the mainframe or not. This gives 4 engineering capabilities × 4 business capabilities × 2 mainframe possibilities = 32 possible alternatives. But one of these has 'no network' for either business or engineering, and then has this connected to the mainframe. That doesn't make sense, so there are really only 31 alternatives. In addition, the engineers agreed that they would not want document processing capabilities without analysis capabilities, and the business operations people agreed that they would not want analysis capabilities without document processing capabilities. When these alternatives are eliminated, there are 17 alternatives left.

"Furthermore, the business operations people agreed that they would not want the network unless it was connected to the mainframe. When you take that into account, there are only 11 alternatives left. Here they are." (See Figure A.4.)

"In this figure," she continued, "the first column has a number that I will use to refer to each of the 11 alternatives. The second and third columns show what capabilities the engineering and business staffs will have for a particular alternative, and the last column shows whether there is a mainframe connection or not. Alternative number 1 is a continuation of the 'status quo' where no network is installed; alternatives 2 through 5 provide a network only to the engineers; while alternatives 6 and 7 provide a network only to the business staff. The remaining alternatives show the various possible combinations where network capabilities are provided to everyone."

"It looks like you have this decision very well specified," Whingard commented. "What is the next step?"

"I have to get numbers for each of the eleven alternatives for all the evaluation measures. Also, I have to obtain a *value function* to combine the various evaluation measures into a single measure of the overall preferability of each alternative. I will go back to the people I have been working with. However, I expect that there will be less agreement on this next part than I have had so far."

A.5 Fourth Meeting: Evaluation of Alternatives

"Before I go and crunch the numbers, I want to report back to you what they look like," Abramson said, as she started her fourth meeting with Whingard. "We had pretty good agreement on the value function, but less agreement on the actual evaluation measure numbers for some of the alternatives. Let's talk about the value function first."

The Value Function

"In value analysis, there are two parts to determining the value function. You need to find *single dimensional value functions* for each of the evaluation measures, and you also need to find *weights* for each evaluation measure. A single dimensional value function takes into account *returns to scale* for a particular evaluation measure. For example, the increment in value that you obtain from decreasing a cost by \$100 may differ, depending on whether you are going from a cost of \$1,000 to a cost of \$900 or going from \$1,000,000 to \$999,900. In the first case, this is a ten percent decrease in cost, while in the second case it is a one-hundredth of a percent decrease.

"The weights show the relative importance of variations in different evaluation measures," she continued. "More specifically, the weight for a particular evaluation measure is equal to the increment in value that is obtained by improving that evaluation measure from its least desirable level to its most desirable level. From this definition you can see that the weight for a particular evaluation measure will depend on the possible range of variation for the evaluation measure. To emphasize this point, the term *swing weight* is sometimes used to indicate that the weight is with respect to the possible 'swing' in the evaluation measure from its least preferred level to its most preferred level.

"Here are the raw data for the single dimensional value functions. (See Figure A.5.) In this figure, the relative value increments are shown for an increase between successive levels of each evaluation measure. For example, the relative value increment for increasing Productivity Enhancement from −1 to 0 is six times as great as the increment for increasing this evaluation measure from 0 to 1. Similarly, the value increment for increasing this evaluation measure from 1

	User Group	Capability	Network Type
1.	Engineering	No network	
	Business	No network	
2.	Engineering	Analysis	No mainframe connection
	Business	No network	
3.	Engineering	Analysis	No mainframe connection
		& document processing	
	Business	No network	
4.	Engineering	Analysis	Networked to mainframe
	Business	No network	
5.	Engineering	Analysis	Networked to mainframe
		& document processing	
	Business	No network	
6.	Engineering	No network	Networked to mainframe
	Business	Document processing	
7.	Engineering	No network	Networked to mainframe
	Business	Analysis	
		& document processing	
8.	Engineering	Analysis	Networked to mainframe
	Business	Document processing	
9.	Engineering	Analysis	Networked to mainframe
		& document processing	
	Business	Document processing	
10.	Engineering	Analysis	Networked to mainframe
	Business	Analysis	
		& document processing	
11.	Engineering	Analysis	Networked to mainframe
		& document processing	
	Business	Analysis	
		& document processing	

Figure A.4   *Alternative networking strategies*

Evaluation Measure	Relative Value Increments				
	$[-2,-1]$	$[-1,0]$	$[0,1]$	$[1,2]$	$[2,3]$
Productivity Enhancement		6	1	4	
User Satisfaction		6	1	4	
Impact on Mainframe Capacity			1	1.25	1.5
Cost	Linear over $0 to $100,000				
Upward Compatibility				1	1.5
Impact on Organizational Structure			2	1	
Computer Security	1.5	1.25	1		

Figure A.5 *Raw data for single dimensional value functions*

to 2 is four times as great as for increasing it from 0 to 1. Notice that all the evaluation measures except Cost have a small number of discrete possible levels. Cost can vary continuously over a range. Because the costs of all the alternatives are small relative to the budget of the division, everyone agreed that the single dimensional value function over cost should be a straight line.

"Notice that there is only one set of numbers for Productivity Enhancement, and also only one set of numbers for User Satisfaction. The engineers and the business operations people agreed on common sets of numbers for these two single dimensional value functions. So the same single dimensional value functions will be used in the evaluation for both groups. As we will see, this is not true for the swing weights for these two evaluation measures.

"Here is the raw data for the swing weights. (See Figure A.6.) The interpretation of this figure is much the same as for the single dimensional value functions. The evaluation measure with the lowest swing weight is Impact on Mainframe Capacity. Engineering Productivity Enhancement has 20 times the weight of Impact on Mainframe Capacity. We had pretty good agreement on most of these relative weights. There was some controversy about the weight for Computer Security. The computer operations people would like to see that higher. Also, there was some discussion about the appropriate weight for Impact on Organizational Structure. I will need to do a sensitivity analysis to determine whether the exact values for these weights affects the ranking of the alternatives."

Evaluation Measure	Range		Relative Weight
	Low	High	
Engineering Productivity Enhancement	−1	2	20
Engineering User Satisfaction	−1	2	20
Business Productivity Enhancement	−1	2	10
Business User Satisfaction	−1	2	5
Impact on Mainframe Capacity	−1	2	1
Cost	$0	$100,000	5
Upward Compatibility	0	2	15
Impact on Organizational Structure	−1	1	10
Computer Security	−2	1	10

Figure A.6 *Raw data for weights*

Evaluation Measure	Networking Strategy										
	1	2	3	4	5	6	7	8	9	10	11
Eng. Prod. Enhance.	0	1	1.5	1.5	2	0	0	1.5	2	1.5	2
Eng. User Satis.	0	1	1	1.5	2	0	0	1.5	2	1.5	2
Bus. Prod. Enhance.	0	0	0	0	0	1	1.5	1	1	1.5	1.5
Bus. User Satis.	0	0	0	0	0	1	1.5	1	1	1.5	1.5
Impact on Main. Cap.	0	0	0	0	0	0	1	0	0	1	1
Cost ($000)	0	20	30	45	55	40	50	60	70	70	80
Upward Compatibility	0	0.5	0.5	2	2	2	2	2	2	2	2
Impact on Org. Struct.	0	0	0	−0.5	−0.5	−0.5	−0.5	−0.5	−0.5	−0.5	−0.5
Computer Security	0	0	0	−1	−1	−1	−1.5	−1	−1	−1.5	−1.5

Figure A.7 *Base case assessments of alternatives*

Assessment of the Alternatives

"There was also some controversy about what evaluation measure levels (scores) are appropriate for the various alternatives," Abramson continued. "Here is a figure with 'base case' numbers (see Figure A.7), but there was quite a bit of discussion about some of these."

Whingard looked over the base case assessments. "When there are negative numbers, something gets worse than the current situation. Right?"

"Yes," Abramson replied. "Those are the cases where there was a lot of discussion. We will need to look at those carefully and see whether variations in them change the results. I will go off and crank through the value analysis calculations."

A.6 Exercises

A.1 Discuss the advantages and disadvantages of conducting a quantitative analysis of the networking decision.

A.2 State whether the diagram in Figure A.1 is a *value hierarchy* as defined in Chapter 2. Give reasons for your answer. If this figure is not a value hierarchy, what changes or additional information are needed to make it a value hierarchy?

A.3 Select one of the evaluation measures listed in Section A.7, other than Cost, and critique the scale definition with respect to how well it meets the clairvoyance test.

A.4 In Section A.4, Abramson discusses the procedure that she used to reduce the number of alternatives to be considered from 32 to 11. List the original 32 alternatives and demonstrate that the conditions given in Section A.4 do, in fact, reduce the number of alternatives to 11.

A.5 Based on the information in Figure A.5, determine the single dimensional values for each of the specified levels (scores) of each of the evaluation measures, except Cost. Determine these values to four decimal places of accuracy. Assuming that piecewise linear single dimensional value functions are used, plot all of the single dimensional value functions, including Cost.

A.6 Based on the information in Figure A.6, determine the swing weights for the nine evaluation measures. Determine these to four decimal places of accuracy. Do these weights sum to 1? If not, explain why not.

A.7 The information in Figure A.7 includes some evaluation measure levels that are between the defined levels for the evaluation measures, as these evaluation measure levels are defined in Section A.7. Select one of these "in between" levels, and explain how such a level might occur.

A.8 Using the information given in the chapter and that you have calculated in preceding questions, determine a correct spreadsheet or other model for the networking decision. Note that the swing weights in a valid model should sum to exactly 1. If the swing weights you determined in Exercise A.6 do not sum to exactly 1, change the weight for "Impact on Mainframe Capacity" to make the total for the weights equal to 1.
  (i) Include a printout of the model with your answer.
  (ii) Determine the values for each of the 11 alternatives, as well as which alternative is most preferred. Include the model output giving this information with your answer.

A.9 Determine, accurate to two decimal places, the range of swing weights for "Computer Security" for which the most preferred alternative identified in Exercise A.8 remains most preferred. As part of your answer, include the model output that you used to determine the answer, and explain which alternative (or alternatives) is most preferred when the swing weight changes enough to change the most preferred alternative. Assume, while you are varying the weight for "Computer Security," that the ratios of the other weights remain the same.

A.7 Supplement: Evaluation Measure Scales

Note: The same scales for Productivity Enhancement and User Satisfaction apply to both the engineering and business staff.

1a Productivity Enhancement

−1) User-group productivity is diminished sufficiently that noticeably longer time or more resources are required to provide the same level of service.

0) No perceivable change in user-group productivity.

1) User-group productivity is enhanced to the extent that group members are perceived by their clients to be providing better service, or somewhat fewer resources are required to provide service at the same level as before the network was installed.

2) Significant and easily perceived increase in user-group productivity. Indicators of this could include significant reduction in staffing level required to carry out user-group activities or considerable improvement in financial performance of the group.

1b User Satisfaction

−1) A significant number of user-group members do not accept use of the network or feel the network is a detraction from their work environment.

0) No noticeable change in user-group satisfaction with their personal computer resources.

1) Many user-group members believe the addition of the network has enhanced their work environment.

2) Virtually all user-group members believe the addition of the network has moderately or significantly enhanced their work environment.

2 Impact on Mainframe Capacity

−1) The addition of the network causes a significant increase in system load on the mainframe, necessitating an acceleration of the system capacity expansion schedule.

0) There is no change in the mainframe load due to implementation of the network.

1) There is a noticeable but relatively minor decrease in the mainframe load due to implementation of the network.

2) The addition of the network causes a significant decrease in mainframe load by supporting a substantial part of the computing load that would otherwise have been on the mainframe.

3 Cost

Net present value of capital and operating costs for the network.

4 Upward Compatibility

0) While the network meets current needs, it has little capability for expanding to meet other needs.

1) The network either currently has the capability of serving all currently projected needs for the network or can be expanded to serve these needs in a straightforward manner.

2) The network could be expanded in a straightforward manner to meet all potential expanded needs, including future uses that do not seem likely but are possible.

5 Impact on Organizational Structure

−1) Significant disruptive and detrimental changes in organizational structure are necessary to effectively utilize the specified network.

0) No changes in organizational structure are necessary to effectively utilize the specified network.

1) While some changes in organizational structure may be necessary to effectively utilize the specified network, these changes are of a minor nature or are generally beneficial, and can be made without disruptive dislocations.

6 Computer Security

−2) The addition of the network causes a substantial decrease in system control and security for the use of data or software.

−1) There is a noticeable diminishing of system control and security.

0) There is no detectable change in system control or security.

1) System control or security is enhanced by addition of a network.

Scenario Planning for Decision Making

Our decision making is often clouded by uncertainty about what will happen in the future. Will those fickle customers buy our new product? Will I be promoted (or perhaps laid off instead)? What will the stock market do next year? Who will win the next election? And so on. . . . Scenario planning addresses this difficulty. Scenarios can help you better understand the potential consequences of your uncertain decisions and design improved alternatives to address the uncertainties that you face.

B.1 The Role of Scenario Planning

The term *scenario* originated in the performing arts, where it means an outline or synopsis of a film or play. As used here, the term means an internally consistent story about how events relevant to your decision might develop over time. A *scenario planning* approach to decision making explicitly considers more than one scenario for possible future events while making a decision.

Good stories transmit ideas in a compelling way that is difficult to replicate with the dry output of a quantitative analysis (Schank 1990). Thus, scenario planning can complement or replace probabilistic decision analysis (Clemen 1996). For example, Wack (1985a, b) discusses scenario planning at Royal Dutch/Shell during the early 1970s. Company planners had concluded that an oil price shock was likely and that it could have serious implications for Shell. They presented their results with the usual tables and charts, and they found they were having difficulty getting Shell's managers to take them seriously. This was because the prediction contradicted the lifetime experience of Shell's managers, who had always seen relatively stable oil prices. Then the planners created scenario "stories" showing what would be necessary to prevent a price shock and also showing the implications of such a price shock. These stories were more compelling than a dry recital with tables and charts, and Shell managers took them seriously. As a result, Shell was better prepared for the 1973 price increases than other oil companies.

In this case, scenarios were used to present the results of a traditional planning analysis. Scenario planning can also serve as an alternative to more traditional approaches. It allows those who are less quantitatively sophisticated to provide input to decision making under uncertainty in a natural way. Only a relatively few people are comfortable with the machinery of probabilistic analysis, but everyone can tell a story. Using the methods presented below, almost anyone can frame a scenario and hence provide valuable input to decision making.

B.2 Structured Decision Making

Scenario planning can form part of a structured decision making process. Such a structured approach to decision making includes the following steps:

1 Define objectives.
2 Specify alternatives.
3 Rate (score) each alternative with regard to how well it achieves each of the objectives.
4 Select the alternative that, on balance, provides the best overall performance with respect to all the objectives.

The degree of formality in implementing these steps can differ greatly. At the informal end, a 1772 letter from Benjamin Franklin to Joseph Priestly (Russo and Schoemaker 1989) discusses how Franklin assesses an alternative (called a "measure" in the letter):

> [I] divide half a sheet of paper by a line into two columns; writing over the one Pro, and over the other Con.... I put down under the different heads short hints of the different motives ... for or against the measure. When I have thus got them all together in one view, I endeavor to estimate their respective weights; and where I find two, one on each side, that seem equal, I strike them both. If I find a reason pro equal to two reasons con, I strike out the three. If I judge some two reasons con, equal to some three reasons pro, I strike out the five; and thus proceeding I find at length where the balance lies.

A somewhat more formal procedure "grades" each alternative in much the same way that a student receives grades in classes. With this procedure, an alternative is given a grade with respect to each of the objectives, and an average grade is calculated for each alternative. (The different objectives may receive different weights in the calculation of this average.) The alternative with the highest average grade is deemed the most preferred. More advanced procedures, such as those presented in the body of this book, use more detailed methods to score the alternatives and select the one that is most preferred. (See also Keeney and Raiffa 1976).

A structured decision making procedure can be very useful in clarifying the elements of a decision and arriving at the best decision. However, a difficulty arises when there is uncertainty—how do you score each of the alternatives? How well a particular alternative will perform if implemented may depend on how the uncertainties resolve. Kirkwood (1982) provides an example. An electric utility company was selecting a site for a nuclear power plant in a region with limited water availability. One possible source of water was considerably cheaper than the other available sources, but there were substantial uncertainties about whether it could be used. Specifically, there were certain political issues that might restrict use. The preferred site for the power plant would change depending on whether or not this site was available. What assumption should be made in the decision analysis?

Scenario planning provides a way to proceed in such a case. The next section discusses how to develop scenarios.

B.3 Developing Scenarios

The following eight-step process to develop scenarios is based on a procedure presented by Russo and Schoemaker (1989). [See also Schoemaker (1995).]

1 Define the key variables in the decision. These definitions should present the time frame over which each variable is important.

2 List the major players in the decision (including companies, governments, consumers, unions, and others). Present each player's role, interests, and power positions.

3 List potentially important economic, political, technological, environmental, and social trends that might affect the decision.

4 Specify key uncertainties: Variables you cannot predict but that can impact the decision outcome positively or negatively.

5 Construct two "extreme" preliminary scenarios by putting all good outcomes in one and all bad outcomes in the other.

6 Assess the internal consistency and plausibility of these extreme scenarios, and rearrange the scenario elements to create at least two internally consistent and plausible scenarios.

7 Evaluate the behavior of the major players (identified in Step 2) in the scenarios created in Step 6 to determine whether the players' actions might change the scenarios.

8 Based on Step 7, create two, three, or four distinctly different scenarios that cover a broad range of possible future conditions, and that are each internally consistent.

As shown in Step 8 of this procedure, experienced scenario planners recommend that you use two, three, or four scenarios in a particular analysis. It is unlikely that the actual course of events will exactly follow one of these scenarios, but using a small number of scenarios allows you to focus on the key issues impacting the success of the selected alternative. As Schwartz (1991) comments, "[T]hose fundamental differences ... must be few in number in order to avoid a proliferation of different scenarios around every possible uncertainty. Many things can happen, but only a few scenarios can be developed in detail, or the process dissipates." Kleiner (1994) further notes, "You don't predict what will happen: you posit several potential futures, none of which will probably come to pass, but all of which, make you more keenly aware of the forces acting on you in the present."

Both experience and scientific research show that we have a strong tendency to underestimate our uncertainty about the future (Dawes 1988). Examples of this abound. Electric utilities who invested in nuclear power plants during the 1960s and 1970s seriously underestimated the financial risks involved. Other examples include Midwest investments (which turned out to be speculations) in farm land during the 1980s, savings and loan mortgages to those farmers and many others for (what turned out to be) shaky real estate investments, the failure by IBM and Digital Equipment Corporation to understand the potential impact of personal computers on their businesses, the failure of the U.S. auto manufacturers to understand the nature of changing consumer preferences and the impact of changing government regulations during the 1970s, and the much studied Challenger space shuttle disaster.

In all these cases, it was possible to foresee what actually happened, but we have a lot of difficulty doing this in actual decisions. Scientific research has shown that the human brain is wired together in such a way that we find it hard to seriously consider a wide enough range of possible futures. We tend to focus in a narrow range of possibilities and quickly come to believe that those are the only things that could happen.

Another difficulty that compounds the problem is that important trends that impact the desirability of different alternatives (Step 3 of the scenario development process) may come from areas that seem distant from the decision. Schwartz (1991) gives an example. He and some associates were considering starting a small business to import high-quality English gardening tools. It turned out that a key factor impacting whether the business would be viable was the future U.S. balance of payments. While this might be obvious to an MBA graduate steeped in international business, how many small business persons interested in selling gardening tools are likely to think of this?

As another example, Kleiner (1994) discusses a workshop where the participants were considering business options related to a national digital network with on-demand video. One participant happened to investigate the semiconductor manufacturing capability required to support such a network and concluded that the United States would have to double this manufacturing capability to meet the need. This established a dramatically different time frame than the participants had previously been assuming for the development of the network. Since none of the participants worked in electronic hardware, it was not natural for them to consider this question.

There are no magic answers to this problem of underestimating the uncertainty about the future, but there are some practices that help. First, simply being aware of the problem seems to help some. Second, the group developing scenarios should be diverse. Kirkwood and Seidman (1985) give a political example:

> Senator Fulbright told the story of the last conference with President Kennedy before the Bay of Pigs invasion of Cuba. The president asked for comments from those present—CIA, military, State Department and the like. All were enthusiastically in favor. Fulbright was last and meekly

asked, "What if it's a failure? What are the consequences?" The president and the others answered, "It will not fail," and the rest is history.

Clearly, the group underestimated the uncertainty in this situation. Perhaps if they had been aware of everyone's tendency to underestimate uncertainty or had included a more diverse group in the decision making, they would have taken Senator Fulbright's question more seriously.

B.4 Common Scenario Plots

Schwartz (1991) states that there are three generic scenario plots that are so common that you should consider whether they are relevant for almost any decision:

- **Winners and Losers (Zero Sum Game):** The idea underlying this plot is that "resources" (natural resources, customers, etc.) are limited, and so if one side gets richer, the other side must get poorer. This can be a particularly appropriate plot for very mature industries or political decision making. Schwartz notes that, in this plot, conflict is inevitable. A balance of power may develop with strange alliances driven by pragmatic considerations. Conspiracy theories are common. In a political setting, war or persistent low-level conflict can result. Such conflict can go on for a long time with no clear winner (for example, the Cold War).
- **Challenge and Response (Race without a Finish Line):** This is a common theme in scripts for adventure stories where the hero faces one unexpected test after another. As a result of each test, the hero is changed. Schwartz notes that this tends to be the way that many Japanese view the world. Also, this may be an appropriate plot for examining environmental/ecological issues.
- **Evolution (Continuous Incremental Change):** Evolution involves slow changes in one direction or another—usually either growth or decline. Because of the slow nature of these changes, they can be hard to spot. However, if they are detected, they are relatively easy to manage because of the slow nature of the change. Schwartz comments that scenarios involving technology are often evolutionary. Social and economic considerations usually slow the pace of technology impacts. Overall business trends also tend to be evolutionary.

Schwartz notes that in addition to these three scenario plots, some others also commonly show up:

- **Revolution (Discontinuity):** Events like the 1929 stock market crash or the Chernobyl nuclear accident can lead to rapid dramatic changes in people's perceptions of what is possible.
- **Cycles:** This type of scenario differs from Evolution in that the direction of motion of some important variable(s) changes. Urban decay and rejuvenation, real estate construction, and other activities tend to follow cyclical patterns. If you think you are facing an Evolution scenario, it is always a good idea to consider whether it might really be cyclical.

- **Infinite Possibility:** The world will expand and improve, infinitely.
- **The Lone Ranger:** This is the "Apple Computer takes on IBM" scenario. The hero rides into town to save the day. Of course, who is the hero and who is the villain may depend on your point of view.
- **My Generation:** To some extent this is more of a background setting for other scenario plots than a plot of its own. The Baby Boom generation has impacted the world in many ways, and any scenario that has not taken the Baby Boomers into account over the last couple of decades has been ignoring an important element of the scenario. There may be a similar background setting that is relevant for the decision you are analyzing.

You should carefully examine any scenario that includes an assumption of some trend continuing in an "unbroken line" without any response. The story above about Royal Dutch/Shell shows that Shell managers in the early 1970s believed that low oil prices would continue indefinitely. However, this unbroken line was unlikely, as is discussed below. Similarly, when a real estate construction boom is on, all the builders can project how much money they will make if it continues. They forget that a lot of other people are also building, and an over-supply will surely develop. If you are in a scenario with competitors, it is likely that they will respond to whatever you do, and hence change the competitive conditions that currently exist.

Schwartz also suggests that you should put considerable thought into naming your scenarios. He notes that if the scenario names are vivid and memorable, they are more likely to be remembered and taken into account in decision making.

B.5 The Royal Dutch/Shell Experience

The Royal Dutch/Shell Group has used scenario planning since the early 1970s. Senge (1990) comments that Shell went from being considered the weakest of the seven largest oil companies in 1970 to being, along with Exxon, the strongest by 1979. The scenario planning work started prior to the 1973 oil price shock (Wack 1985a, b). At that time, oil prices had been relatively steady since World War II. Royal Dutch/Shell planners concluded that a number of contingencies made this unlikely to continue, including (1) the prospect of exhausting U.S. oil reserves while U.S. demand for oil continued to grow, and (2) resentment among Islamic countries about Western support for Israel after the 1967 six-day Arab-Israeli war. The planners wrote two sets of scenarios, each with projected oil price figures. One set of scenarios presented the conventional expectation that prices would remain stable, while the other showed a price crisis sparked by the Organization of Petroleum Exporting Countries.

Initially, these were presented as dry recitals of facts, from which a careful analysis could conclude that stable oil prices were unlikely to continue. However, there was relatively little reaction from Shell management. The projected price crisis was so contradictory to the managers' many years of experience that they paid little attention to it.

In early 1973, the scenarios were rewritten to show a story about how the projected future could come about and what it would mean. It was much clearer from reading this set of scenarios that the "business as usual" future was unlikely, and furthermore, that the impacts on the oil business would be very significant. As a result, Shell managers began to prepare for the possibility of substantial change. Schwartz (1991) comments, "In October 1973, after the 'Yom Kippur' war in the Middle East, there *was* an oil price shock. The 'energy crisis' burst upon the world. Of the major oil companies, only Shell was prepared emotionally for the change. The company's executives responded quickly. During the following years, Shell's fortunes rose."

Later, in 1983, Shell was moving toward a decision whether to develop the Troll gas field, a deposit of natural gas under water one thousand feet deep in the North Sea. The field would supply natural gas to Europe after a projected development cost of six billion dollars. The profitability of this investment depended on selling the natural gas at a relatively high price, and Shell planners realized that the Soviet Union could potentially provide natural gas to Europe at a much lower price. However, the Europeans had an informal agreement not to take more than thirty-five percent of their natural gas from the Soviet Union. In addition, it would take a development effort in the Soviet Union by the multinational oil companies to provide significantly more natural gas, and the Soviets had never allowed this.

An end to the Cold War was almost inconceivable in 1983, but Shell planners realized that a significantly larger Soviet gas supply to Europe could seriously reduce the value of the Troll field, and thus they developed scenarios investigating this possibility. They came to realize that economic conditions in the Soviet Union, driven by long-term demographic trends, could force a change in Soviet policies. Schwartz (1991) comments that when a possible scenario for this was presented to government agencies, virtually every Soviet expert told them they were crazy. "Our insight came solely from asking the right questions. From having to consider more than one scenario. If we had to pick only one, we might have been just as wrong as the CIA.... But having more than one scenario allowed us to anticipate [Gorbachev's] arrival and understand its significance when he ascended to the leadership."

B.6 Planning for the Next Ten Thousand Years

In 1990, I served on a four-person team that developed scenarios for possible inadvertent intrusion into the Waste Isolation Pilot Project (WIPP) over the next ten thousand years (Benford et al., 1991). The WIPP, located twenty-six miles east of Carlsbad, New Mexico, is a defense activity of the Department of Energy that is to serve as a research and development facility to demonstrate the safe disposal, in natural bedded salt formations, of radioactive waste resulting from the defense activities and programs of the U.S. government. By late 1989, over 10 miles of underground structures had been excavated. This included four deep shafts extending 2,150 feet below the surface, and horizontal tunnels and rooms at that depth. Underground rooms and connecting passageways were 13 feet high and 33 feet wide.

Previous studies had identified inadvertent intrusion as a significant possible cause of radioactivity release over the ten-thousand-year period that had been selected for study. The team I served on was one of four charged with investigating the likelihood of this inadvertent intrusion and how it might occur. Our team was varied: An astrophysicist who also writes science fiction, a decision analyst, a physical scientist turned social scientist who had once trained to be an astronaut, and a geographer. We differed from the other three teams in that we were from the Southwest. We commented in the introduction to our report that

> While reviewing the material on markers [to prevent inadvertent intrusion into the WIPP] provided by U.S. Department of Energy personnel and contractors, we were struck by the fact that these recommendations regarding markers implicitly assume that future potential inadvertent intruders will look basically like Twentieth Century archaeologists (except, perhaps, that they will not understand English very well). We hope our report gives images of how truly different the future is likely to be.... What will be the worldview of someone contemplating the WIPP site in 12,000 A.D.?

The truly awe-inspiring task of trying to write scenarios for the next 10,000 years may seem distant from business planning, but it places in sharp focus the issues involved in effective scenario planning. Is predicting the next ten years really that much easier than trying to say something useful about the next ten thousand years? In 1985, almost no one gave serious attention to the possibility of the Soviet Union breaking up. Yet within ten years this event was having an impact on businesses and governments all around the world. Whether you are worrying about a business decision with impacts for the next ten years or storing nuclear waste for the next ten thousand years, uncertainties about what will happen are a reality.

Our approach to writing scenarios for the next ten thousand years was not that different from writing scenarios for more routine business decision making. We followed essentially the eight-step process outlined earlier in this appendix. We realized that the key variables impacting inadvertent intrusion were (1) the identity of possible intruders, (2) their motivations for intruding, and (3) their technological capabilities. We soon came to the conclusion that a potentially important political/social trend (Step 3 of the scenario development process) was what would happen to political control of the area around the WIPP. Loss of control by the United States, or any other advanced nation, is certainly possible. As we commented,

> Those who travel Interstate Route 8 between Arizona and San Diego are familiar with the agricultural inspection and immigration(!) checkpoints. This is more control on transit than there is between some Western European nations, and it provides an appropriate image of the place of the Southwest in U.S. history. Antonio de Espejo crossed the WIPP region in 1582. This is, as the saying goes, an ancient land, and one where the impact of U.S. control is light and, possibly, transient.

The characteristics of the final set of scenarios we developed are indicated by their names:

1 Technological knowledge increases
- Mole miner scenario
- Nanotechnology scenario

2 Technological knowledge decreases: Doom and gloom scenario

3 Decline and rebuilding of technological knowledge: Seesaw scenario

4 Altered political control: The Free State of Chihuahua

5 Stasis: 10,000 years of solitude

This violates the "no more than four scenarios" rule. The "technological knowledge decreases" scenario doesn't really add much that is useful to the discussion because it doesn't have much in the way of implications for preventing inadvertent intrusion. Thus, there are really four significant scenarios, one of which (technological knowledge increases) has two variations. As we presented each scenario, we told a story with specific dates and activities to give a concrete picture of what it would feel like to live through the scenario and how it could come about.

In retrospect, this exercise wasn't as useful as I had hoped. Follow-on work by others resulted in mundane proposals for marking the WIPP to forestall inadvertent intrusion. The proposed markers were basically along the lines of the pyramids or Stonehenge. These sorts of markers had been proposed before our scenario work, and it is difficult to detect that we had much of an impact on the new work. A significant problem in much of the work on public health and safety risks is the narrowness of the group involved. It is basically the "Bay of Pigs" problem—everyone involved, including those who oppose projects with technological risks, has essentially the same view of what is possible. Even our modestly daring scenarios for the WIPP were too outlandish to be taken seriously.

Perhaps we can hope that the following vignette from our report comes to pass:

> TDY1142 released her sleeping cocoon and mumbled to her dressing robot, "Something blue." Then "news on." The announcer's image materialized above the kitchen table. "Good morning. In the top of the news today: The City Builders have discovered some prehistoric ruins at 2100 feet while moving south toward the Mexican isthmus. Following the disastrous release of the common cold last year from other ruins, they are proceeding with caution ..."

B.7 Using Scenarios for Assessment of Alternatives

Using scenarios to assess alternatives is straightforward. Each scenario should be defined in enough detail so that all important considerations needed to assess alternatives are specified. Therefore, the assessment procedure can be applied to all alternatives successively using the different assumptions from each scenario. If fate is kind, the same alternative will be preferred under all of the different scenarios, and you can proceed to implement that alternative.

However, that is not likely to happen unless you work to develop alternatives that are designed to provide good results under all the scenarios. In fact, a key use of scenarios is to design alternatives that are resilient in the face of different future events.

Schwartz (1991) comments on this in his discussion of the scenario work at Royal Dutch/Shell on the Troll gas field. While Schwartz's management was skeptical that the Soviet Union would be a significant factor in European gas markets, he still proceeded to develop a scenario investigating this possibility and soon realized that the Soviet economy was facing a major crisis driven by declining productivity and a long-term decline in birthrate. This might lead them to open up their economy. Shell proceeded with the Troll field, but made substantial efforts to keep the cost down and also to avoid investing in new oil fields. Thus, the scenario planning exercise helped guide them in developing new alternatives. This process is discussed in more detail in the next section.

B.8 Using Scenarios for Design of Alternatives

Crawford, Huntzinger, and Kirkwood (1978) provide a business planning example relevant to using scenarios to aid design of alternatives. An electric utility was planning to build a high-voltage transmission line and was attempting to decide on a size for the electrical conductor to be used in the transmission line. A smaller conductor would cost less to build than a larger conductor, but it would cost more to operate because of the higher electrical resistance of the smaller conductor. Thus, in selecting the size for the conductor there was a tradeoff between immediate construction costs and longer-term operating costs.

The future operating costs would depend on what it cost to generate the electricity that would be sent over the transmission line. This generation cost was dependent on the cost of the oil used to fuel the generating plant. Initially, the choice seemed to be between a small conductor (with low capital costs, but higher operating costs) and a large conductor (with higher capital costs, but lower operating costs). When this analysis was conducted in the mid 1970s, it was clear from recent experience that there was substantial uncertainty about the cost of oil. Which size conductor would be best depended on the future price of oil to generate the electricity that would be sent over the transmission line, and this price was highly uncertain at the time of the decision.

By explicitly examining different scenarios, this tradeoff was made clear, and a new alternative was developed: The towers to hold the transmission line could be built strong enough to hold a large conductor, but only a small conductor would be installed. That way, if the price of oil later turned out to be high, a larger transmission line could be installed without having to completely rebuild the entire transmission system. Of course, stronger towers would be more expensive, but not as expensive as building an entire system with larger transmission lines. (Further analysis indicated that this flexible option was not worth the extra cost, but it illustrates the idea behind using scenarios as a guide to developing alternatives.)

If you have developed an initial set of alternatives, some of which are better under some scenarios and some of which are better under other scenarios, then there may be ways to combine these alternatives to design new alternatives that are good under a wider range of scenarios. Here are four approaches to doing this: hedging, sequencing, risk sharing, and insuring.

The traditional meaning of *hedging* in financial transactions is to protect yourself from the risk of a loss by a countervailing transaction. More generically, hedging means to develop a new alternative that is a combination of existing alternatives, each of which performs well under a different scenario. The intent is for the combined alternative to perform well under a wider range of scenarios. Thus, in the electrical transmission line decision, a hedging alternative would combine a large and a small conductor, which results in a medium-sized conductor. This might not be the cheapest alternative with either low or high future oil prices, but it would not do too badly in either case. By selecting this hedging alternative, you ensure that you will not do very poorly regardless of what happens. Of course, you also give up the opportunity of doing as well as you would have if you happened to guess right about the future.

It may also make sense to design an alternative that allows you to *sequence* your decisions. That is, if you can break up a decision into a sequence of smaller decisions that occur over a period of time, you will have more time to determine what scenario for the future is actually occurring before you have to make some of the later decisions. The alternative with "overbuilt" towers in the electrical transmission line decision is an example of this type of alternative. This example illustrates that with this type of alternative, you must generally pay more in the early stages to retain the flexibility to select among multiple options later. As another example, I once did work for an electric utility that was considering building a new power plant. At the time of this work, it was becoming clear that nuclear power might have some difficulties. Thus, the company chose to select good sites for both nuclear and coal power plants. Site selection for either type of plant is expensive, and so there was a cost associated with retaining the flexibility to later choose either type of plant in a sequential decision process.

Another type of alternative that addresses uncertainties about the future is *risk sharing*. With this approach, you take a partner. When you do this, you lose the opportunity to make all the profit if the uncertainty turns out well, but you also share the loss if things don't turn out so well. For example, property hazard insurance companies will often sell portions of their portfolio of policies if they have a large number of policies of a particular type (for example, earthquake insurance in California). In this way, if there is a serious natural disaster in a particular area, they will not have such a large loss. Of course, they will also not make as large a profit if there are no disasters in the area.

Finally, you may be able to take out *insurance* against the risks in some scenarios. Of course, we are all familiar with personal health or life insurance, which we take out even though we do not plan to use it. In the electrical transmission line example, the company might have been able to find someone else willing to build the line and rent capacity to the company for a fixed cost. As with the other alternatives discussed in this section, the cost for this alternative would probably have been higher than what it would cost to build and operate the line under some scenarios, but it would have avoided the risk of high costs under some other scenarios.

B.9 Scenario Planning as a Philosophy for Decision Making

At its core, scenario planning for decision making is a philosophy for making decisions. It says that you should not "bet the house" on a particular future and leave yourself vulnerable if unexpected events occur. With a scenario approach to decision making, you explicitly consider different possible futures and design alternatives that allow you to face these futures without undue worries about the risks associated with your alternatives if things don't turn out as you expected.

Schwartz (1991) summarized the scenario work at Royal Dutch/Shell in the 1980s by commenting

> [W]e were certainly pleased with our anticipation of perestroika. But the test of our scenarios, the justification for their practice, was the fact that Shell prospered because of them. They cut costs on the Troll gas fields. They did not buy oil fields when oil was thirty dollars, when everyone else did; they bought them after the price fell to fifteen. Scenarios gave them a huge long-term advantage and allowed them to think in long-term strategies. They could act with the confidence that comes from saying, "I have an understanding of how the world might change. I know how to recognize it when it is changing, and if it changes, I know what to do."

B.10 References

G. Benford, C. W. Kirkwood, H. Otway, M. J. Pasqualetti, "Ten Thousand Years of Solitude? On Inadvertent Intrusion into the Waste Isolation Pilot Project Repository," Report LA-12048-MS, Los Alamos Laboratory, Los Alamos, New Mexico 87545, March 1991.

R. T. Clemen, *Making Hard Decisions: An Introduction to Decision Analysis*, Second Edition, Duxbury Press, Belmont, California, 1996

D. M. Crawford, B. C. Huntzinger, and C. W. Kirkwood, "Multiobjective Decision Analysis for Transmission Conductor Selection," *Management Science*, Vol. 24, pp. 1700–1709 (1978).

R. M. Dawes, *Rational Choice in an Uncertain World*, Harcourt Brace Jovanovich, San Diego, 1988.

R. L. Keeney and H. Raiffa, *Decisions with Multiple Objectives: Preferences and Tradeoffs*, Wiley, New York, 1976.

C. W. Kirkwood, "A Case History of Nuclear Power Plant Site Selection," *Journal of the Operational Research Society*, Vol. 33, pp. 353–363 (1982).

C. W. Kirkwood and L. W. Seidman, "Avoiding Decision-Making Errors," *Pace*, Vol. 12, No. 9, pp. 65–69 (September 1985).

A. Kleiner, "Creating Scenarios," in P. M. Senge, C. Roberts, R. B. Ross, B. J. Smith, and A. Kleiner, *The Fifth Discipline Fieldbook: Strategies and Tools for Building a Learning Organization*, Doubleday Currency, New York, 1994.

J. E. Russo and P. J. H. Schoemaker, *Decision Traps: Ten Barriers to Brilliant Decision-Making and How to Overcome Them*, Simon and Schuster, New York, 1989.

R. C. Schank, *Tell Me a Story: A New Look at Real and Artificial Memory*, Scribner, New York, 1990.

P. J. H. Schoemaker, "Scenario Planning: A Tool for Strategic Thinking," *Sloan Management Review*, Vol. 36, No. 2, pp. 25–40 (Winter 1995).

P. Schwartz, *The Art of the Long View*, Doubleday Currency, New York, 1991.

P. M. Senge, *The Fifth Discipline: The Art and Practice of the Learning Organization*, Doubleday Currency, New York, 1990.

P. Wack, "Scenarios: Uncharted Waters Ahead," *Harvard Business Review*, September/October 1985a, pp. 73–89.

P. Wack, "Scenarios: Shooting the Rapids," *Harvard Business Review*, November/December 1985b, pp. 139–150.

B.11 Exercise

B.1 Project (Part D′). This is a continuation of the project from Chapters 1, 2, and 3.[1] In this part of the project, you determine a value function, collect data about your alternatives, develop scenarios to address uncertainties, and conduct a value analysis for your alternatives.

    (i) Assess a value function for your decision. Present the general procedure used for the assessment but not a blow-by-blow description.

---

[1] This is an alternate assignment to the Part D project assignment presented in Exercise 4.8.

Include assessed "raw data" used to determine the value function, perhaps in a table. Show the math used to obtain the final value function from the assessed raw data (perhaps in a figure), as well as the parameters for the final value function.

(ii) Present the scenarios used to analyze the impact of uncertainties. Describe the process used to develop the final set of scenarios, as well as other scenarios that seem relevant and the reasons these were not included.

(iii) Present the procedure used to determine the evaluation measure scores (levels) for each alternative under each scenario. Present the evaluation measure scores for each scenario, perhaps in one or more tables. Reference data sources, including interviews with experts, in standard bibliographic style.

(iv) Present the value calculations for the alternatives for each scenario, as well as a sensitivity analysis. Briefly describe how computations were done, but you do not have to present the actual computations if you use a spreadsheet program to do these calculations. Include a display of the equations for any spreadsheet model that you use. Conduct and present a systematic sensitivity analysis to investigate how variations in key assumptions impact the analysis results.

(v) Present your conclusions based on the analysis in the preceding parts, including a qualitative discussion of the reasons that the preferred alternative is best. The goal of this discussion is that someone who does not understand the details of decision analysis methods will find your discussion to be a convincing argument for the preferred alternative. That is, the analysis should not be a mysterious procedure, but rather a way of developing insight about the key factors in the decision and how these lead to selection of the preferred alternative.

APPENDIX C

Probability Elicitation Interview

This appendix[1] presents an annotated transcript of an interview in which a decision analyst elicited a judgmental (subjective) probability distribution from a senior aerospace executive. This interview was conducted using the elicitation protocol presented in Sections 5.4 through 5.8.

In practice, an analyst must balance the need for careful and detailed elicitation of probabilities against other requirements in the analysis process. Many real-world analyses are conducted under (often severe) time and resource constraints. Several probability distributions may be needed for an analysis, and the analyst must consider the total available time and budget for all the elicitations (as well as other parts of the analysis) when deciding how much effort to apply to determining each distribution. In addition, the analyst must judge the tolerance of a particular manager for participating in a detailed elicitation—it may be better to obtain a small amount of information about the probability distribution rather than persevere with a formal protocol until the manager is no longer willing to participate.

In other words, managing the probability elicitation process requires that the analyst make *tradeoffs* among a variety of objectives that are important to completing a decision or risk analysis in a timely manner within a reasonable budget. Learning to make such tradeoffs, which are common to many analyst/manager interactions, is an important part of becoming a good analyst. The analyst must learn to take into account not only manager verbal statements but nonverbal cueing as well. This appendix contributes to the learning process by documenting an actual elicitation session. The session is annotated to show the thought process of the analyst while managing the process, and these annotations illustrate both the elicitation protocol and interview management issues that an analyst must address while interacting with a manager.

---

[1] This appendix is adapted from Shephard and Kirkwood (1994). Copyright ©1994 IEEE. Reprinted, with permission, from *IEEE Transactions on Engineering Management*; Vol. 41, No. 4, pp. 414–425; November 1994.

C.1 Judgmental Probability Elicitation Protocol

When eliciting judgmental probabilities for continuous uncertainty quantities, the cumulative distribution function is usually the most convenient summary of the probability distribution, as discussed in Section 5.8. Several fractiles of the probability distribution are elicited from the subject, and the remainder of the cumulative distribution function is estimated, either by drawing the curve in by hand or by more formal curve-fitting techniques. Figure C.3 below shows an example of a cumulative distribution function $F(X)$ determined in this manner. In this figure, the 0.01, 0.25, 0.50, 0.75, and 0.99 fractiles are marked by the symbol $\otimes$. (For example, the 0.50 fractile shown in Figure C.3 is 44. Thus, there is a 0.50 probability that the actual value of the uncertain quantity will be less than or equal to 44.)

The probability elicitation protocol presented in Sections 5.4 through 5.8 includes five stages: (1) motivating, (2) structuring, (3) conditioning, (4) encoding, and (5) verifying. The annotated transcript in this appendix demonstrates the use of this protocol.

C.2 Data Collection Procedure

The probability elicitation interview presented here was conducted as part of research (Shephard 1990) investigating the feasibility of guiding experts to self-elicit their own knowledge or expertise. The author led a judgmental probability elicitation interview with a senior executive of a large aerospace company. The participating executive had major planning responsibilities for his corporation. His educational background was in industrial engineering with training in probability theory, and he had also taught the subject.

Videotape recording and postproduction processing was used as a basis for investigating the probability elicitation process used by the decision analyst. Figure C.1 shows the interview setup for the elicitation interview. Two video cameras were used to record the analyst's perceptions, one focusing on the elicitation interview subject (the aerospace executive) and one focusing on the notepad used by the decision analyst.

Following completion of the elicitation interview, the two recorded videotapes were merged into a single tape containing a large picture of the elicitation subject, a small insert of the decision analyst's notepad, and an elapsed time code. The sound track from the original tape of the elicitation interview was also inserted into this tape. The decision analyst then reviewed this composite videotape to reconstruct the thought processes he used during the interview. He labeled the identifiable "chunks" of organized information that he used while leading the elicitation process, and these labels were then overlaid onto the previously constructed composite tape.

The research reported in Shephard (1990) used this final videotape as a basis for developing rule conditions and actions in a cognitive information processing system model (Gilmartin, Newell, and Simon 1976; Newell and Simon 1972) to

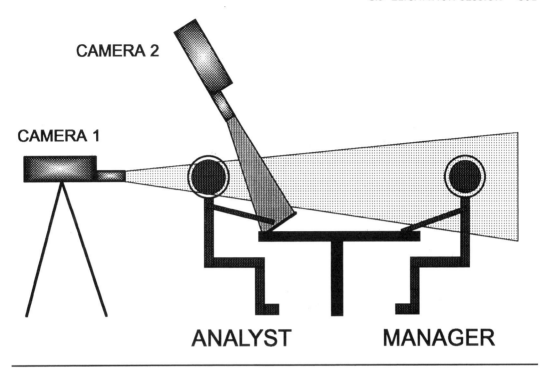

Figure C.1 *Probability elicitation interview*

explain the analyst's recorded elicitation activities. In this appendix, we present the elicitation interview together with explanatory material that describes the thought process of the analyst as he managed the interview following the elicitation protocol in Sections 5.4 through 5.8.

C.3 Elicitation Session

The following interview transcript was taken from the videotape made using the procedure reviewed in the preceding section. The actual spoken conversation included grammatical mistakes as well as various verbal conversation delimiters ("Ah, heh." "Okay."), which would distract the reader of an unedited written transcript. These have been corrected or removed to create a more meaningful written dialogue. A few proprietary items have been removed or masked. Some long sections have been condensed where these do not aid in understanding the analyst/manager interaction. References in the conversation to the equipment and procedures used to record the elicitation session have also been deleted.

The interview dialogue provides only a partial description of the interaction between the analyst and manager. The interaction also includes nonverbal visual and aural cues plus the analyst's interpretation of what is said (or not said) by the manager. The thought process of the analyst, as reconstructed and reported by the analyst himself while reviewing the interview recording, is presented to assist the reader in better understanding how the analyst managed the interview as it progressed. Comments concerning the interview management process are shown in slightly larger type and a different typeface than the actual interview transcript. The dialogue has been marked to show the cognitive information "chunks" identified by the analyst to describe external interview cue interpretations (marked **Cue:**) and his own actions (marked **Action:**) performed during the interview process. Dialogue corresponding to each **Cue** or **Action** immediately follows the label that applies. The dialogue is indented from the discussion of the analyst's thought process as he manages the interview.

Preliminary Discussion

The analyst established a strategy for conducting the interview based on the context within which it was being conducted. The manager was a senior executive with a busy schedule. He had agreed to participate because of his interest in the technical subject area as well as a desire to assist university research that might be relevant to industrial practice. The subject matter of the elicitation was of substantial interest to the executive, but there was no immediate decision to be made based on the results of the elicitation. Section 5.8 notes that up to a half day may be required for a probability elicitation, but in this interview less time was available because of the executive's schedule. Furthermore, because the probability elicitation did not address an immediate decision, the analyst had to be continually aware of the need to keep the manager interested in the elicitation.

These issues are also significant in elicitations being conducted to support an immediate decision or risk analysis. Managers and other subjects often have busy schedules, and they frequently have other, more pressing concerns than the analysis. Thus, the issues of available time and manager interest are frequently concerns that an analyst must address while conducting an elicitation interview.

In a meeting prior to the interview reported below, the purpose of the elicitation had been already discussed. In addition, the equipment being used had also been discussed, as well as the process of working with the equipment. The analyst starts the interview below with some seemingly idle chitchat, which is used to judge the interest level of the manager and to determine the extent to which the recording equipment is likely to intrude into the elicitation process.

> ANALYST: Let me start with a little example. One of my colleagues relates a story. He was working with a firm that was thinking of manufacturing some new widget. They made a decision to go ahead and market it, and then they decided, after the fact, to call in the consultants to look at some of the risks associated with this decision to see what could go wrong. So he [the colleague] went around and talked to the chief marketing guy and a few other people and did a little bit of financial analysis. When he came in to give his final report, the CEO of the firm and the chief marketing guy and a few other people were sitting around the table.
>
> Summarizing, he said, "The conclusion out of all of this, after working with your marketing people, is that there is a forty percent chance this product will be a success." The CEO leans back in his chair, gasps a little bit, looks over at the

chief marketing guy and says, "I don't understand. You told me it was *very likely* that this would be a success, and now I hear it's a forty percent chance!" The marketing guy says, "Yeah, it's very likely. This is a very competitive market, and a forty percent chance *is* very likely to be successful. The CEO says, "There's no way I would have approved this thing if I had realized that's what you meant by very likely!"

This is a real problem in a lot of business planning. We try to communicate to each other what are the risks associated with different activities. We use terms like "very likely" or "not a very high chance" or "very good chance," and we all have different meanings for those terms.

You're an IE [industrial engineer] by background ...

MANAGER: Yes.

ANALYST: So you use probabilities as a language for talking about these issues.

MANAGER: I've even taught elementary probability courses. I recognize the problem; we've all faced it.

ANALYST: In many situations—like the marketing case—you don't have very good objective data. You have data that's relevant, but the situation you face is not exactly like the situation where the data was collected. So, if you're going to use probabilities, they are going to come out of somebody's head, and, of course, have the deficiencies and the strengths that such probabilities have. We've found over the years in investigating this that it's not easy to get numbers that really represent properly people's information about uncertainty. The result is that you need somebody who's reasonably well trained to get the numbers from people.

From observing the manager, it is evident to the analyst by this point that the manager is either experienced with cameras or does not find them distracting. He is not paying any visible attention to the cameras, nor is he giving any indication of nervousness about his appearance before the cameras. It is also apparent from his reactions that the manager is fully participating in the interview. He is visibly concentrating on what the analyst is saying, and is responding promptly with relevant comments to what the analyst is saying. Based on these favorable indications, the analyst begins the actual elicitation process.

Stage 1: Motivating (Elapsed Interview Time: 0 Minutes)

During the motivating stage, the analyst continues to establish rapport with the manager and determines whether there is a significant potential for motivational bias in the elicitation.

ANALYST: We're going to work on the probability distribution for the number of engines that you will ship this year.

MANAGER: Right. It has definite pay value for us if we forecast that accurately, and it definitely has uncertainty.

ANALYST: I understand that your customers place "supposedly" firm orders.

MANAGER: Well stated. "Supposedly" they are firm orders by contract, requesting certain delivery. Yet, in the contract it also has provisions for being able to slide those deliveries out to a certain point in time. When you get into the reality of how the world works, if they don't have the money to pay for the engines, a firm contract doesn't mean too much. You're tied to their needs, and they don't forecast very far out. I should say, they may forecast far out, but it's not very accurate.

The following sequence of questions checks to see if there are reasons that the manager might have a motivational bias. That is, is there a possibility for a conscious or unconscious adjustment in the manager's probabilities because of perceived personal rewards?

Action: *Check whether there is an official forecast.*

ANALYST: Do you make forecasts of the number of engines that you will ship?

Cue: *Acknowledges that an official forecast exists, and that the short-term forecast is a goal.*

MANAGER: Yes. We make two flavors: a five-year forecast that is recognized to have quite a bit of variability in it, and then for next year we make a commitment as to what we will ship. We make that commitment to our [parent] corporation. So we have what we consider to be a very tight forecast at that point, but yet a year out in our business is still a long way out. This is a lot of lead time for the kind of OEMs [original equipment manufacturers] we deal with.

By this point, the analyst is convinced that the manager is fully engaged in the interview. The manager is looking directly at the analyst, and is responding forcefully with relevant answers to the analyst's questions. (This is in contrast to some other experiences of the analyst where the subjects have evidenced confusion at this point in the interview about what this line of questioning has to do with the subject matter of the interview. Such confusion is generally indicated by puzzled expressions or comments that are not directly relevant to the interview.) Hence, the analyst proceeds to address the potentially delicate issue of motivational biases.

Action: *Check for personal motivational bias.*

ANALYST: Are things like compensation for people tied to shipping the forecasted number?

The analyst notes that the manager evidences some discomfort at the mention of compensation—perhaps this is going to be a difficult issue to address.

Cue: *Acknowledges that personal rewards are tied to the forecast.*

MANAGER: At the executive level. The working people in the company in general are not on an incentive plan, although we do have a bit of profit sharing and a yearend bonus that goes to everybody.

Thus, the manager acknowledges that personal compensation is tied to "making" the numbers shown in the forecasts. Hence, there is a basis for a possible motivational bias.

Action: *Begin "debiasing" for personal motivational bias.*

ANALYST: The reason I raise that is that it's sometimes a problem in this type of assessment. Some people's bonuses are tied to the forecast, so there's hesitation to forecast you'll be below.

Cue: *Weakly responsive to debiasing.*

MANAGER: Well, the incentives are very much for making the forecast you put in. That's what I meant by a commitment.

From both the vocal tone in which the manager responds and his somewhat stilted body movements and position, the analyst develops some concern that what he had intended as a general comment appears to be striking the manager personally—perhaps as a comment that he (the manager) can't keep his thinking straight in the face of this reward structure. Thus, the analyst proceeds with some caution to check if the manager considers the uncertain quantity to be a goal, rather than an uncertainty. If it is considered to be a goal, then the manager might not be willing to admit to significant uncertainty about its value ("management bias").

Action: *Reinforce potential for personal motivational bias.*

ANALYST: So there's a commitment both that you can deliver them and that you can in fact market this number, also.

Cue: *Still somewhat weakly responsive to debiasing.*

MANAGER: That we can actually manufacture them, and that we can produce the revenues that go with that—the margins and so forth.

Action: *Reinforce potential for personal motivational bias.*

ANALYST: Is it tied also to finding another customer if one of yours decides not to buy this year?

Cue: *Still weakly responsive to debiasing.*

MANAGER: You can do that to an extent, although each engine model is customized to that particular customer. You have to be very knowledgeable about specific models to know if you can make these changes.

Action: *Continue to debias for personal motivational bias.*

ANALYST: Do you make incentives to your customers if it looks like they won't take delivery, and you might be below your goal?

MANAGER: We try and work with them to assure that we get the shipments that we're looking for during the year. We don't have year-end sales, but it's usually in their best interest to take the engines by the end of the year that they signed up for. Otherwise there will a penalty because there is generally price escalation from year to year.

The manager's responses seem to be avoiding the issue of whether motivational biases might impact the forecasts that he makes. At this point, the analyst decides that it is better not to press this issue further. There is a need to proceed so as to not spend excessive time on this issue, and also a need to keep the manager participating fully in the interview.

Now the analyst switches to a bias question that reflects less directly on the manager—whether the manager considers the forecasts made by experts to be accurate. If so, the manager might consider the uncertainty to be minimal. This is "expert" bias.

Action: *Begin to check for expert bias.*

ANALYST: Who actually does these forecasts?

Cue: *Indicates understanding of expertise needed for forecasting.*

MANAGER: The forecasts rest with our directors of programs. For the turbofan that we're talking about here, the responsibility rests with the director of turbofan programs. He draws on sales and contracts people, along with people from field sales, and has to balance this with what manufacturing can produce. The lead time for our engines is more than a year, so when you talk about the forecast for a year, you already have something in work.

Action: *Begin debiasing for expert bias.*

ANALYST: Traditionally, how good have these forecasts been?

Cue: *Indicates problems with experts dealing with uncertainty.*

MANAGER: There are times that they're really close. There's times that you exceed them like you'd like to, or you exceed them quite a bit and it can be a real strain. You have trouble producing enough for what the market wants. Other times, we don't make the forecasts. That's a real concern. If you forecast more than you are actually able to ship, of course you create inventory.

It appears that the manager understands that there is uncertainty in the quantity of interest.

ANALYST: I imagine that with these kinds of items it's very expensive [to create inventory] . . .

MANAGER: Oh, yes.

The manager's comments leave the analyst with some concern that the manager either does not understand the potential for the various motivations to bias his probability distributions or is not willing to admit that possibility for the record. However, there is little more that can be done to address this point now without appearing to "beat a dead horse" and possibly either lose the interest of the manager or even get him irritated. Thus, the analyst proceeds with the structuring part of the interview.

Stage 2: Structuring (Elapsed Interview Time: 6 Minutes)

The purposes of the structuring stage are (1) to make sure the quantity of interest is well defined, (2) to determine if this quantity should be disaggregated into multiple different variables before assessment, (3) to find out what assumptions the manager is making during the assessment, and (4) to determine an appropriate measuring scale. While these tasks are being done, the potential for cognitive biases is explored, so these biases can be addressed during the conditioning stage.

Action: *Check for possible disaggregation.*

ANALYST: Now we're talking about turbofan engines. Are there different variations within that category?

Cue: *Acknowledges potential for disaggregation.*

MANAGER: There are different models. At the first level, we say there are different series. We call these different dash numbers. Dash two, dash three, and dash five are the three series of the turbofan that we actively have. Most of our production now is in dash five type engines. But within a dash number, different customers may get slightly different engines. That's the model number. Each customer has its own model number.

The discussion of the differences between the models continues at some length. The manager speaks with considerable enthusiasm about this material. It is also important information for deciding exactly how to conduct the probability elicitation. Thus, the analyst continues to encourage the discussion. (Also, the analyst has an engineering background, and this material is interesting to him. Practitioners in operations research and systems analysis note that successful analysts often get interested in the actual substance of the problems they are analyzing. This helps both in developing a sense in the manager that the analyst is "one of us" and also in avoiding naive mistakes in the models that are built.)

Since there are significant differences between different dash numbers and even between different models, it appears that the uncertain quantity should be disaggregated before the elicitation is conducted. The analyst now works with the manager to determine an appropriate disaggregation.

Action: *Ask whether to disaggregate forecast.*

ANALYST: When you do your forecasts, do you break it down by dash number?

Cue: *Acknowledges that disaggregation is normal practice.*

MANAGER: We even build up the forecast by model number, and then to an extent we look within dash numbers to certain things that are common.

Action: *Suggest disaggregation for elicitation.*

ANALYST: So for our elicitation, we are looking at a single customer?

Cue: *Agrees to specific variable.*

MANAGER: Yes, "Customer A."

Now that it has been established that the probability elicitation should focus on Customer A, the analyst explores what is the most natural way for the manager to think about the number of engines sold to Customer A.

Action: *Clarify variable definition so as to pass the clairvoyance test.*

ANALYST: We're somewhat into the year already. So are you going to include any engines already shipped this year to this customer?

MANAGER: Yes. I'm talking about the calendar year. That's the easiest.

ANALYST: So you have some information already. You know if they've accepted any orders yet this year. •

MANAGER: Yes.

Action: *Make sure variable passes clairvoyance test.*

ANALYST: Sometimes you run into problems in communicating because you don't have the quantity you are talking about very well defined. The technical jargon for this is to ask if the quantity passes the *clairvoyance test.* That is, after the year is over, could I in fact figure out what the number was? I think this definition looks good. You can look at that customer and see how many engines were shipped.

Cue: *Agrees that variable passes clairvoyance test.*

MANAGER: Yes, very easily. Every month we have a meeting and publish a report about manufacturing plans, which is our projection by customer. We can compare the yearend report to actual numbers shipped and say how we did versus what we had in the plan.

This is a more technical discussion of the clairvoyance test than would be conducted with most managers. However, the manager is actively participating in this discussion. Therefore, the analyst pursues the topic in some detail—both to ensure that the variable is well defined and to keep the interest level of the manager up. Now the analyst investigates whether there are unspoken assumptions that the manager is making about the uncertain quantity.

Action: *Probe for assumptions underlying variable probability elicitation.*

ANALYST: Another question is whether there are any assumptions we ought to make for purposes of doing this elicitation—perhaps about the general economy or any unusual circumstances in your manufacturing capabilities.

Cue: *Discusses assumptions underlying elicitation.*

MANAGER: Basically, what's happened with the market in the last ten years is that it peaked out about 1979–80. Everybody was shipping two to three to four times as much as they are today. [Additional discussion of this follows. It is clearly of interest to the manager from both the tone of his voice and the extended nature of his comments.]

We think 1988 was kind of the bottom of the tub, if you will. We see a fair increase for this year, but not a "leaps and bounds" increase. [Some additional discussion of the general industry follows.]

The analyst now works to bring some closure to this discussion. As in many interviews, the available time is limited, and there is a need to move on in order to ensure that there is enough time available to determine a probability distribution.

Action: *Reinforce basic structuring assumption.*

ANALYST: So your underlying assumption is that the bottom of the trough has been reached.

Cue: *Acknowledge the structuring assumptions to be made.*

MANAGER: There will be general growth [in the overall market] but in the range of two to six percent.

ANALYST: Now if we were doing a very detailed model here, we might break out this probability elicitation based on the percent growth for the industry and look at what this means for this specific customer. For the purposes here, we will sort of eyeball that one and go ahead.

Stage 3: Conditioning (Elapsed Time: 15 Minutes)

The purpose of the conditioning stage is to bring into the manager's immediate consciousness the relevant knowledge about the uncertain quantity. This helps combat the cognitive biases of anchoring/adjustment and availability by helping ensure that *all* relevant information is considered, and not just that which immediately comes to mind.

Action: *Describe the general nature of probability elicitation issues.*

ANALYST: Let me talk about some of the problems that tend to arise when you do these elicitations. There are two that are by far the biggest problems that you run into. These problems are so reliable that I will routinely, when I teach an introductory class, do a quick and dirty test. And I've never, when I've had ten or fifteen students, not had the effects show up very strongly. They're that strong.

Action: *Describe overconfidence bias.*

ANALYST: The first is that people are overconfident about what they know. Their bounds on what they think can happen are too tight. This shows up routinely, time and time again. Now you may say, "What do you mean by too tight; this is a number coming out of my head." But, for example, suppose we came up with a distribution that said there's a ninety-five percent chance the quantity is in such-and-such a range. Suppose you do that ten or fifteen times, and every time the number is outside the range. You would start to get concerned that the person is not representing his true uncertainty in the problems. This effect shows up very strongly, time and time again. It's even got a name, overconfidence, because it shows up so widely.

Cue: *The manager acknowledges understanding with a shake of his head, but no other visible signs.*

Action: *Describe availability bias.*

ANALYST: The other thing that shows up is that people concentrate on particular cases. For example, we say that we already know what is happening for the first two months of the year, and it's either looking grim or great. . . .

Cue: *The manager now breaks in with a comment that demonstrates an understanding of the difficulty.*

MANAGER: You tend to extrapolate that for the whole year. . . .

ANALYST: But it's only two months out of an ongoing five-, ten-, or twenty-year history.

Cue: *Strongly acknowledges understanding of availability bias with relevant comments about his experiences.*

MANAGER: We have seen both of those phenomena. We've been forecasting about four engines a month for Customer A. Forty-eight engines for the year. If you talk about that long enough, everyone homes in and says, "Boy, that's it." We've heard it and talked about it. I think the repeated discussion probably narrows that perception, too. [The discussion of cognitive biases continues.]

The analyst was concerned about the lack of much reaction from the manager in the early part of Stage 3: Conditioning. The analyst was starting to develop some concern that, as with the motivational biases, the manager seemed to think that these difficulties didn't apply to *him* or that the manager was losing interest in the entire interview. However, the strong reaction of the manager to the discussion of the availability bias is encouraging, and the analyst moves on to the actual encoding of the probability distribution.

Stage 4: Encoding (Elapsed Interview Time: 21 Minutes)

The analyst starts encoding the probability distribution by working first on the extreme fractiles. As discussed in Section 5.3, research results and practical experience both indicate that determining these before the more central fractiles helps to combat the biases that lead to distributions that are too narrow.

Action: *Introduce encoding task.*

ANALYST: This is particularly an issue when you have official forecasts, such as you have talked about. So it is something to keep in mind as we go on. In fact, because of this problem, rather than talking about the forecast, we'll talk about how good things could be—maybe "good" is not the right word here—how *high* things could be, and how low things could be. And what kind of conditions could lead to each of those [results], before we talk about what sort of middle number makes sense. We look at the extremes first because of the fact that people tend to "underthink" about those extremes.

So let's do that. You talked about a number of things that affect the number of engines that will be shipped to Customer A. Presumably there's some general business effects—general across the whole economy—that would affect the customer. There are also presumably some general industry—at least to their segment of the aviation industry—type effects. They may have a particular kind of client that tends to buy. There may be effects due to the business of that kind of client.

The analyst is starting to work with the manager to identify conditions that could lead to extreme values for the uncertain quantity. By discussing these conditions, they become more available in the manager's thinking, and hence will be judged more likely. This combats overconfidence. Also, working with both the high and low extreme values before considering the central fractiles of the distribution will provide two very different cognitive anchors, which helps to offset the tendency to anchor on a middle value of the uncertain quantity.

Cue: *Discusses conditions leading to extreme cases.*

MANAGER: Very much so. The manufacturers in this segment tend to be relatively small compared to the airplane manufacturers we usually think of. The particular product we're talking about, turbofans, tends to be very responsive to corporate profits as a measure. They're generally used as executive jets and in business aviation. When times are tight and corporate profits are down, people will not typically go out and buy a new turbofan. The economy determines a lot of what their [the airplane manufacturer's] business will be like.

The analyst now introduces a probability wheel (discussed in Section 5.1) as a visual aid to assist with the probability elicitation. Because of the manager's formal probability theory background, he is comfortable with thinking directly in numerical terms about probabilities. As a result, the wheel plays less of a role in the elicitation than with some less technically sophisticated subjects.

ANALYST: I'll bring out my trusty little probability wheel. The consultants have fancy versions of these. . . .

MANAGER: I'm familiar with those.

ANALYST: [Demonstrating the operation of the probability wheel.] So you know that it adjusts. The fraction of each color adjusts. [The wheel has pink and blue sectors.] The idea is that this is something that you would spin, and you can pick either pink or blue. Let's say you win with blue. The idea here—and this is really the way in which you define probabilities for this kind of situation—is that if you had to put your own money down, would you rather bet on blue coming up or some quantity out in the world? You're familiar with that?

Cue: *Indicates understanding of probability concepts.*

MANAGER: Right.

Action: *Describe meaning of 0.99 fractile.*

ANALYST: Okay. Focusing in on the top number [of possible engine sales], we tend to think of something like a one-in-a-hundred chance. In other words, a number that's high enough so that it's not very likely to be exceeded, but it could happen. If you were to look over a hundred years of data, which you don't have, occasionally something very wild happens and you get very high numbers. Right now, I'm trying to get an upper bound that's a realistic bound on what could happen, although occasionally you might be surprised at that. Again, if you want to think in quantitative terms, you can think of something like a one-in-a-hundred chance, or a one percent probability. [Shows this as a small sliver of blue on the probability wheel.]

The familiarity of the manager with probability wheels indicates to the analyst that it is not necessary to give a further explanation of the concept of the 0.99 fractile. The analyst now works to bring to the manager's conscious reasoning the various conditions that could lead to a high value for the uncertain quantity.

Action: *Debias against overconfidence and begin to elicit 0.99 fractile.*

ANALYST: From what you just said, the kind of situation you'd have that would lead to that [high value] is an economy that's very strong in general; [an economy that] would lead to more corporate customers taking delivery of jets. So, if the economy is good, then shipments would be their highest.

Thinking in terms of the rest of the year, let's try and get a feel for what might be a realistic upper bound, where realistic means, again, as I showed on the wheel, something that could be exceeded, but you probably wouldn't plan on it.

Cue: *Describes conditions leading to bounds on possible values and gives initial 0.99 fractile.*

MANAGER: A couple of realities enter in. An engine manufacturer can only accelerate so much from what we laid out as a plan. The same thing is true of the aircraft manufacturer and his other suppliers. So, regardless of what the economy does, I might say double. I want to say twice as many engines and airplanes.

ANALYST: You can't make more?

MANAGER: You can't get there from here. They're currently producing two airplanes a month. So that's four engines a month—there are two engines per aircraft. So, if you could dream and said we could really do everything we can, and they can do everything they can, you're probably up in the range of seventy engines.

Action: *Probe initial number.*

ANALYST: So your base is now four a month, which is thirty-six for the year.

MANAGER: No. four a month is . . .

ANALYST: Forty-eight. I can't do my arithmetic.

MANAGER: Yeah. Forty-eight, and I'm saying that somewhere up in the range of sixty to seventy would be my first guess at what could be done.

Having already spent quite a bit of time on the high end, the analyst decides not to pin down an exact number for the 0.99 fractile at this time because of a concern that the manager might start to anchor on the high number and not provide a realistic low number. This concern is based on the general research findings about anchoring, rather than any particular response of the manager. The analyst now shifts to looking at the low end of the range for the uncertain quantity.

Action: *Shift to low end of distribution, and debias with scenarios for what could go wrong.*

ANALYST: And that's assuming that the economy is going very well and all those other things we talked about. I assume on the low end, then, to focus on that for a minute, that the economy falls apart, or something like that. Are there any catastrophes that are within reason, like this customer goes out of business?

Cue: *Describes scenarios leading to low values.*

MANAGER: Well, they come pretty close to that at times. We've literally had engines—again, this is a product with an eighteen-month to two-year lead time— in shipping, ready to go to the customer, and then be stopped from shipping them because of credit problems. So it's not as speculative as you might think for that kind of product.

Yeah, if you assumed a real disaster. Another stock market Black Monday, and everything really clamped up....

Action: *Reinforce possibility of low result.*

ANALYST: Which is certainly, I would say, within a one-in-a-hundred chance.

Cue: *Concurs with realism of bad scenarios.*

MANAGER: I would ...

ANALYST: What do you think of that?

MANAGER: I would say you could end up essentially shutting them down. We've shipped eight engines. We're two months underway. So you can't get below eight. They've taken title to those, and there's probably some going in March. So the lower bound would be somewhere in the low teens probably.

ANALYST: So maybe ten to ...

MANAGER: In the range of ten to twenty if you want to get the same kind of lower bound [as the upper bound].

While the manager is still actively discussing the lower bound, the analyst decides to proceed without obtaining an exact number for the 0.01 fractile because of some concern about the growing length of the interview. The analyst is genuinely surprised at the breadth of the range and now expresses this surprise.

Action: *Probe to see if he believes range from 0.01 to 0.99.*

ANALYST: Well, that's a pretty impressive range.

The research literature indicates that almost everyone *underestimates* the amount of uncertainty in the forecasts they make. In light of this, it was probably a mistake for the analyst to make this comment about the breadth of the range. As the discussion presented below at the very end of the interview shows, the comments by the analyst may have triggered some second thoughts by the manager about the breadth of the range from the 0.01 to the 0.99 fractiles.

Cue: *Confirms range.*

MANAGER: Yes. And we've seen both extremes. We've had times where the shipments dropped by seventy to eighty percent within a three-month time period.

ANALYST: Is that right!

MANAGER: At other times—back in the late seventies, anything we could make they would take. Just strictly a capacity limitation on what we could produce.

ANALYST: I don't think this is consumer products.

MANAGER: No, that's true!

Actually, you *can* get variations like this in consumer product sales. This discussion is mostly idle chitchat by the analyst to show interest in the subject matter, and the manager appears to accept it as such since he does not make any further comments. The analyst now shifts to working on the center of the probability distribution for the uncertain quantity. He starts by specifying values for the quantity and asking for probabilities.

Action: *Begin assessing the middle of the probability range.*

ANALYST: These are extremes. Let's look a little bit at more realistic numbers. Realistic is not the right word, because I certainly think these are realistic. But rather the kind of numbers you tend to plan on. In fact, let me throw a few numbers out and see if we can attach probabilities to those. You said somewhere in the range of ten to twenty [for the lower limit]. Let me go up a little bit from that. Say, thirty. Now suppose I said, "What's the probability that you'd exceed thirty? Is that something you can think about? If that's not easy to think about, we can look at it a different way.

Cue: *Confirms understanding.*

MANAGER: Yeah, I can think about it. It's a quite high probability that we will exceed thirty.

Action: *Probe for specific probability number.*

ANALYST: [Displays the probability wheel with equal segments of the two colors.] More than fifty-fifty, to put my wheel out here again?

MANAGER: Oh, yes. Definitely more than fifty-fifty.

ANALYST: [Adjusts the wheel so that blue covers three-quarters of the circle.] Is it as high as three-quarters, do you think?

Cue: *Responds with number.*

MANAGER: I'd say it's in that range. Around three-quarters.

Having obtained an initial value for the 0.25 fractile, the analyst now shifts to the upper part of the probability distribution, rather than spend more time finding a definite number for the 0.25 fractile. The interview is now starting to get long, and the analyst is concerned that he will run out of time. The manager does not appear to be losing interest, but the interview is now approaching the time limit that had been indicated when it was set up. The analyst is concerned that the manager may have other commitments, or that the video equipment might be scheduled for other use.

Action: *Probe for another intermediate probability number.*

ANALYST: Moving up to the other end now and coming a bit off of that [i.e., the upper limit of sixty to seventy], say around fifty. You can think about shipping more than that or shipping fewer than that, whichever way you want to think about it.

MANAGER: Meeting fifty or more. I'd say your probability is less than fifty percent [of shipping fifty or more engines].

ANALYST: [Adjusts the probability wheel so that blue covers one-quarter of the circle.] Okay, it's less than fifty percent. Think it's less than twenty-five percent?

MANAGER: No. It's higher than that, but probably not a whole lot. And it tails off pretty fast there.

Action: *Verify number.*

ANALYST: So its somewhere in that range.

MANAGER: Yeah. Twenty-five to thirty percent. Somewhere in there probably.

Now having preliminary values for the 0.25 and 0.75 fractiles, the analyst works to clarify these. He shifts from specifying values of the uncertain quantity and asking for probabilities, to specifying probabilities and asking for values. He starts by asking for the median (0.50 fractile). The manager continues to be actively participating in the interview as evidenced by his ready responses to questions.

Action: *As a check, shift to specifying probabilities and asking for levels.*

ANALYST: All right, let me try now to look at this a different way to do some checks. Suppose, rather than giving you a number and saying, "give me a probability," that I give you a probability—do it the other way around. Let's try to find a number where you think there's a fifty-fifty chance you'll either be above or below that number. Let's try some numbers. For example, let me pick a number in the middle, say around forty. Do you think you're more likely to ship more than forty or fewer than forty?

MANAGER: I'd say a little more likely to ship more that forty.

ANALYST: A little more likely to ship more than forty. How about ...

MANAGER: But the fifty-fifty number would probably be in the forties, I would guess. It would be somewhere between forty and the forty-eight rate we're running now.

Action: *Solidify number.*

ANALYST: Okay. To pick a number to have, say around forty-five.

MANAGER: I'd pick an even number, since there's two per airplane.

ANALYST: Ha, ha, ha. Forty-four!?

MANAGER: Forty-four would probably be my pick off the top of my head.

The analyst acquiesces to the manager's statement about needing an even number, although there really isn't anything wrong with the median being odd. If the median were forty-five, then this would mean there was a fifty percent chance that the uncertain quantity would be forty-six or higher and a fifty percent chance that it would be forty-four or lower. During the entire preceding section while the probability elicitation was going on, the manager has appeared to be intensely concentrating. The analyst interpreted this as indicating that these questions were both difficult and interesting for him.

Action: *Check consistency of number just obtained with earlier numbers.*

ANALYST: Now let's compare that with the numbers from before and see if we're consistent. You say there's a seventy-five percent chance of shipping more than thirty, so that agrees. And there's about a twenty-five percent chance of shipping more than fifty. So that certainly agrees.

Cue: *Confirms numbers.*

MANAGER: It tails off pretty fast there, I think. [Discussion of the consistency continues.] It definitely tails off faster on the upside. There are a lot more things that have to happen for them [the airplane manufacturer] to increase their rates [than to decrease their rates]. It's easy to stop things.

Stage 5: Verifying (Elapsed Interview Time: 36 Minutes)

The analyst now moves on to verify the 0.25 fractile by examining the conditional probability, given that the actual value of the uncertain quantity is below the median.

Action: *Verify lower quartile.*

ANALYST: Let's try to get a little bit more detail. Of course, you can go on forever on this and get detail that you don't really have knowledge about, but let's try for a little more. Suppose you knew for a fact that they were going to ship fewer than forty-four. So you're not in a great situation. If you had to split the interval below forty-four into two equally likely pieces, where would you split?

MANAGER: Well, somewhere in the thirties probably. No, that's probably more than fifty percent. Somewhere in the twenty-five to thirty range.

ANALYST: It's got to be even, right? So how about twenty-eight, as a first cut?

MANAGER: As a construction. Sure. [This approximately agrees with the value of thirty that was previously determined for the 0.25 fractile.]

In a similar way, the analyst moves to confirm the 0.75 fractile.

Action: *Verify upper quartile.*

ANALYST: How about if we're on the up end—if we're above forty-four? If you knew you were going to ship more than forty-four? Again splitting it into two equally likely pieces.

MANAGER: Probably somewhere around fifty or fifty-two. Say fifty-two. [This approximately agrees with the manager's earlier statement that the probability of shipping fifty or more engines is twenty-five to thirty percent.]

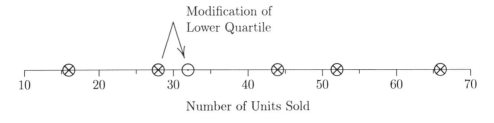

Figure C.2   *Fractiles for probability distribution*

The verification continues with a graphical aid. First, the 0.01 and 0.99 fractiles are more exactly specified. The manager still appears to be paying close attention to the interview.

Action: *Plot numbers to see how they look.*

ANALYST: Now let me plot where we are at this point and see if it looks reasonable. [Draws an axis with labeled ticks. See Figure C.2.] You said the lower end was somewhere around ten to twenty. I'll pick a number, say sixteen.

MANAGER: Okay. [The analyst inserts a ⊗ at sixteen.]

ANALYST: Up on the high end, it was somewhere from sixty to seventy. I'll be a little optimistic—how about sixty-six?

MANAGER: Okay. [The analyst inserts a ⊗ at sixty-six.]

Now the analyst works to confirm that the specified interquartile range has a probability of 0.5.

Action: *Begin checks for consistency.*

ANALYST: [Inserts ⊗ at the median number of forty-four, and then also at the lower and upper quartiles of twenty-eight and fifty-two.] These are the numbers [for the various fractiles]. Let me ask you a couple of questions on this. First, let's just confirm what I think you said. You said it's equally likely to be above or below forty-four. Okay?

Cue: *Confirms median.*

MANAGER: I think it's in the ballpark. Right.

Action: *Check for middle interquartile range.*

ANALYST: [Pointing to the interquartile range from twenty-eight to fifty-two on Figure C.2.] How about to be inside this range versus outside? Is one of those more likely?

MANAGER: It's probably more likely that it's going to be in there [pointing to the inside of the interquartile range]. Your construction says it should be fifty percent between those two numbers, and it's probably a little more likely than that. [The manager realizes that his response is inconsistent with earlier responses.]

ANALYST: So we should pull these numbers in a little bit.

MANAGER: Right. Yes.

Action: *Determine which number should be changed.*

ANALYST: Think there's one side or the other that . . .

MANAGER: Probably the down side. I'd come up a little bit on that, I think.

ANALYST: So you'd pull it up two or four.

MANAGER: Yeah, about thirty-two. That's probably better.

The manager appears to be thinking intensely about the modifications he is making. This gives the analyst confidence that the modified distribution is appropriate.

Action: *Check modified interquartile range.*

ANALYST: Okay. [Inserts a ⊙ at thirty-two in Figure C.2.] Then, as you pointed out, we want to get this so that fifty percent of the time you're going to be inside and fifty percent you're going to be out.

MANAGER: That's why I adjusted it. That's probably closer.

The inconsistency has been resolved. Now, the analyst confirms that the probabilities of being below the 0.25 fractile and above the 0.75 fractile are the same.

ANALYST: Going back and looking at this, do you think you're still about equally likely to be in this range [pointing to range above the upper quartile of fifty-two] or below that [pointing to range below the lower quartile of thirty-two]?

MANAGER: Yes.

The manager understands probability theory. Thus, it is appropriate to use probability terminology in some final consistency checks. The analyst now plots the cumulative probability distribution, and then uses this to determine an approximate probability density function to use in a further consistency check.

Action: *Plot curve as another consistency check.*

ANALYST: Okay, you know some probability—in fact, you have taught it. Let's do a quickie plot of the cumulative distribution for this data. [The analyst draws the axes for a cumulative probability distribution, and labels the x-axis at intervals of ten and the y-axis at intervals of 0.25. He then inserts ⊗ at the assessed values for the 0.01, 0.25, 0.50, 0.75, and 0.99 fractiles. Finally, he draws a cumulative distribution function through these values freehand. See Figure C.3.]

Let's do a density function for this. I've got to do a derivative in my head. [Draws an approximate density function $f(X)$. See Figure C.4.] It comes up pretty fast, stays flat for a while, and then drops off pretty fast. I interpret this as saying there's really quite a bit of uncertainty. There's a fairly even probability of it being anywhere over a fairly broad range.

Cue: *Confirms distribution.*

MANAGER: That's what it would reflect. It's probably a little tighter than that.

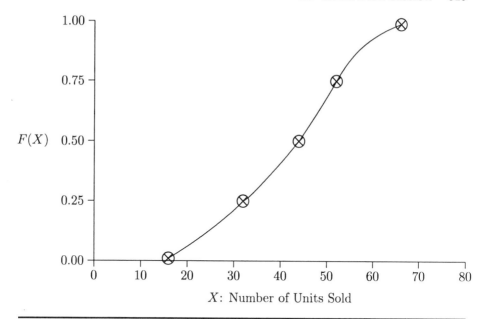

Figure C.3   *Cumulative distribution function*

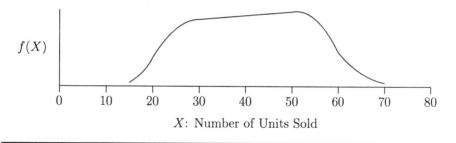

Figure C.4   *Approximate probability density function*

This comment shows that the manager is starting to be concerned that his probability distribution is too wide. The analyst begins to debias for overconfidence.

Action: *Re-emphasize overconfidence bias.*

ANALYST: Although, remember what I said. People tend to understate. . . .

MANAGER: I've probably erred on the other side.

ANALYST: Well, we'll come back next year and find out.

MANAGER: Yes, it's easy to check.

[Completion Time: 44 Minutes]

C.4 Concluding Comments

The elicitation session presented above is typical of those seen in decision analysis practice, with the exception that the manager was more sophisticated about probability concepts than is often true. For example, he recognized inconsistencies in his responses in cases where that might not be true for managers with less knowledge of probability. Because of this, it was easier than is sometimes the case to resolve inconsistencies in responses.

The interview transcript and annotations presented above show that the analyst must continually balance the desire for a probability distribution that more accurately represents the manager's knowledge against the need to retain the interest and attention of the manager, as well as the need to complete the elicitation in a time-efficient manner.

C.5 References

A. A. Gilmartin, A. Newell, and H. A. Simon, "A Program Modeling Short-term Memory under Strategy Control." In C. N. Cofer (Editor), *The Structure of Human Memory*, Freeman, San Francisco, 1976.

A. Newell and H. A. Simon, *Human Problem Solving*, Prentice Hall, Englewood Cliffs, New Jersey (1972).

G. G. Shephard, *Guided Self-Elicitation of Decision Analyst Expertise*, Ph.D. dissertation, Arizona State University, Tempe, AZ, 1990.

G. G. Shephard and C. W. Kirkwood, "Managing the Judgmental Probability Elicitation Process: A Case Study of Analyst/Manager Interaction," *IEEE Transactions on Engineering Management*, Vol. 41, pp. 414–425 (1994).

APPENDIX D

Interdependent Uncertainties

This appendix introduces probability analysis methods for decisions where the uncertainties are *interdependent*. In such situations, the probability distribution for one uncertain quantity depends on the level of other uncertain quantities. These situations are not emphasized in this book, but some understanding of the material in this appendix is required to follow Section 7.6.

The issues of interest are illustrated by the following example: An aircraft engine manufacturing company is considering changing its production facilities. As part of the analysis, a probability distribution must be elicited for the number of a particular type of engine that will be sold during the next year. Discussions with the marketing specialist for the engine make it clear that the sales will differ, depending on the state of the economy. This is because the engine is used in corporate aircraft, and companies tend to purchase fewer airplanes when the economy is poor.

Suppose, to keep the example simple, that if the economy is good, then there is a thirty percent chance that 40 engines will be sold and a seventy percent chance that 30 engines will be sold. On the other hand, if the economy is poor, there is a twenty-five percent chance that 35 engines will be sold and a seventy-five percent chance that 25 engines will be sold. The marketing specialist declines to predict the state of the economy for the next year, but a member of the corporate planning department says the chance is sixty percent that the economy will be good and the chance is forty percent that it will be poor. Given this information, what is the probability distribution for the number of engines that will be sold in the next year?

D.1 Conditional Probabilities

The difficulty in determining the probability distribution for the number of engines sold arises from the fact that this probability distribution differs, depending on the state of the economy. Thus, it is necessary to combine the probability information about the state of the economy with the information about the number of engines that will be sold. Figure D.1 displays the probability information given above in a *probability tree*. Each circle (called a *chance node*) represents an uncertain quantity, and the lines emanating from a chance node represent the possible results when the uncertainty represented by that node is resolved. The tree is read from left to right, and it starts on the left from a single node (called the *root node*). Each line emanating from a chance node has a number in parentheses, which is the probability that the event represented by that line will occur.

For a probability tree to be valid, the lines emanating from each chance node must constitute an *event space*, as defined in Section 6.1. That is, one and only one of the lines emanating from a chance node can occur when the uncertainty is resolved. Hence, the probabilities on the lines emanating from a particular chance node must sum to 1.

The final required characteristic for a probability tree to be valid is that the probabilities assigned to the lines emanating from each chance node must take into account the particular path that leads from the root node to that chance node. Thus, the probabilities for the upper-right chance node in Figure D.1 are for the number of engines sold, *given that* the state of the economy is good. Similarly, the probabilities for the lower-right chance node are for the number of engines sold, *given that* the state of the economy is poor. Probabilities that depend on the results of other uncertain events are called *conditional probabilities*. Thus, the conditional probability that 40 engines will be sold, given that the state of the economy is good, is 0.3. Similarly, the conditional probability that 35 engines will be sold, given that the state of the economy is poor, is 0.25.

This example shows that a probability tree represents situations where there are *interdependent uncertainties*. Specifically, the tree in Figure D.1 shows that the probability distribution for the number of engines sold depends on the state of the economy. In this example, we are directly interested only in the probability distribution for the number of engines sold. The state of the economy is considered only because the marketing specialist believes that the state of the economy influences the number of engines that will be sold.

Given the information in Figure D.1, how can the probability distribution be determined for the number of engines sold regardless of the state of the economy? This question can be answered by returning to the definition of a probability in terms of a probability wheel, which was given in Section 5.1. To illustrate this, consider the probability that 40 engines will be sold. In Figure D.1, this outcome occurs if the state of the economy is good, and then 40 engines are sold, given that the state of the economy is good. The probability that the state of the economy will be good is 0.6, and the probability that forty engines will be sold, given that the state of the economy is good, is 0.3.

Consider this in terms of a probability wheel. Since there are two different uncertainties (the state of the economy and the number of engines sold), it is necessary to use two spins of a probability wheel occurring in sequence. First, a wheel that represents the state of the economy is spun, where the "good" sector of the wheel covers sixty percent (0.6 portion) of the wheel. If the good sector comes up when the wheel is spun, then a second wheel is spun for which the "40" sector covers thirty percent (0.3 portion) of the wheel. If the 40 sector comes up, then 40 engines will be sold. Thus, we see that to sell 40 engines, it is necessary

State of
Economy

Number of
Engines Sold

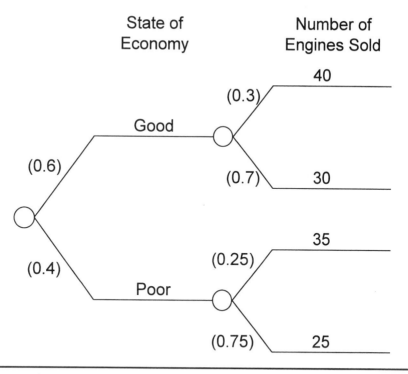

Figure D.1   *Probability tree for engine sales*

to obtain *both* a "good" result on the first wheel spin *and* a "40" result on the second wheel spin.

Some thought shows that these two probability wheel spins in sequence have the same overall probability of yielding a "40" as a single wheel spin when the wheel has a "40" sector equal to a $0.6 \times 0.3 = 0.18$ portion of the wheel. That is, make thirty percent of the "good" sector (which is sixty percent of the initial probability wheel) into a "40" (that is, where you obtain engine sales of 40). The probability tree in Figure D.1 shows that the 0.6 and 0.3 numbers, which need to be multiplied together, are the two probabilities that are on the path from the root node of the tree to the branch where engine sales are equal to 40. Reasoning by analogy, we would say that the probability of 35 sales must be $0.4 \times 0.25 = 0.1$. That is, multiply the probabilities along the path from the root node to the branch where engine sales are equal to 35.

In fact, this procedure works in general, as long as probabilities are assigned to branches taking into account all the events along the path leading from the root node to the event of interest. Thus, the complete probability distribution for the number of engines sold is

- Probability of 40 sold $= 0.6 \times 0.3 = 0.18$
- Probability of 35 sold $= 0.4 \times 0.25 = 0.10$
- Probability of 30 sold $= 0.6 \times 0.7 = 0.42$
- Probability of 25 sold $= 0.4 \times 0.75 = 0.30$

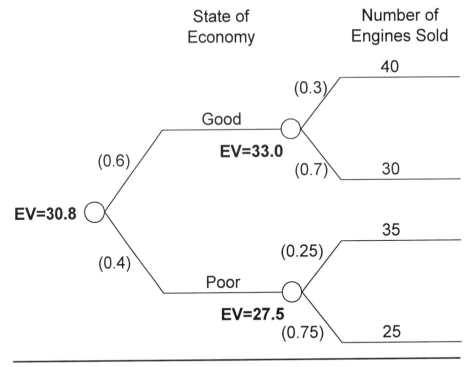

Figure D.2   *Expected value rollback for engine sales*

D.2 Calculating Expected Values on Probability Trees

The use of expected values as a criterion for making decisions was considered in Section 6.2. Using the probabilities determined in the preceding section, the expected number of engine sales $= 40 \times 0.18 + 35 \times 0.10 + 30 \times 0.42 + 25 \times 0.30 = 30.8$. It can be shown that the expected value can also be calculated by a procedure that does not require that you first combine the probabilities on the tree as we did in the preceding section. This procedure, called *rolling back* the tree, is illustrated in Figure D.2, and is discussed further in Clemen (1996), Chapter 4.

The procedure is as follows: Start by calculating the expected values for each of the rightmost chance nodes in the tree. In Figure D.2, the upper-right chance node has an expected value of $0.3 \times 40 + 0.7 \times 30 = 33.0$, and the lower-right chance node has an expected value of $0.25 \times 35 + 0.75 \times 25 = 27.5$. Write the calculated expected values on the tree next to the appropriate nodes, as shown in Figure D.2. Now move to the left in the tree and calculate the expected value for the root node using the previously calculated expected values for the right chance nodes. Thus, the expected value for the root node is $0.6 \times 33.0 + 0.4 \times 27.5 = 30.8$. This agrees with the expected value calculated above, and the procedure works in general.

D.3 Decision Trees

This section introduces the use of decision trees by expanding the engine sales example to consider a possible modification to the current engine production process. Assume the company currently uses a production process that involves mostly purchasing components from subcontractors. As a result, there are no fixed costs of production. The marginal production cost per engine depends on the state of the economy. When the economy is good, the marginal production cost per engine is $125,000, while when the economy is poor, the marginal production cost per engine is $110,000.

The proposed modified production process involves more in-house manufacturing, and as a result there would be a fixed production cost of $1,000,000 per year regardless of the number of engines produced. Marginal production costs would still depend on the state of the economy, and would be $90,000 per engine when the economy is good and $85,000 per engine when the economy is poor.

The sale price for the engines does not depend on which production process is used, but does depend on the state of the economy. Specifically, it is $150,000 per engine when the economy is good and $120,000 when the economy is poor. Using expected net profit as the criterion for making a decision, should the company stick with the current production process or switch to the proposed modified process? (*Net profit* is used here to mean the difference between sales revenue and production costs.)

Before proceeding to solve this problem, it is worth noting that the answer is not immediately apparent. In fact, it is confusing to even try and sort out all the pieces. With one process, there are fixed production costs, while with the other there are not. The various costs and revenues depend on both the state of the economy and the production process that is chosen, as well as the number of engines sold.

A *decision tree* can be used to organize the information for this decision, and such a decision tree is shown in Figure D.3. In this figure, the square box on the left side of the decision tree is called a *decision node*, and it represents the decision to be made between the current and proposed production processes. The lines emanating from a decision node represent the various alternatives that are available. In the same way that the lines emanating from a chance node in a probability tree must represent a collectively exhaustive and mutually exclusive listing of the possible uncertain outcomes, the lines emanating from a decision node must represent a collectively exhaustive and mutually exclusive listing of the available decision alternatives.

The remainder of the Figure D.3 decision tree represents the uncertainties that follow the selection of a decision alternative. At the right-hand side of the decision tree, the net profits for each path through the tree are given on the endpoints of the tree. Determining each of these endpoint profits requires some calculation. For example, the topmost number (1,000) is calculated in two steps. First, determine the net profit for each engine sold when the current production process is used and the state of the economy is good, which is $150 - $125 = $25 (in thousands of dollars). Then multiply this by the number of engines sold:

$40 \times \$25 = \$1,000$. A similar calculation is done to determine the net profit for each of the other three possible endpoints for the current production process.

When an endpoint is analyzed that involves the proposed production process, the fixed production cost must also be taken into account. Thus, the very bottommost endpoint value ($-\$125$) is calculated in three steps: First, determine the marginal net profit for each engine sold when the proposed production process is used and the state of the economy is poor. This is $\$120 - \$85 = \$35$. Then multiply this by the number of engines sold to yield $25 \times \$35 = \$875$. Finally, subtract off the fixed production cost: $\$875 - \$1,000 = -\$125$. The net profits for each of the other three endpoints that can result from the proposed production process are calculated in a similar manner.

Once the net profit figures are determined for all of the endpoints, the rollback procedure discussed above can be used to determine the expected net profit for each of the two production processes. As an example of the calculations, the expected net profit of $\$825$ shown for the upper-rightmost chance node is calculated by $0.3 \times \$1,000 + 0.7 \times \$750 = \$825$. The results of the rollback calculations of expected net profit are shown in Figure D.3. This shows that using the current production process has an expected net profit of $\$605$, while using the proposed production process has an expected net profit of $\$573$. Thus, the current process has a higher expected net profit by $\$605 - \$573 = \$32$ (in thousands of dollars).

D.4 Influence Diagrams

The decision tree in Figure D.3 clarifies this production decision; however, it may occur to you that the decision tree will turn into a bushy mess if the decision problem is much more complex than the one shown in the figure.

More complex decisions can be addressed with *influence diagrams*. These diagrams allow you to specify qualitative information about the interdependencies in a decision model in a way that is more compact than a decision tree. An influence diagram for the production decision is shown in Figure D.4. In this diagram, the rectangular box represents the production process decision, and thus corresponds to the decision node in the Figure D.3 decision tree. Circles or ovals represent uncertainties, and thus correspond to chance nodes in a decision tree. (These two types of symbols in an influence diagram are called *decision nodes* and *chance nodes*, respectively, just like their counterparts in a decision tree.) A double-lined circle or oval (for example, "Total Cost" in Figure D.4) is called a *deterministic node*, and it represents a quantity whose value is known for certain once the inputs to the node are specified. Finally, a double-lined, rounded-corner rectangle (for example, "Net Profit" in Figure D.4) represents the quantity whose value is of concern for making the decision. Thus, this node represents all of the endpoints in a decision tree taken together. This is called the *value node* for the influence diagram.

The arrows pointing into a chance node, deterministic node, or value node represent the influences that determine the value of the quantity represented by the node. Thus, Figure D.4 shows that the number of engines sold depends on

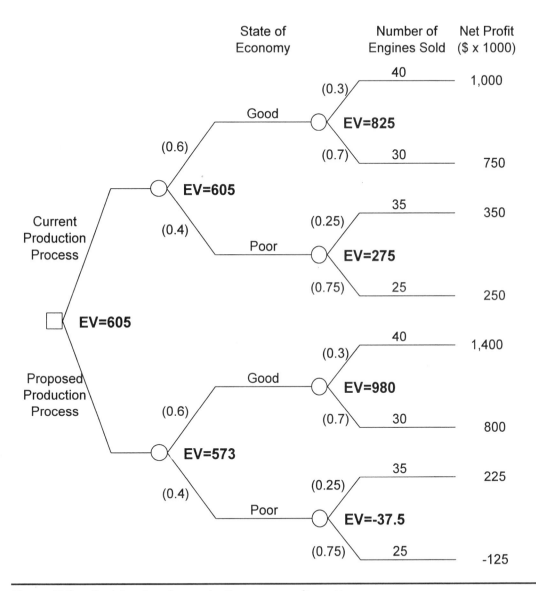

Figure D.3 *Decision tree for production process alternatives*

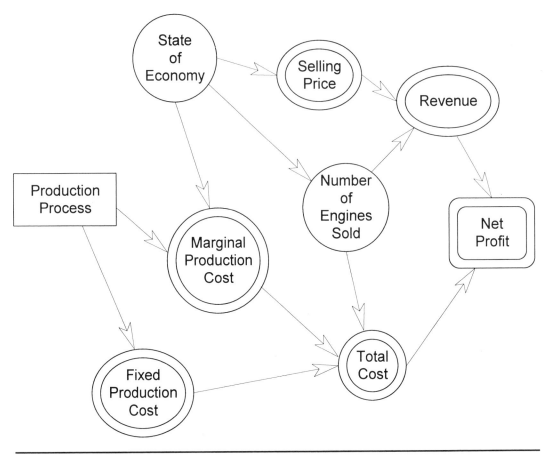

Figure D.4   *Influence diagram for production alternatives*

the state of the economy, while the total cost depends on the number of engines sold, the marginal production cost, and the fixed production cost.

In some respects, the Figure D.4 influence diagram shows more information about the structure of the decision problem than the Figure D.3 decision tree. The influence diagram shows general qualitative relationships among the variables in the production decision that cannot be directly determined from the decision tree. On the other hand, there are no numbers on the influence diagram. Thus, it is necessary to specify these numbers in addition to the influence diagram to conduct expected value calculations.

Influence diagrams can compactly represent the structure of a large decision model. For example, Kirkwood (1993) shows an influence diagram on one page that represents a decision model for which a decision tree would have 25,272 endpoints. While many realistic decision analysis problems do not require decision trees that large, it is frequently necessary to have decision trees with several dozen to several hundred endpoints. It is generally not feasible to conduct a paper-and-pencil analysis for such sized models. There are a variety of software packages available for personal computers that automate the construction and solution of decision tree and influence diagram models (Buede 1993). More details on using decision trees and influence diagrams are included in Clemen (1996) and McNamee and Celona (1990).

D.5 Drug Testing: "Backwards" Probabilities

The athletic governing board of Santa Clara University had to decide whether to recommend implementing a drug-testing program for its intercollegiate athletes. Charles D. Feinstein, a faculty member on the board, conducted a decision analysis to support the board's decision-making process (Feinstein 1990). This analysis illustrates an important issue that can arise when a decision situation involves the use of information that has uncertainty.

While drug testing raises a number of potentially contentious issues related to privacy rights and the legalities of the testing, Professor Feinstein restricted his analysis to the issue of the accuracy of the test results. His analysis convinced the board of governance to recommend against implementing a drug-testing program. This was true even though the director of athletics had previously recommended that such a program be started.

The accuracy issue with regard to drug testing is illustrated by a study conducted by the Centers for Disease Control (Hansen et al., 1985). The results of that study were summarized as follows:

> In response to questions about the reliability of the results of screening urine for drugs, we evaluated the performance of 13 laboratories, which serve a total of 262 methadone treatment facilities, by submitting preferenced samples through the treatment facilities as patient samples (blind testing). Error rates for the 13 laboratories on samples containing barbiturates, amphetamines, methadone, cocaine, codeine, and morphine ranged from 11% to 94%, 19% to 100%, 0% to 33%, 0% to 100%, 0% to 100%, and 5% to 100%, respectively. Similarly, error rates on samples not containing these drugs (false-positives) ranged from 0% to 6%, 0% to 37%, 0% to 66%, 0% to 6%, 0% to 7%, and 0% to 10%, respectively.

Thus, errors can be made in drug testing, and these errors include both *false positives* (that is, samples not containing drugs that are incorrectly identified as containing drugs) and *false negatives* (that is, samples containing drugs that are incorrectly identified as not containing drugs).

The error rates identified in the Centers for Disease Control study differed from drug to drug, as well as from laboratory to laboratory. To illustrate the analysis procedure for such information, we will use the following representative numbers:

1 For any specified test involving a sample that contains a drug, the probability of a false negative is 8%. That is, the probability that the sample will incorrectly be identified as drug free is 0.08.

2 For any specified test involving a sample that does not contain a drug, the probability of a false positive is 4%. That is, the probability that the sample will incorrectly be identified as containing drugs is 0.04.

3 The probability that any randomly selected Santa Clara University athlete is a drug user is 10% (0.1).

A brief note on terminology: Some literature about testing errors uses the terms *sensitivity* and *specificity* for a test. The sensitivity is the probability that a test result will be positive for a sample that contains drugs, while the specificity is the probability that a test result will be negative for a sample that does not contain drugs. That is, the sensitivity is 1 minus the probability of a false positive, and the specificity is 1 minus the probability of a false negative. Hence, for the example we are analyzing, the sensitivity is $1 - 0.08 = 0.92$, and the specificity is $1 - 0.04 = 0.96$.

The interrelations among the probabilities given above are shown in the Figure D.5a probability tree. On first examination, it appears that this test is fairly good. It is true that there is an 8% chance it will not identify a drug user, but there is only a 4% chance it will incorrectly identify a nonuser as a user. Since this latter error is probably more serious, it is encouraging to see the low probability associated with it occurring.

However, further thought leads to the conclusion that the probability tree in Figure D.5a is "backward" from what we are really interested in. Figure D.5b shows what we really want to know about. In this tree, the order of the "Drug User?" and "Test Result" nodes has been reversed from those in Figure D.5a. This is because we are interested in the conditional probability that someone is a drug user, *given that* a test result is positive, and also the probability that the person is not a drug user, *given that* a test result is negative. From the information shown in Figure D.5a, you might initially expect that these two probabilities are very high. However, we shall show below that the probability someone is a drug user, given that he or she tests positive, is lower than you might expect.

Before proceeding with the analysis, it is worth thinking about this situation for a while to see if you can identify why the probability that someone is a drug user is relatively low even if there is a positive test result. Here is the reason: The difficulty with using the test is that there are so many more nonusers than users. Thus, even though the test is fairly accurate at identifying nonusers, the small percentage of errors in identifying nonusers is made on a much larger group of people than the user population. Hence, the percentage of positive test results for nonusers is higher than you might initially expect, as is demonstrated below.

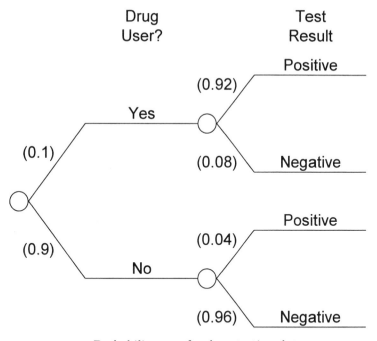

a. Probability tree for drug-testing data

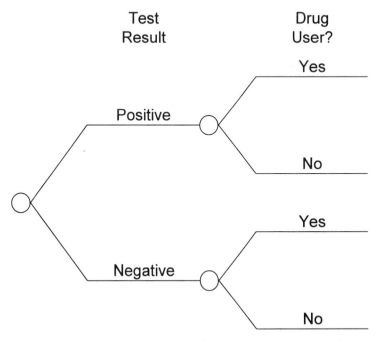

b. "Flipped" probability tree for interpreting test results

Figure D.5   *The drug-testing problem*

"Flipping" the Tree

The probabilities for the tree in Figure D.5b can be obtained from those in Figure D.5a by straightforward calculation. The process is illustrated in Figure D.6. In part a of that figure, the tree in Figure D.5a is reproduced with the addition of the probabilities for each path through the tree. For example, the 0.092 shown at the upper-right endpoint of the tree is the probability that an athlete is a drug user and also tests positive. As was shown earlier in this appendix, this is calculated by multiplying the probabilities along the branches leading to this endpoint: $0.1 \times 0.92 = 0.092$. The other three endpoint probabilities are calculated in an analogous manner.

Comparing the tree in Figure D.6b with the one in Figure D.6a shows that the endpoints of both trees represent the same set of events. However, these events are arranged in a different order in the two trees. This is shown in the figure by marking each path probability with a letter A, B, C, or D. By comparing the two trees, we see that in both trees, the uppermost path through the tree represents the case where an athlete is a drug user and tests positive. However, the next lower path in the Figure D.6a tree represents an athlete who is a drug user and tests negative, while the next lower path in the Figure D.6b tree represents an athlete who is not a drug user but who tests positive. (The corresponding path in the Figure D.6a tree is the third path from the top.)

Since the various endpoints for the Figure D.6b tree represent the same events as the endpoints in Figure D.6a (but in a different order), the probabilities for the Figure D.6b endpoints can be determined from Figure D.6a. These probabilities are shown on the right side of Figure D.6.

Once the path probabilities are determined for the "flipped" tree in Figure D.6b, it is possible to determine the various branch probabilities in this tree. To aid the discussion of this process, these (not yet known) branch probabilities have been marked on Figure D.6b by the letters E, F, G, H, I, and J.

Because the endpoints of a probability tree are collectively exhaustive and mutually exclusive, they are an *event space*. Therefore, the probability for any collection of endpoints can be calculated by summing the probabilities of the endpoints in the collection. Specifically, the two endpoints for which an athlete tests positive are the upper two endpoints in the Figure D.6b tree, and hence the probability that an athlete will test positive is $A + C = 0.092 + 0.036 = 0.128$. However, the probability that an athlete will test positive is just what is supposed to be on the upper-left branch of the tree. Hence, $E = A + C = 0.128$. In the same way, $F = B + D = 0.008 + 0.864 = 0.872$. (Note that E and F sum to 1, as we know they must.)

Once E and F have been determined, it is straightforward to calculate the conditional probabilities G, H, I, and J. We know that the product of the probabilities along any path through a probability tree must equal the path probability. For example, it must be true that $E \times G = A$. But since we know the values for E and A, we can determine the value for G ($0.128 \times G = 0.092$, and therefore $G = 0.092/0.128 = 0.719$).

In the same way, $H = C/E = 0.036/0.128 = 0.281$, $I = B/F = 0.008/0.872 = 0.009$, and $J = D/F = 0.864/0.872 = 0.991$. (Note that $G + H = 1$ and $I + J = 1$,

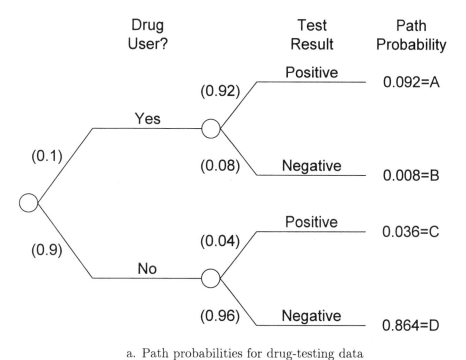

a. Path probabilities for drug-testing data

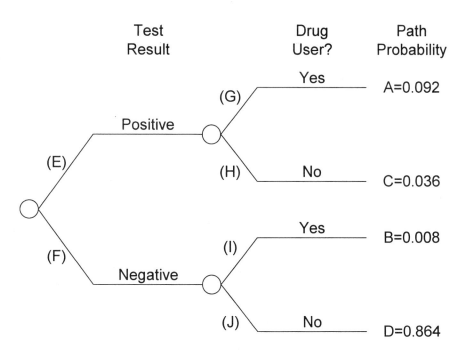

b. Path probabilities for interpreting test results

Figure D.6   *The drug-testing problem path probabilities*

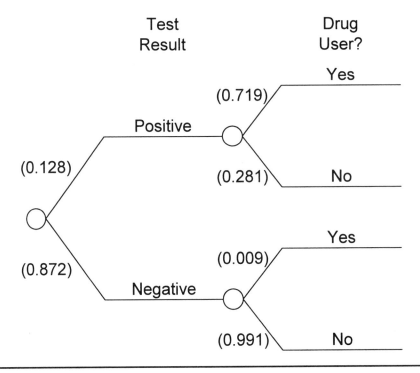

Figure D.7   *Probability tree for interpreting drug-testing results*

as we know they must.) The Figure D.6b tree with probabilities inserted is shown in Figure D.7.

Figure D.7 shows that if an athlete tests positive, there is only a 0.719 probability (approximately 72% chance) that he or she is a drug user, and hence a 0.281 probability (approximately 28% chance) that he or she is not a drug user. While it is gratifying to see from the figure that there is only a 0.009 probability (approximately 1% chance) that someone who tests negative is a drug user, the high probability of a false positive test is discouraging.

In his analysis, Professor Feinstein considered the impact of variations in the probability numbers used in the analysis. Over reasonable ranges of variation, his basic conclusion remained the same: There is a relatively high probability that an athlete who tests positive is actually not a drug user. Another member of the board of governance noted that Professor Feinstein's "decision analytic approach provided the framework that the board needed to debate the costs and benefits [of drug testing].... The board concluded that the incidence of drug use on our campus was so low and the cost of a false accusation so high that drug-testing was not the correct decision for us." The president of the university adopted that recommendation.

D.6 Concluding Comments

The calculation procedure given in the preceding section was first discovered by Thomas Bayes around 1760, but the confusion about this matter still continues today. Dawes (1988) provides a number of examples. Here is one: Studies show that a high proportion of users of other illegal drugs also use marijuana. (That is, the conditional probability is high that someone smokes marijuana, given that he or she uses other illegal drugs.) This has been misinterpreted to mean that a high proportion of marijuana users also smoke other illegal drugs. (That is, the conditional probability is high that someone uses other illegal drugs, given that he or she uses marijuana.) Note the unjustified "flipping" of the direction of the conditional probability. The information that a high proportion of users of other illegal drugs also smoke marijuana does not by itself tell you what proportion of marijuana smokers use other illegal drugs.

D.7 References

D. Buede, "Aiding Insight: Decision Analysis Survey," *OR/MS Today*, Vol. 20, No. 2, pp. 52–60 (April 1993).

R. T. Clemen, *Making Hard Decisions: An Introduction to Decision Analysis*, Second Edition, Duxbury Press, Belmont, California, 1996

R. M. Dawes, *Rational Choice in an Uncertain World*, Harcourt Brace Jovanovich, San Diego, 1988.

C. D. Feinstein, "Deciding Whether to Test Student Athletes for Drug Use," *Interfaces*, Vol. 20, No. 3, pp. 80–87 (May–June 1990).

H. J. Hansen, S. P. Caudill, and J. Boone, "Crisis in Drug Testing: Results of CDC Blind Study," *Journal of the American Medical Association*, Vol. 253, No. 16, pp. 2382–2387 (April 26, 1985).

C. W. Kirkwood, "An Algebraic Approach to Formulating and Solving Large Models for Sequential Decisions Under Uncertainty," *Management Science*, Vol. 39, pp. 900–913 (1993).

P. McNamee and J. Celona, *Decision Analysis with Supertree*, Second Edition, The Scientific Press, South San Francisco, California, 1990.

D.8 Review Questions

RD-1 Define *conditional probability*.

RD-2 Describe the characteristics of a probability tree. Include in this description definitions for *chance node* and *root node*, as well as the two required properties for a chance node to be valid.

RD-3   Define *sensitivity* and *specificity*. Explain the relationship of these to false negatives and false positives.

RD-4   Explain how to "flip" a probability tree.

RD-5   Describe the characteristics of a decision tree. Include in this description the definition of a *decision node*, as well as the required properties for a decision node to be valid.

RD-6   Describe the purpose of an *influence diagram*. In particular, explain the meaning of arrows pointing into a chance node in an influence diagram.

D.9 Exercises

D.1   When I started college, all entering freshman were given a test for tuberculosis (TB). This consisted of pricking us with a short pin treated with a test substance. If the area around the prick became reddened within the next three days, the test was positive. One of my friends had a positive test and became quite concerned. When he returned to the doctor, he was told not to worry because he probably didn't have TB, although they would take a chest X-ray to confirm this. In fact, my friend did not have TB. At the time I wondered why they gave the pin prick test if most people who tested positive didn't have TB.

For the analysis below, assume that both the sensitivity and specificity of this test are 90%, and that one-tenth of one percent (0.001 proportion) of the population has TB.

(i) Determine the probability that a randomly selected individual will test positive on the pin prick test.

(ii) Determine the conditional probability that someone has TB, given that the person tests positive on the pin prick test.

(iii) Determine the conditional probability that someone has TB, given that the person tests negative on the pin prick test.

(iv) In light of the relative costs of the pin prick test (cheap) and X-rays (more expensive), explain why it might make sense to use the pin prick test and give X-rays only to those who test positive, rather than give everyone an X-ray.

D.2   Consider a variation of the athlete drug-testing data shown in Figure D.5a. Specifically, assume that all the probabilities in that figure still hold, except that the probability an athlete is a drug user is reduced from 0.1 to 0.05 (and correspondingly, the probability that an athlete is not a drug user is increased to 0.95).

(i) Carry out the required tree flip and calculate the conditional probability that an athlete uses drugs, given that he or she tests positive.

(ii) Give an intuitive explanation of the direction of change in your answer to part (i) relative to the corresponding probability shown in Figure D.7.

D.3 One reader of this appendix objected to the conclusion reached in the drug-testing example in Section D.5 by saying that it was "really unsophisticated.... A sophisticated and widely-used approach is to send the 'positives' on for further, albeit more expensive, testing." Discuss the pros and cons of this approach.

D.4 The same reader quoted in the preceding exercise also commented that the marijuana example in Section D.6 is "a real straw man. Simple study of conditional probabilities and Bayes' Theorem will bring a stop to that fast!" (Bayes' Theorem is the mathematical result that justifies the tree flipping procedure presented in Section D.5.) Discuss the pros and cons of teaching conditional probabilities and Bayes' Theorem as a solution to the problem of incorrect reasoning about uncertainty.

Index